Selected Titles in This Subseries

35 **G. I. Olshanski, Editor,** Kirillov's Seminar on Representation Theory (TRANS2/181)
34 **A. Khovanskiĭ, A. Varchenko, and V. Vassiliev, Editors,** Topics in Singularity Theory (TRANS2/180)
33 **V. M. Buchstaber and S. P. Novikov, Editors,** Solitons, Geometry, and Topology: On the Crossroad (TRANS2/179)
32 **R. L. Dobrushin, R. A. Minlos, M. A. Shubin, and A. M. Vershik, Editors,** Topics in Statistical and Theoretical Physics (F. A. Berezin Memorial Volume) (TRANS2/177)
31 **R. L. Dobrushin, R. A. Minlos, M. A. Shubin, and A. M. Vershik, Editors,** Contemporary Mathematical Physics (F. A. Berezin Memorial Volume) (TRANS2/175)
30 **A. A. Bolibruch, A. S. Merkur'ev, and N. Yu. Netsvetaev, Editors,** Mathematics in St. Petersburg (TRANS2/174)
29 **V. Kharlamov, A. Korchagin, G. Polotovskiĭ, and O. Viro, Editors,** Topology of Real Algebraic Varieties and Related Topics (TRANS2/173)
28 **L. A. Bunimovich, B. M. Gurevich, and Ya. B. Pesin, Editors,** Sinai's Moscow Seminar on Dynamical Systems (TRANS2/171)
27 **S. P. Novikov, Editor,** Topics in Topology and Mathematical Physics (TRANS2/170)
26 **S. G. Gindikin and E. B. Vinberg, Editors,** Lie Groups and Lie Algebras: E. B. Dynkin's Seminar (TRANS2/169)
25 **V. V. Kozlov, Editor,** Dynamical Systems in Classical Mechanics (TRANS2/168)
24 **V. V. Lychagin, Editor,** The Interplay between Differential Geometry and Differential Equations (TRANS2/167)
23 **Yu. Ilyashenko and S. Yakovenko, Editors,** Concerning the Hilbert 16th Problem (TRANS2/165)
22 **N. N. Uraltseva, Editor,** Nonlinear Evolution Equations (TRANS2/164)

Published Earlier as Advances in Soviet Mathematics

21 **V. I. Arnold, Editor,** Singularities and bifurcations, 1994
20 **R. L. Dobrushin, Editor,** Probability contributions to statistical mechanics, 1994
19 **V. A. Marchenko, Editor,** Spectral operator theory and related topics, 1994
18 **Oleg Viro, Editor,** Topology of manifolds and varieties, 1994
17 **Dmitry Fuchs, Editor,** Unconventional Lie algebras, 1993
16 **Sergei Gelfand and Simon Gindikin, Editors,** I. M. Gelfand seminar, Parts 1 and 2, 1993
15 **A. T. Fomenko, Editor,** Minimal surfaces, 1993
14 **Yu. S. Il'yashenko, Editor,** Nonlinear Stokes phenomena, 1992
13 **V. P. Maslov and S. N. Samborskiĭ, Editors,** Idempotent analysis, 1992
12 **R. Z. Khasminskiĭ, Editor,** Topics in nonparametric estimation, 1992
11 **B. Ya. Levin, Editor,** Entire and subharmonic functions, 1992
10 **A. V. Babin and M. I. Vishik, Editors,** Properties of global attractors of partial differential equations, 1992
9 **A. M. Vershik, Editor,** Representation theory and dynamical systems, 1992
8 **E. B. Vinberg, Editor,** Lie groups, their discrete subgroups, and invariant theory, 1992
7 **M. Sh. Birman, Editor,** Estimates and asymptotics for discrete spectra of integral and differential equations, 1991
6 **A. T. Fomenko, Editor,** Topological classification of integrable systems, 1991
5 **R. A. Minlos, Editor,** Many-particle Hamiltonians: spectra and scattering, 1991
4 **A. A. Suslin, Editor,** Algebraic K-theory, 1991
3 **Ya. G. Sinaĭ, Editor,** Dynamical systems and statistical mechanics, 1991
2 **A. A. Kirillov, Editor,** Topics in representation theory, 1991
1 **V. I. Arnold, Editor,** Theory of singularities and its applications, 1990

Kirillov's Seminar on Representation Theory

American Mathematical Society

TRANSLATIONS

Series 2 • Volume 181

Advances in the Mathematical Sciences — 35

(*Formerly Advances in Soviet Mathematics*)

Kirillov's Seminar on Representation Theory

G. I. Olshanski
Editor

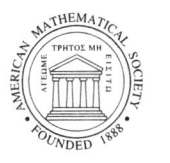

American Mathematical Society
Providence, Rhode Island

ADVANCES IN THE MATHEMATICAL SCIENCES
EDITORIAL COMMITTEE

V. I. ARNOLD

S. G. GINDIKIN

V. P. MASLOV

Translation edited by A. B. Sossinsky

1991 *Mathematics Subject Classification*. Primary 05Exx, 17Bxx, 22E15;
Secondary 53C35.

ABSTRACT. The book is a collection of papers written by students of A. A. Kirillov and participants of his seminar on Representation Theory at Moscow University. The papers deal with various aspects of representation theory for Lie algebras and Lie groups and its relations to algebraic combinatorics, theory of quantum groups, and geometry. The book is useful for researchers and graduate students working in representation theory and its applications.

Library of Congress Card Number 91-640741
ISBN 0-8218-0669-6
ISSN 0065-9290

Copying and reprinting. Material in this book may be reproduced by any means for educational and scientific purposes without fee or permission with the exception of reproduction by services that collect fees for delivery of documents and provided that the customary acknowledgment of the source is given. This consent does not extend to other kinds of copying for general distribution, for advertising or promotional purposes, or for resale. Requests for permission for commercial use of material should be addressed to the Assistant to the Publisher, American Mathematical Society, P. O. Box 6248, Providence, Rhode Island 02940-6248. Requests can also be made by e-mail to reprint-permission@ams.org.

Excluded from these provisions is material in articles for which the author holds copyright. In such cases, requests for permission to use or reprint should be addressed directly to the author(s). (Copyright ownership is indicated in the notice in the lower right-hand corner of the first page of each article.)

© 1998 by the American Mathematical Society. All rights reserved.
The American Mathematical Society retains all rights
except those granted to the United States Government.
Printed in the United States of America.

∞ The paper used in this book is acid-free and falls within the guidelines
established to ensure permanence and durability.
Visit the AMS home page at URL: http://www.ams.org/

10 9 8 7 6 5 4 3 2 1   03 02 01 00 99 98

Contents

Preface	ix
Screenings and a universal Lie–de Rham cocycle Victor Ginzburg and Vadim Schechtman	1
Interlacing measures Sergei Kerov	35
Quasicommuting families of quantum Plücker coordinates Bernard Leclerc and Andrei Zelevinsky	85
Factorial supersymmetric Schur functions and super Capelli identities Alexander Molev	109
Yangians and Capelli identities Maxim Nazarov	139
Hinges and the Study–Semple–Satake–Furstenberg–De Concini–Procesi–Oshima Boundary Yurii A. Neretin	165
Multiplicities and Newton polytopes Andreĭ Okounkov	231
Shifted Schur functions II. The binomial formula for characters of classical groups and its applications Andreĭ Okounkov and Grigori Olshanksi	245

Preface

The present volume was prepared for publication by students and friends of Alexandr Alexandrovich Kirillov in connection with his 60th anniversary.

A. A. Kirillov's numerous students (and not only his students) studied at his seminar on representation theory at Moscow State University. This seminar functioned for nearly 30 years, beginning in the early sixties when A. A. began teaching at the chair of function theory and functional analysis of the Mechanics and Mathematics Department of MSU, and continuing until A. A. started working at the University of Pennsylvania in Philadelphia. I first came to A. A.'s seminar in the winter of 1964–65 as a freshman, so I am one of his first students and one of the oldest participants of the seminar.

For many years Kirillov's seminar was one of the best known and popular Moscow mathematical seminars, and for me as well as for Kirillov's other students, the most customary and comforting one. It took place on Mondays, two hours before the Gelfand seminar. On Thursdays, A. A. also conducted a seminar for beginners (first and second year students), which was especially well attended. Active students of the latter would eventually move on to the Monday seminar, intended for older undergraduates, graduate students, and professional research mathematicians.

The topics discussed at the seminar ranged quite widely, reflecting Kirillov's broad research interests.[1] It included finite-dimensional representation theory; unitary representations of reductive, solvable, and general Lie groups; representations of infinite-dimensional groups. Of course, the *orbit method*; the universal formula for characters; symplectic geometry. The fractional fields of enveloping algebras and other noncommutative rings; identities in noncommutative rings. Infinite-dimensional Lie algebras. Superalgebras. C^*-algebras. Combinatorics. Quantum groups. Mathematical physics ... (I am afraid that I have missed many topics.)

The Kirillov seminar was neither primarily intended to inform on various topics, with experts taking turns to lecture on them, nor was it a working group concentrating on a specific cycle of papers,[2] although to some extent it performed

---

[1] When first talking to a new student, A. A. would usually ask what the latter would like to study under him—algebra, geometry, or analysis.

[2] In general, the organization of the seminar did not involve any rigid planning: lecturers and titles were not written out and displayed in advance, and everything seemed to take place spontaneously.

both functions. Above all, it was a place where one learned to do mathematics "according to Kirillov".

Participants of the seminar would usually assemble in advance, and while waiting for A. A. (who often came a bit late), they would conduct animated conversations.[3] When Kirillov appeared at the doorstep, all the participants would rise. If no talk was planned and A. A. did not intend to lecture himself, he would conduct a poll: who had done something new? He would then call someone to the blackboard and ask to state the result "in five minutes". In fact, few succeeded in complying with this sacramental time interval, but if the topic was interesting, A. A. would often forget this constraint, and "five minutes" could easily become a detailed account with a subsequent discussion.[4]

The atmosphere of the seminar was very free, relaxed, and informal. The lecturer was often interrupted by questions, and whenever A. A. felt that the listeners were losing track, he would explain the difficult parts in his own way or discuss improvised examples.

I am convinced that for us, just beginning to do mathematics, the main profit from participating in the seminar had to do with the impact of A. A. Kirillov's personality, his manner of explaining things simply, his light irony concerning an overly "scientific" style of exposition, his sharp remarks, and strong dislike of artificial constructions. All this contributed to form a proper taste in mathematical style, such an important component of one's mathematical education. And, of course, a crucial role was played by the problems that Kirillov systematically produced during the seminar. Some were prepared in advance, others arose spontaneously during discussions. A good result leading to new problems was particularly praised by Kirillov.[5]

One of the specific traits of A. A. Kirillov's style as a teacher was that he never liked to impose research topics for *kursovye* (term papers), *diplomnye* (MS theses), or *kandidatskie* (PhD dissertations). It was assumed that each student must find a topic himself on the basis of problems set at the seminar. Of course, this was not an absolute rule, but to the students that he rated among the best, Kirillov always gave complete freedom in the choice of a research topic.

Now, when Kirillov's seminar in Moscow no longer functions, while his students have dispersed all around the world and mostly communicate by e-mail,[6] I piercingly realize how much I owe to the seminar. I have no doubts that similar feelings are experienced by Kirillov's other students.

* * *

I shall briefly review the contents of the contributions.

1. In the paper "Screenings and a universal Lie–de Rham cocycle" by V. Ginzburg and V. Schechtman, a generalization of the classical Feigin–Fuchs construction

---

[3]Many Moscow seminars were, to some extent, something like clubs (this was especially true of I. M. Gelfand's famous seminar) and the discussions before they formally began, as well as the positive influence of the late arrival of the seminar's head, deserve special analysis.

[4]Such a poll would invariably take place at the first session after vacations. I vividly remember the feeling of frustration that arose if my turn to be polled was not reached.

[5]In assessing mathematical achievements, A. A. half-jokingly used "economic" terminology, distinguishing results that destroy "workplaces" for mathematicians from those that create them.

[6]Recently A. A. told me that he can invite people to his new seminar in Philadelphia from within a radius of $300 (that is the amount that can be allocated for travel expenses). Unfortunately, typical distances are now measured by larger amounts.

is presented. It provides canonical mappings from the homology of one-dimensional local systems on the configuration spaces appearing in conformal field theory to the *Ext*-spaces between modules of semi-infinite forms over the Virasoro algebra or Wakimoto modules over affine Lie algebras. An analog of this construction for finite-dimensional semisimple Lie algebras is given.

2. The paper "Interlacing measures" by S. Kerov deals with the asymptotic behavior of pairs of interlacing sequences,

$$x_1 < y_1 < x_2 < \cdots < x_{n-1} < y_{n-1} < x_n.$$

A typical example of such pairs is provided by roots of polynomials of adjacent degrees in a family of orthogonal polynomials. The author introduces and studies a more general object, a pair of interlacing measures. As a matter of fact, to each pair of interlacing measures with difference τ there corresponds a unique probability distribution μ such that

$$\exp \int \ln \frac{1}{z-u} \tau(du) = \int \frac{\mu(du)}{z-u}, \qquad \text{Im } z \neq 0.$$

This equation has a number of interesting applications, including

(1) the connection between additive and multiplicative integral representations of analytic functions of negative imaginary type;
(2) the Markov moment problem;
(3) distributions of mean values of Dirichlet random measures;
(4) the theory of spectral shift function in scattering theory;
(5) the Plancherel measure of the infinite symmetric group.

Apparently, the paper gives the first unified survey of all these topics. A special emphasis is on the combinatorial connections between the moments of the measures τ and μ in the above formula. One of the new results is an explicit formula for the multiplicative integral representation of the Gaussian measure on the real line.

3. The paper "Quasicommuting families of quantum Plücker coordinates" by B. Leclerc and A. Zelevinsky is devoted to the study of q-deformations of Plücker coordinates on the flag variety. The authors give a criterion for quasi-commutativity of two such coordinates and study their maximal quasi-commuting families (here "quasi-commutativity" means "commutativity up to a power of q"). The results have applications to the description of canonical bases for the quantum group GL_n, the geometry of Bott–Samelson desingularizations of Schubert varieties, and combinatorics of the "second Bruhat order" due to Manin and Schechtman.

4. The paper "Factorial supersymmetric Schur functions and super Capelli identities" by A. Molev is devoted to super generalization of a remarkable class of combinatorial functions—the so-called factorial Schur polynomials. These polynomials, introduced by the mathematical physicists L. C. Biedenharn and J. D. Louck and further studied by I. G. Macdonald and other authors, are certain multidimensional inhomogeneous polynomials whose highest degree terms are ordinary Schur polynomials. They have numerous applications in algebraic combinatorics and representation theory. The author develops the super counterpart of the theory. The main applications of his results are "factorial" analogs of the Jacobi–Trudi and Sergeev–Pragacz formulas; construction of a distinguished linear basis in the center of the universal enveloping algebra of $\mathfrak{gl}(m|n)$; a super analog of the higher Capelli identities. Related topics are discussed in the papers by M. Nazarov and by A. Okounkov and G. Olshanski (see below).

5. In the paper "Yangians and Capelli identities" by M. Nazarov, the R-matrix formalism is applied to higher Capelli identities. Recall that the classical Capelli identity (which is discussed in H. Weyl's famous book on classical groups) provides remarkable determinantal expressions for canonical generators of the center of the universal enveloping algebra $U(\mathfrak{gl}(n))$. The higher Capelli identities are stated for a much wider family of central elements, which form a distinguished linear basis in the center of $U(\mathfrak{gl}(n))$. Note that under the Harish–Chandra isomorphism, these basis elements turn into the factorial Schur polynomials mentioned above. The methods of the paper are inspired by quantum group theory. The author studies the image of the universal R-matrix for the Yangian $Y(\mathfrak{gl}(n))$ with respect to the evaluation homomorphism of $Y(\mathfrak{gl}(n))$ to $U(\mathfrak{gl}(n))$. The fusion procedure as defined by I. Cherednik is used. The higher Capelli identities are obtained as a corollary of this machinery. Although the Yangian techniques used in the paper may first seem rather sophisticated, the Yangians are actually a very natural and powerful tool for handling many problems concerning classical Lie algebras. Note that in another paper by the same author, the same approach is carried over the "true" super analog of $\mathfrak{gl}(n)$, the queer Lie superalgebra $\mathfrak{q}(n)$, and in the recent paper by A. Molev and M. Nazarov, the Yangian techniques are used to obtain new Capelli-type identities (for the orthogonal and symplectic Lie algebras). Different approaches to the higher Capelli identities for $\mathfrak{gl}(n)$ were developed by A. Okounkov.

6. The aim of the paper "Hinges and the Study–Semple–Satake–Furstenberg–De Concini–Procesi–Oshima boundary" by Yu. Neretin is to propose a unified elementary geometric description for various boundaries and completions of groups and symmetric spaces—the Satake–Furstenberg boundary, the Martin boundary, the Karpelevich boundary, complete symmetric varieties in the sense of De Concini and Procesi, compactifications of Bruhat–Tits buildings, etc. The key element of the author's constructions is the new concept of a "hinge" (a finite collection of points of a Grassmann manifold subject to certain conditions).

7. The paper "Multiplicities and Newton polytopes" by A. Okounkov deals with Newton polytopes associated in the author's recent paper (Invent. Math. **125** (1996), 405–411) to G-spaces X, where G is a connected reductive group,

$$(*) \qquad X \subset \mathbb{P}(V), \qquad X \text{ is closed, irreducible and } G\text{-stable},$$

and V is a finite-dimensional representation of G. The first result of the paper is the explicit computation of the polytope for the case when G is the symplectic group, $G = Sp(2n)$, and X is the flag variety. The polytope thus obtained coincides with the Gelfand–Zetlin-type polytope that appears in the well-known description (due to Zhelobenko) of weight multiplicities for the reduction scheme $Sp(2n) \downarrow Sp(2n-2) \downarrow \cdots$. This gives yet another proof and a geometric interpretation of Zhelobenko's theorem. The second result is that the polytopes corresponding to all the different G-equivariant embeddings $(*)$ of X can be arranged into a convex cone. This gives a strengthening of the theorem from the author's paper cited above: the semi-classical limit of weight multiplicities for the action $(*)$ is a log-concave function of both the weight and the G-linearized invertible sheaf that defines the embedding $(*)$.

8. The paper "Shifted Schur functions II. The binomial formula for characters of classical groups and its applications" by A. Okounkov and G. Olshanski continues the authors' previous work (referred to as Part I) but can be read independently.

Note that the shifted Schur functions are a modification (of a special case) of factorial Schur polynomials. The results of Part I have a direct relationship to the groups $GL(n)$; the aim of Part II was to find their counterparts for the orthogonal and symplectic groups. The paper starts with the binomial formula, which is a kind of Taylor expansion for finite-dimensional characters. This is a simple result, which has a number of important consequences. For instance, it suggests the definition of a distinguished linear basis in $Z(\mathfrak{g})$, the center of the universal enveloping algebra $U(\mathfrak{g})$, where $\mathfrak{g}$ stands for an orthogonal or symplectic Lie algebra. The basis elements can then be characterized in several different ways. Note that their images under the Harish–Chandra isomorphism can be expressed through certain factorial Schur polynomials. A natural basis in $I(\mathfrak{g})$, the subalgebra of invariants in the symmetric algebra $S(\mathfrak{g})$, is also examined. Both bases turn out to be related via the "special symmetrization map" $S(\mathfrak{g}) \to U(\mathfrak{g})$, an equivariant linear isomorphism, which differs from the usual symmetrization map. More involved versions of the binomial formula and the combinatorics of "generalized symmetrization maps" are studied in subsequent works by the same authors (cited in Part II).

G. Olshanski
Moscow, 1997

Screenings and a Universal Lie–de Rham Cocycle

Victor Ginzburg and Vadim Schechtman

To A. A. Kirillov on the occasion of his 60th birthday

Contents

1. Introduction
Chapter 1. Toy examples
  2. The toy example
  3. Generalization of the toy example
Chapter 2. Virasoro
  4. Cartan cocycle
  5. Bosonization for the Virasoro algebra
  6. Bosonic vertex operators
  7. Screening charges
Chapter 3. $\widehat{\mathfrak{sl}}(2)$
  8. Wakimoto realization
  9. Screening current
Chapter 4. Affine Lie algebras (general case)
  10. Bosonization
  11. Screening currents
References

§1. Introduction

In the pioneering paper [**FF**], Feigin and Fuchs have constructed intertwining operators between "Fock-type" modules over the Virasoro algebra via contour integrals of certain operator-valued one dimensional local systems over top homology classes of a configuration space. Similar constructions exist for affine Lie algebras. Key ingredients in such a construction are the so called "screening operators". The main observation of the present paper is that the screening operators contain more information. Specifically, at the chain level, the screening operators provide a certain canonical cocycle of the Virasoro (resp. affine) Lie algebra with coefficients in the de Rham complex of an operator-valued local system on the configuration space.

1991 *Mathematics Subject Classification.* Primary 17B56, 17B67, 17B68.

©1998 American Mathematical Society

This way we obtain canonical morphisms from *higher* homology spaces of the above local systems to appropriate higher *Ext*'s between the Fock space representations.

Our construction is motivated by, and in a special case reduces to, the construction of [**BMP1**], [**BMP2**]. In fact, as the results of *loc. cit* and [**FS**] suggest, the explanation of our results should lie in some equivalence of (derived?) categories of representations of quantum groups, and the corresponding affine algebras, sending (contragradient) Verma modules to Wakimoto modules.

On the other hand, we believe that the cocycles we study can be most adequately interpreted as de Rham cohomology classes of the chiral algebras considered by Beilinson–Drinfeld [**BD**].

Recently, Sebbar [**S**] was able to obtain a "*q*-deformation" of most of the constructions of this paper, where the de Rham cohomology is deformed to Aomoto-Jackson *q*-de Rham cohomology. It is interesting to note that, as was expected, the operators from Lemma 3.3 are deformed to Kashiwara operators (see [**L**], Part III).

The main results of this paper were obtained back in 1991, and written up in February 1996. We are deeply indebted to Ed Frenkel for the numerous enlightning discussions. We thank Jim Stasheff for useful remarks.

Chapter 1. Toy examples

§2. The toy example

In this work, everything will be over the base field $\mathbb{C}$.

2.1. Let $\mathfrak{g}$ be the Lie algebra $\mathfrak{sl}(2)$, with the standard generators E, F, H. For $\lambda \in \mathbb{C}$, let $M(\lambda)$ denote the Verma module over $\mathfrak{g}$ generated by the vacuum vector v_λ subject to the relations $Ev_\lambda = 0$, $Hv_\lambda = \lambda v_\lambda$. We shall use the formulas

$$EF^a v_\lambda = a(\lambda - a + 1) F^{a-1} v_\lambda, \qquad HF^a v_\lambda = (\lambda - 2a) F^a v_\lambda.$$

(In the sequel, in our formulas we shall use the agreement $F^b v_\lambda = 0$ for $b < 0$.)

Let us pick $\lambda, \lambda' \in \mathbb{C}$. For an integer $n \geqslant 0$, let us consider the $\mathbb{C}$-linear operator

$$V_n \colon M(\lambda' - 1) \to M(\lambda - 1), \qquad V_n(F^a v_{\lambda'-1}) = F^{a+n} v_{\lambda-1}.$$

These operators satisfy the following commutation relations.

(a) $\qquad [E, V_n](F^a v_{\lambda'-1}) = (-n^2 + (\lambda - 2a)n + a(\lambda - \lambda')) F^{a+n-1} v_{\lambda-1},$

(b) $\qquad [H, V_n](F^a v_{\lambda'-1}) = (\lambda - \lambda' - 2n) F^a v_{\lambda-1},$

(c) $\qquad [F, V_n] = 0.$

2.2. Consider the operator

$$V(z) = \sum_{n \geqslant 0} V_n z^{-n-1} dz \colon M(\lambda' - 1) \to M(\lambda - 1)[[z^{-1}]] \frac{dz}{z}$$

(here z is a formal variable). Let us try to find a number $\alpha \in \mathbb{C}$ and an operator

$$V(E; z) = \sum_{n \geqslant 0} V_n(E) z^{-n} \colon M(\lambda' - 1) \to M(\lambda - 1)[[z^{-1}]]$$

such that

(a) $\qquad\qquad [E, V(z)] = (d + \alpha \, dz/z) V(E; z).$

The equation (a) is equivalent to the system of equations

(b) $$[E, V_n] = (-n + \alpha) V_n(E) \qquad (n \geq 0).$$

So, we have

$$(-n + \alpha) V_n(E)(F^a v_{\lambda'-1}) = (-n^2 + (\lambda - 2a)n + a(\lambda - \lambda')) F^{a+n-1} v_{\lambda-1};$$

therefore $V_n(E)(F^a v_{\lambda'-1}) = (n + \beta(a)) f^{a+n-1} v_{\lambda-1}$ for some function $\beta(a)$ such that

$$(-n + \alpha)(n + \beta(a)) = -n^2 + (\lambda - 2a)n + a(\lambda - \lambda'),$$

that is, $\beta(a) - \alpha = -\lambda + 2a$, i.e., $\beta(a) = 2a + \alpha - \lambda$, and $\beta(a)\alpha = a(\lambda - \lambda')$ for all a.

Suppose that $\lambda \neq \lambda'$. Then we must have $\beta(a) = 2a$, $\alpha = \lambda$, hence from the second equation we obtain $\lambda' = -\lambda$.

2.3. From now on we suppose that $\lambda' = -\lambda$. Thus, we have

(a) $$[E, V_n] = (-n + \lambda) V_n(E),$$

where $V_n(E)(F^a v_{-\lambda-1}) = (n + 2a) F^{a+n-1} v_{\lambda-1}$.

From 2.1 (b) we obtain

(b) $$[H, V_n] = (-n + \lambda) V_n(H),$$

where $V_n(H) = 2V_n$. Finally,

(c) $$[F, V_n] = 0.$$

Therefore, we come to the following conclusion.

2.4. The operator

$$V(z) = \sum_{n \geq 0} V_n z^{-n-1} dz \colon M(-\lambda - 1) \to M(\lambda - 1)[[z^{-1}]] \frac{dz}{z}$$

defined by $V_n(F^a v_{-\lambda-1}) = F^{a+n} v_{\lambda-1}$, satisfies the following relation

(a) $$[X, V(z)] = (d + \lambda\, dz/z) V(X; z) \qquad (X \in \mathfrak{g}),$$

where the operators

$$V(X; z) = \sum_{n \geq 0} V_n(X) z^{-n} \colon M(-\lambda - 1) \to M(\lambda - 1)[[z]],$$

linearly depending on $X \in \mathfrak{g}$, are defined by

$$V_n(E)(F^a v_{-\lambda-1}) = (n + 2a) F^{a+n-1} v_{\lambda-1}, \quad V_n(H) = 2V_n, \quad V_n(F) = 0.$$

2.5. We have

$$[H, V_n(E)](F^a v_{-\lambda-1}) = -2(n + 2a)(n - \lambda - 1) F^{a+n-1} v_{\lambda-1},$$
$$[F, V_n(E)] = -V_n(H),$$
$$[E, V_n(H)](F^a v_{-\lambda-1}) = -2(n + 2a)(n - \lambda) F^{a+n-1} v_{\lambda-1},$$
$$[F, V_n(H)] = 0.$$

It follows that for any $X, Y \in \mathfrak{g}$ and $n \geq 0$, we have

(a) $$V_n([X, Y]) = [X, V_n(Y)] - [Y, V_n(X)].$$

2.6. Let us consider the complex (of length 1)

(a) $$\Omega^{\cdot}: 0 \to \Omega^0 \xrightarrow{d_\lambda} \Omega^1 \to 0,$$

where $\Omega^0 = \mathbb{C}[[z^{-1}]]$, $\Omega^1 = \mathbb{C}[[z^{-1}]]\, dz/z$, $d_\lambda = d + \lambda\, dz/z$.

For any two integers $i, j \geqslant 0$, set

$$C^{ij}(\mathfrak{g}; M(-\lambda-1), M(\lambda-1)) = \mathrm{Hom}\,(\Lambda^i \mathfrak{g} \otimes M(-\lambda-1), M(\lambda-1) \otimes \Omega^j)$$
$$= \mathrm{Hom}\,(\Lambda^i \mathfrak{g}, \mathrm{Hom}\,(M(-\lambda-1), M(\lambda-1) \otimes \Omega^j)).$$

The bigraded space $C^{\cdot\cdot}(\mathfrak{g}; M(-\lambda-1), M(\lambda-1))$ has the natural structure of a bicomplex. The first differential

$$d': C^{ij}(\mathfrak{g}; M(-\lambda-1), M(\lambda-1)) \to C^{i+1,j}(\mathfrak{g}; M(-\lambda-1), M(\lambda-1))$$

is induced by the standard Chevalley–Eilenberg differential in the cochain complex of the Lie algebra $\mathfrak{g}$ with coefficients in the module $\mathrm{Hom}\,(M(-\lambda-1), M(\lambda-1))$.

The second differential

$$d'': C^{ij}(\mathfrak{g}; M(-\lambda-1), M(\lambda-1)) \to C^{i,j+1}(\mathfrak{g}; M(-\lambda-1), M(\lambda-1))$$

is induced by the differential in the complex $\Omega^{\cdot}$.

Let $C^{\cdot}(\mathfrak{g}; M(-\lambda-1), M(\lambda-1))$ denote the associated total complex.

The operator $V(z)$ is an element of the space $C^{01}(\mathfrak{g}; M(-\lambda-1), M(\lambda-1))$. Let us denote this element by $V^{01}(z)$. The operators $V(X; z)$ ($X \in \mathfrak{g}$) define an element $V^{10}(z)$ of the space $C^{10}(\mathfrak{g}; M(-\lambda-1), M(\lambda-1))$.

Property 2.4 (a) means that $d'(V^{01}(z)) = d''(V^{10}(z))$. Property 2.5 (a) means that $d''(V^{10}(z)) = 0$. Therefore, the element $(V^{01}(z), V^{10}(z))$ is a 1-*cocycle* of the complex $C^{\cdot}(\mathfrak{g}; M(-\lambda-1), M(\lambda-1))$.

2.7. Suppose that λ is a nonnegative integer. In this case, the complex $\Omega^{\cdot}$ has two one-dimensional cohomology spaces: $H^0(\Omega^{\cdot})$ generated by the function $z^{-\lambda}$ and $H^1(\Omega^{\cdot})$ generated by the class of the form $z^{-\lambda}\, dz/z$. (If $\lambda \notin \mathbb{N}$, the complex $\Omega^{\cdot}$ is acyclic.)

Let us consider the dual spaces $H_i = H^i(\Omega^{\cdot})^*$. The space H_1 is generated by the functional $\Omega^1 \to \mathbb{C}$ which assigns to a form ω the residue $\mathrm{res}_{z=0}(\omega z^\lambda)$. The space H_0 is generated by the (restriction of) the functional $\Omega^0 \to \mathbb{C}$ which assigns to a function $f(z)$ the residue $\mathrm{res}_{z=0}(f(z)\, z^\lambda\, dz/z)$.

The previous discussion implies the following.

(a) The operator

$$\mathop{\mathrm{res}}_{z=0} (V^{01}(z)\, z^\lambda) \in \mathrm{Hom}\,(M(-\lambda-1), M(\lambda-1))$$

is an intertwiner. It is the unique $\mathfrak{g}$-homomorphism $M(-\lambda-1) \to M(\lambda-1)$ sending $v_{-\lambda-1}$ to $F^\lambda v_{\lambda-1}$.

(b) The operator

$$\mathop{\mathrm{res}}_{z=0} (V^{10}(z)\, z^\lambda\, dz/z) \in \mathrm{Hom}\,(\mathfrak{g}, \mathrm{Hom}\,(M(-\lambda-1), M(\lambda-1)))$$

is a 1-cocycle of the Lie algebra $\mathfrak{g}$ with coefficients in the $\mathfrak{g}$-module

$$\mathrm{Hom}\,(M(-\lambda-1), M(\lambda-1)).$$

Therefore, this operator defines a certain element of the space

$$\operatorname{Ext}_{\mathfrak{g}}^1(M(-\lambda-1), M(\lambda-1)).$$

§3. Generalization of the toy example

3.1. Let $A = (a_{ij})_{i,j=1}^r$ be a symmetrizable generalized Cartan matrix, and let $\mathfrak{g}$ be the corresponding Kac–Moody Lie algebra defined by the Chevalley generators E_i, F_i, H_i ($i = 1, \ldots, r$) and relations

$$[H_i, H_j] = 0, \quad [H_i, E_j] = a_{ij} E_i, \quad [H_i, F_i] = -a_{ij} F_i, \quad [E_i, F_j] = \delta_{ij} H_i,$$
$$\operatorname{ad}(E_i)^{-a_{ij}+1}(E_j) = \operatorname{ad}(F_i)^{-a_{ij}+1}(F_j) = 0$$

(see [**K, 0.3**]). Let $\mathfrak{h} \subset \mathfrak{g}$ be the Cartan subalgebra spanned by the elements $H_1, \ldots, H_r$. For $i = 1, \ldots, r$, let $\alpha_i \in \mathfrak{h}^*$ be the corresponding simple root; let $r_i \colon \mathfrak{h}^* \to \mathfrak{h}^*$ be the corresponding simple reflection,

$$r_i \lambda = \lambda - \langle H_i, \lambda \rangle \alpha_i.$$

Let $\rho \in \mathfrak{h}^*$ be the element defined by $\langle H_i, \rho \rangle = 1$ ($i = 1, \ldots, r$).

For $\lambda \in \mathfrak{h}^*$, let $M(\lambda)$ denote the Verma module over $\mathfrak{g}$, defined by one generator v_λ and relations $E_i v_\lambda = 0$, $H_i v_\lambda = \langle H_i, \lambda - \rho \rangle v_\lambda$.

3.2. Let us pick an element $\lambda \in \mathfrak{h}^*$ and $i \in \{1, \ldots, r\}$. Set $\lambda' = r_i \lambda$. For each integer $n \geq 0$, consider the operator

$$V_{i;n} \colon M(\lambda') \to M(\lambda), \qquad V_{i;n}(x v_{\lambda'}) = x F_i^n v_\lambda$$

where $x \in U\mathfrak{n}_-$, $\mathfrak{n}_- \subset \mathfrak{g}$ being the Lie subalgebra generated by the elements $F_1, \ldots, F_r$.

3.3. LEMMA-DEFINITION. *There exists a unique linear operator $\partial_i \colon U\mathfrak{n}_- \to U\mathfrak{n}_-$ such that $\partial_i(F_j) = \delta_{ij} \cdot 1$ ($j = 1, \ldots, r$) and for all $x, y \in U\mathfrak{n}$ we have $\partial_i(xy) = \partial_i(x) y + x \partial_i(y)$.*

PROOF. The uniqueness is clear. Let us prove the existence. Let A be the free associative $\mathbb{C}$-algebra with generators θ_j, $j = 1, \ldots, r$. It is clear that there exists a unique linear operator $\partial_i \colon A \to A$ such that $\partial_i(\theta_j) = \delta_{ij} \cdot 1$ and for any $x, y \in A$, $\partial_i(xy) = \partial_i(x) y + x \partial_i(y)$.

For an integer $a \geq 1$ and $j \neq k$ in $\{1, \ldots, r\}$, define the following element in A:

$$C(j, k; a) = \operatorname{ad}(\theta_j)^a(\theta_k) = \sum_{p=0}^{a} (-1)^p \binom{a}{p} \theta_j^{a-p} \theta_k \theta_j^p.$$

We claim that

(a) for any a, j, k, $C(j, k; a) \in \operatorname{Ker}(\partial_i)$.

Indeed, the claim is clear for (j, k) such that $i \neq j$ and $i \neq k$. We have

$$\partial_i(C(j, i; a)) = \sum_{p=0}^{a} (-1)^p \binom{a}{p} \theta_i^a = (1-1)^a \theta_j^a = 0.$$

Let us prove that $\partial_i(C(i, k; a)) = 0$ by induction on a. For $a = 1$ this is obvious.

We have
$$\partial_i(C(i,k;a)) = \partial_i(\theta_i C(i,k;a-1) - C(i,k;a-1)\theta_i)$$
(by induction)
$$= \partial_i(\theta_i)C(i,k;a-1) - C(i,k;a-1)\partial_i(\theta_i)$$
$$= C(i,k;a-1) - C(i,k;a-1) = 0.$$

The claim is proved.

It follows from (a) that the operator $\partial_i \colon A \to A$ induces the operator $\partial_i \colon \mathbf{Un}_- \to \mathbf{Un}_-$ since $\mathbf{Un}_-$ is the quotient of A by the two-sided ideal generated by the elements $C(j,k;-a_{jk}+1)$. The lemma is proved. $\square$

3.4. Let us define the linear operators $V_{i;n}(E_j) \colon M(\lambda') \to M(\lambda)$ $(n \geqslant 0)$ by setting

(a) $\qquad V_{i;n}(E_j)(xv_{\lambda'}) = a_{ji}\partial_j(x) F_i^n v_\lambda + \delta_{ij} n x F_i^{n-1} v_\lambda \qquad (x \in \mathbf{Un}_-).$

3.5. LEMMA. *For any j, n,*
$$[E_j, V_{i;n}] = (-n + \langle H_i, \lambda \rangle) V_{i;n}(E_j).$$

This is checked by a direct computation.

3.6. Let us define the operators $V_{i;n}(H_j) \colon M(\lambda') \to M(\lambda)$ by

(a) $\qquad\qquad\qquad V_{i;n}(H_j) = a_{ji} V_{i;n}.$

One easily checks that

(b) $\qquad\qquad [H_j, V_{i;n}] = (-n + \langle H_i, \lambda \rangle) V_{i;n}(H_j).$

Finally, it is evident that

(c) $\qquad\qquad [F_j, V_{i;n}] = 0 \quad \text{for all } j.$

Set $V_{i;n}(F_j) = 0$.

3.7. LEMMA-DEFINITION. *For each $n \geqslant 0$, there exists a unique element*
$$V_{i;n}(\,\cdot\,) \in \mathrm{Hom}\,(\mathfrak{g}, \mathrm{Hom}\,(M(\lambda'), M(\lambda)))$$
such that $V_{i;n}(E_j), V_{i;n}(H_j), V_{i;n}(F_j)$ are defined above and the following cocycle condition holds:

(a) *For any $X, Y \in \mathfrak{g}$, $V_{i;n}([X,Y]) = [X, V_{i;n}(Y)] - [Y, V_{i;n}(X)]$.* $\square$

3.8. Let us consider the operators
$$V_i(z) = \sum_{n=0}^{\infty} V_{i;n} z^{-n-1} dz \in \mathrm{Hom}\,(M(r_i\lambda), M(\lambda)[[z^{-1}]]\,dz/z),$$
$$V_i(X;z) = \sum_{n=0}^{\infty} V_{i;n}(X) z^{-n} \in \mathrm{Hom}\,(M(r_i\lambda), M(\lambda)[[z^{-1}]]) \qquad (X \in \mathfrak{g}).$$

The previous considerations may be reformulated as the following theorem.

3.9. Theorem. *For any $X, Y \in \mathfrak{g}$, we have*
$$[X, V_i(z)] = (d + \langle H_i, \lambda \rangle \, dz/z) V_i(X; z),$$
$$V_i([X, Y]; z) = [X, V_i(Y; z)] - [Y, V_i(X; z)]. \quad \Box$$

3.10. Suppose an element w of the Weyl group of $\mathfrak{g}$ together with its reduced decomposition $w = r_{i_a} \cdots r_{i_1}$ and an element $\lambda \in \mathfrak{h}^*$ is given. For $p = 1, \ldots, a$, set $\lambda_p = r_{i_{p-1}} \cdots r_{i_1} \lambda$.

Let us define the complex
$$\Omega^{\cdot} : 0 \to \Omega^0 \to \cdots \to \Omega^a \to 0$$
as follows. Set $A = \mathbb{C}[[z_1^{-1}, \ldots, z_a^{-1}]]$. By definition, Ω^p is the free A-module with basis $\{(dz_{j_1}/z_{j_1}) \wedge \cdots \wedge (dz_{j_p}/z_{j_p}), 1 \leqslant j_1 < \cdots < j_p \leqslant a\}$. The differential is defined by
$$d\eta = d_{DR}(\eta) + \left(\sum_{p=1}^{a} \frac{\langle H_{i_p}, \lambda_p \rangle}{z_p} dz_p \right) \wedge \eta,$$
where d_{DR} is the de Rham differential.

3.11. For each $p = 1, \ldots, a$, consider the operators
$$\omega_p = V_{i_p}(z_p) \colon M(\lambda_{p+1}) = M(r_{i_p} \lambda_p) \to M(\lambda_p) z_p^{-1}[[z_p^{-1}]],$$
$$\tau_p(X) = V_{i_p}(X; z_p) \colon M(\lambda_{p+1}) \to M(\lambda_p)[[z_p^{-1}]], \qquad X \in \mathfrak{g},$$
defined above.

For each m, $0 \leqslant m \leqslant a$, let us define the operators
$$V^{m,a-m} \in \mathrm{Hom}\,(\Lambda^m \mathfrak{g}, \mathrm{Hom}\,(M(w\lambda), M(\lambda) \otimes \Omega^{a-m}))$$
as follows. We set
$$V^{m,a-m}(X_1 \wedge \cdots \wedge X_m)$$
$$= \sum_{1 \leqslant p_1 < \cdots < p_m \leqslant a} (-1)^{\mathrm{sgn}(p_1, \ldots, p_m)}$$
$$\times \left(\sum_{\sigma \in \Sigma_m} (-1)^{\mathrm{sgn}(\sigma)} \omega_1 \cdots \tau_{p_1}(X_{\sigma(1)}) \cdots \tau_{p_m}(X_{\sigma(m)}) \cdots \omega_a \right)$$
$$\times dz_1 \wedge \cdots \wedge \widehat{dz_{p_1}} \wedge \cdots \wedge \widehat{dz_{p_m}} \wedge \cdots \wedge dz_a.$$

Here, for each sequence $1 \leqslant p_1 < \cdots < p_m \leqslant a$, the corresponding summands are obtained by replacing the operators ω_{p_j} in the product $\omega_1 \cdots \omega_a$, by $\tau_{p_j}(X_{\sigma(j)})$. Here Σ_m denotes the group of all bijections $\sigma \colon \{1, \ldots, m\} \xrightarrow{\sim} \{1, \ldots, m\}$. The sign $\mathrm{sgn}(p_1, \ldots, p_m)$ is defined by induction on m as follows. We set
$$\mathrm{sgn}(\ ) = 0, \quad \mathrm{sgn}(p_1, \ldots, p_m) = \mathrm{sgn}(p_1, \ldots, p_{m-1}) + p_m + m.$$

For example,
$$V^{0a} = V_{i_1}(z_1) \cdots V_{i_a}(z_a) \, dz_1 \wedge \cdots \wedge dz_a.$$

Let us consider the double complex $C^{\cdot}(\mathfrak{g}; \mathrm{Hom}\,(M(w\lambda), M(\lambda) \otimes \Omega^{\cdot}))$. Here the first differential is the Chevalley–Eilenberg differential in the complex of cochains of the Lie algebra $\mathfrak{g}$ with coefficients in the complex of $\mathfrak{g}$-modules $\mathrm{Hom}\,(M(w\lambda), M(\lambda) \otimes \Omega^{\cdot})$, the action of $\mathfrak{g}$ being induced by the standard action (by the commutator) of $\mathfrak{g}$ on $\mathrm{Hom}(M(w\lambda), M(\lambda))$. The second differential is induced by the differential in $\Omega^{\cdot}$.

By definition,
$$V^{m,a-m} \in C^m(\mathfrak{g}; \mathrm{Hom}(M(w\lambda), M(\lambda) \otimes \Omega^{a-m})).$$

3.12. THEOREM. *The element $V = (V^{0a}, \ldots, V^{a0})$ is an a-cocycle in the total complex associated with the double complex $C^{\cdot}(\mathfrak{g}; \mathrm{Hom}(M(w\lambda), M(\lambda) \otimes \Omega^{\cdot}))$.* □

3.13. COROLLARY. *The cocycle V induces linear maps*
$$f_m \colon H^m(\Omega^{\cdot})^* \to \mathrm{Ext}_{\mathfrak{g}}^{a-m}(M(w\lambda), M(\lambda)), \qquad m = 0, \ldots, a.$$

3.14. EXAMPLE. Assume that all numbers $\langle H_p, \lambda_p \rangle$, $p = 1, \ldots, a$, are nonnegative integers. The highest homology space $H^a(\Omega^{\cdot})$ is one-dimensional, generated by the (image of) the functional $r \in \Omega^{a*}$ defined by
$$r(\eta) = \mathop{\mathrm{res}}_{z_a=0} \cdots \mathop{\mathrm{res}}_{z_1=0} (z_1^{\langle H_{i_1}, \lambda_1 \rangle} \cdots z_a^{\langle H_{i_a}, \lambda_a \rangle} \eta).$$
The image $f_0(r) \in \mathrm{Hom}_{\mathfrak{g}}(M(w\lambda), M(\lambda))$ is the unique (up to proportionality) intertwiner sending $v_{w\lambda}$ to $F_{i_a}^{\langle H_{i_a}, \lambda_a \rangle + 1} \cdots F_{i_1}^{\langle H_{i_1}, \lambda_1 \rangle + 1} v_\lambda$.

Chapter 2. Virasoro

§4. Cartan cocycle

4.1. Let $\mathfrak{g}$ be a Lie algebra. Consider the differential graded (dg) Lie algebra $\mathfrak{g}^{\cdot} = \mathfrak{g}^{-1} \oplus \mathfrak{g}^0$ defined as follows. We set $\mathfrak{g}^{-1} = \mathfrak{g}^0 = \mathfrak{g}$; the differential $d \colon \mathfrak{g}^{-1} \to \mathfrak{g}^0$ is the identity map; the bracket $\Lambda^2 \mathfrak{g}^0 \to \mathfrak{g}^0$ coincides with the bracket in $\mathfrak{g}$; the bracket $\mathfrak{g}^{-1} \otimes \mathfrak{g}^0 \to \mathfrak{g}^0$ coincides with the bracket in $\mathfrak{g}$.

For $x \in \mathfrak{g}$, let us denote by the same letter x the corresponding element of $\mathfrak{g}^0$, and by i_x the corresponding element of $\mathfrak{g}^{-1}$.

As a dg Lie algebra, the algebra $\mathfrak{g}^{\cdot}$ is generated by the Lie algebra $\mathfrak{g}$ (in degree 0) and by the elements $i_x \in \mathfrak{g}^{-1}$ ($x \in \mathfrak{g}$) subject to the relations (a)–(c) below.

(a) $\qquad d(i_x) = x \qquad (x \in \mathfrak{g}),$

(b) $\qquad [x, i_y] = i_{[x,y]} \qquad (x, y \in \mathfrak{g}),$

(c) $\qquad [i_x, i_y] = 0 \qquad (x, y \in \mathfrak{g}).$

4.2. Let $M^{\cdot}$ be a complex of vector spaces. A $\mathfrak{g}^{\cdot}$-module structure on $M^{\cdot}$ is the same as a collection of data (a), (b) below satisfying the properties (c)–(e) below.

(a) Morphisms of complexes $x \colon M^{\cdot} \to M^{\cdot}$ ($x \in \mathfrak{g}$) that define an action of the Lie algebra $\mathfrak{g}$.

(b) Morphisms of graded spaces $i_x \colon M^{\cdot} \to M^{\cdot}[-1]$ ($x \in \mathfrak{g}$).

(c) (*Cartan formula*) $[d, i_x] = x$ ($x \in \mathfrak{g}$).

Here (and below) the commutators are understood in the graded sense, i.e., $[d, i_x] = d \circ i_x + i_x \circ d$.

(d) $[x, i_y] = i_{[x,y]}$ ($x, y \in \mathfrak{g}$).

(e) $[i_x, i_y] = 0$ ($x, y \in \mathfrak{g}$).

4.3. Let us consider the enveloping algebra $U\mathfrak{g}^{\cdot}$. It is a dg associative algebra. We have the canonical embedding

$$\Lambda^{\cdot}(\mathfrak{g}^{-1}) \hookrightarrow U\mathfrak{g}^{\cdot}.$$

We shall use the notation $i_{x_1\ldots x_a}$ for the elements $i_{x_1}\cdots i_{x_a} \in (U\mathfrak{g}^{\cdot})^{-a}$, $x_1,\ldots,x_a \in \mathfrak{g}$.

4.4. LEMMA. *For any $x_1,\ldots,x_a \in \mathfrak{g}$, we have*

$$di_{x_1\ldots x_a} = \sum_{p=1}^{a}(-1)^{p-1}x_p i_{x_1\ldots\hat{x}_p\ldots x_a} + \sum_{1\leqslant p<q\leqslant a}(-1)^{p+q}i_{[x_p,x_q]x_1\ldots\hat{x}_p\ldots\hat{x}_q\ldots x_a}$$

$$= \sum_{p=1}^{a}(-1)^{p-1}i_{x_1\ldots\hat{x}_p\ldots x_a}x_p + \sum_{1\leqslant p<q\leqslant a}(-1)^{p+q+1}i_{[x_p,x_q]x_1\ldots\hat{x}_p\ldots\hat{x}_q\ldots x_a}.$$

PROOF. Induction on a. For $a = 1$ it is the Cartan formula. Suppose that $a > 1$. We have

$$di_{x_1\ldots x_a} = d(i_{x_1}\cdot i_{x_2\ldots x_a}) = di_{x_1}\cdot i_{x_2\ldots x_a} - i_{x_1}\cdot di_{x_2\ldots x_a}$$

(by induction)

$$= x_{i_1}\cdot i_{x_2\ldots x_a} - i_{x_1}\cdot\left(\sum_{p=2}^{a}(-1)^p x_p i_{x_2\ldots\hat{x}_p\ldots x_a} + \sum_{2\leqslant p<q\leqslant a}(-1)^{p+q}i_{[x_p,x_q]x_2\ldots\hat{x}_p\ldots\hat{x}_q\ldots x_a}\right)$$

(we use $i_{x_1}x_p = x_p i_{x_1} + i_{[x_1,x_p]}$ and anticommutation of the various elements i_x)

$$= x_{i_1}\cdot i_{x_2\ldots x_a} + \sum_{p=2}^{a}(-1)^{p-1}x_p i_{x_1\ldots\hat{x}_p\ldots x_a} + \sum_{p=2}^{a}(-1)^{1+p}i_{[x_1,x_p]x_2\ldots\hat{x}_p\ldots x_a}$$

$$+ \sum_{2\leqslant p<q\leqslant a}(-1)^{p+q}i_{[x_p,x_q]x_1\ldots\hat{x}_p\ldots\hat{x}_q\ldots x_a}$$

$$= \sum_{p=1}^{a}(-1)^{p-1}x_p i_{x_1\ldots\hat{x}_p\ldots x_a} + \sum_{1\leqslant p<q\leqslant a}(-1)^{p+q}i_{[x_p,x_q]x_1\ldots\hat{x}_p\ldots\hat{x}_q\ldots x_a}.$$

This proves the first equality. The second equality is proved in the same manner, or may be deduced from the first one. $\square$

4.5. Let M be a $\mathfrak{g}$-module. Recall that the complex $C^{\cdot}(\mathfrak{g};M)$ of cochains of $\mathfrak{g}$ with coefficients in M is defined by $C^a(\mathfrak{g};M) = \text{Hom}(\Lambda^a\mathfrak{g},M)$, the differential $d\colon C^{a-1}(\mathfrak{g};M) \to C^a(\mathfrak{g};M)$ acts as

$$d\phi(x_1\wedge\cdots\wedge x_a) = \sum_{p=1}^{a}(-1)^{p-1}x_p\phi(x_1\wedge\cdots\wedge\hat{x}_p\wedge\cdots\wedge x_a)$$

$$+ \sum_{1\leqslant p<q\leqslant a}(-1)^{p+q}\phi(x_1\wedge\cdots\wedge\hat{x}_p\wedge\cdots\wedge\hat{x}_q\wedge\cdots\wedge x_a).$$

The differential in $C^{\cdot}(\mathfrak{g};M)$ is called the *Chevalley–Eilenberg differential*.

4.6. REMARK. Let us consider $U\mathfrak{g}^{\cdot}$ as a $\mathfrak{g}$-module by means of the left multiplication, where $\mathfrak{g}$ is identified with $\mathfrak{g}^0$. Let us consider the complex of cochains of $\mathfrak{g}$ with coefficients in $U\mathfrak{g}^{\cdot}$, $C^{\cdot}(\mathfrak{g}, U\mathfrak{g}^{\cdot})$. It is a double complex in which the first differential is the Koszul differential, and the second one is induced by the differential in $U\mathfrak{g}^{\cdot}$. Let $C^{\cdot}$ be the associated total complex.

For each $a \geq 0$, let us define an element $c^{a,-a} \in C^a(\mathfrak{g}, U\mathfrak{g}^{-a})$ by

$$c^{a,-a}(x_1 \wedge \cdots \wedge x_a) = i_{x_1\ldots x_a}, \qquad c^{00} = 1.$$

One can reformulate the previous lemma, as in the following statement.

(a) The element $c = \sum_{a \geq 0} c^{a,-a}$ is a 0-cocycle in $C^{\cdot}$.

4.7. Suppose we are given the collection of data (a)–(d) below.

(a) A Lie algebra $\mathfrak{g}$;

(b) a $\mathfrak{g}$-module M;

(c) a dg $\mathfrak{g}^{\cdot}$-module $\Omega^{\cdot}$;

(d) an element $\omega \in M \otimes \Omega^n$, for some $n \geq 0$.

Assume that the element ω satisfies the properties (i), (ii) below.

(i) $d\omega = 0$.

Here $d = \mathrm{Id}_M \otimes d_{\Omega^{\cdot}} : M \otimes \Omega^n \to M \otimes \Omega^{n+1}$.

(ii) $\omega \in (M \otimes \Omega^n)^{\mathfrak{g}}$.

Here the superscript $(\cdot)^{\mathfrak{g}}$ denotes the subspace of $\mathfrak{g}$-invariants. We consider $M \otimes \Omega^n$ as a $\mathfrak{g}$-module equal to the tensor product of the two $\mathfrak{g}$-modules M and Ω^n, the last one being the $\mathfrak{g}$-module obtained via the identification $\mathfrak{g} = \mathfrak{g}^0$.

Let us consider the double complex $C^{\cdot}(\mathfrak{g}; M \otimes \Omega^{\cdot})$ defined by

$$C^a(\mathfrak{g}; M \otimes \Omega^b) = \mathrm{Hom}(\Lambda^a \mathfrak{g}, M \otimes \Omega^b).$$

The first differential is the Koszul differential in the standard cochain complex of $\mathfrak{g}$ with coefficients in $M \otimes \Omega^{\cdot}$. Here the action of $\mathfrak{g}$ on $M \otimes \Omega^{\cdot}$ is defined *through the first factor* (i.e., $x \cdot (m \otimes \alpha) = xm \otimes \alpha$). The second differential is induced by the differential in $\Omega^{\cdot}$. Let $C^{\cdot}(\mathfrak{g}; M, \Omega^{\cdot})$ denote the associated total complex.

For $0 \leq a \leq n$, let us define the elements $w^a \in C^a(\mathfrak{g}; M \otimes \Omega^{n-a})$ by

$$w^0 = 0, \qquad w^a(x_1 \wedge \cdots \wedge x_a) = i_{x_1\ldots x_a}\omega.$$

Here the action of $U\mathfrak{g}^{\cdot}$ is defined through the second factor. Let us consider the element $\widehat{\omega} = \sum_{a=0}^n w^a \in C^n(\mathfrak{g}; M, \Omega^{\cdot})$.

4.8. LEMMA. *The element $\widehat{\omega}$ is an n-cocycle in $C^{\cdot}(\mathfrak{g}; M, \Omega^{\cdot})$.*

We shall call $\widehat{\omega}$ the *Cartan cocycle* associated with ω.

PROOF. Let d' (resp., d'') denote the first (resp., second) differential in the complex $C^{\cdot}(\mathfrak{g}; M \otimes \Omega^{\cdot})$. We have $d''(w^0) = 0$ by property (i). We have

$d'w^{a-1}(x_1 \wedge \cdots \wedge x_a)$

$$= \sum_{p=1}^{a}(-1)^{p-1} x_p \cdot w^{a-1}(x_1 \wedge \cdots \wedge \hat{x}_p \wedge \cdots \wedge x_a)$$

$$+ \sum_{1 \leq p < q \leq a}(-1)^{p+q} w^{a-1}([x_p, x_q] \wedge x_1 \wedge \cdots \wedge \hat{x}_p \wedge \cdots \wedge \hat{x}_q \wedge \cdots \wedge x_a)$$

(by Lemma 4.4, taking into account the fact that

$$x_p \cdot (i_{x_1\ldots\hat{x}_p\ldots x_a}\omega) = i_{x_1\ldots\hat{x}_p\ldots x_a}(x_p^{(1)}\omega)$$
$$= \text{(by (ii))} = -i_{x_1\ldots\hat{x}_p\ldots x_a}(x_p^{(2)}\omega) = -(i_{x_1\ldots\hat{x}_p\ldots x_a}x_p)\omega\,.$$

Here $x^{(1)}\cdot?$ (resp., $x^{(2)}\cdot?$) denotes the action of $\mathfrak{g}$ on $M\otimes\Omega^n$ through the first (resp., second) factor.)

$$= (di_{x_1\ldots x_a})\omega = \text{(by (i))} = d(i_{x_1\ldots x_a}\omega) = (d''\omega^a)(x_1 \wedge \cdots \wedge x_a).\square$$

4.9. Let us consider the dual complex $\Omega^{\cdot *}$. Note that the cocycle $\widehat{\omega}$ defines a morphism of complexes

$$\widehat{\omega}\colon (\Omega^{n-\cdot})^* \to C^{\cdot}(\mathfrak{g};M)\,.$$

Let us denote $H_i(\Omega) := H^{-i}(\Omega^*)$. The morphism $\widehat{\omega}$ induces the maps

$$H^{-i}(\omega)\colon H_i(\Omega^{\cdot}) \to H^{n-i}(\mathfrak{g};M) \qquad (0 \leqslant i \leqslant n)\,.$$

4.10. The construction in this section should be compared with [**Br**].

§5. Bosonization for the Virasoro algebra

Our account of the bosonization for the Virasoro algebra essentially follows the paper [**F**].

5.1. Heisenberg algebra. The *Heisenberg algebra* is the Lie algebra $\mathcal{H}$ defined by the generators b_n ($n \in \mathbb{Z}$) and $\mathbf{1}$, and relations

(a) $[b_n, b_m] = 2n\delta_{n+m,0}\mathbf{1}$, $[\mathbf{1}, b_n] = 0$ $(n,m \in \mathbb{Z})$.

The elements $\mathbf{1}$ and b_0 lie in the center of $\mathcal{H}$. The elements b_n, $n > 0$, are called *annihilation operators*.

We shall denote by $\mathcal{H}^+$ (resp., $\mathcal{H}^-$, $\mathcal{H}^0$) the Lie subalgebra of $\mathcal{H}$ generated by the elements b_n, $n > 0$ (resp., by b_n, $n < 0$, by b_0 and $\mathbf{1}$).

5.2. Given two generators b_n, b_m, their *normal ordered product* $:b_nb_m: \in \mathrm{U}\mathcal{H}$ is defined as follows. If $n > 0$ and $m \leqslant 0$, we set $:b_nb_m: = b_mb_n$; otherwise $:b_nb_m: = b_nb_m$. We define

$$[b_nb_m] = b_nb_m - :b_nb_m:$$

(not to be confused with the Lie bracket!).

Similarly, the normal ordering of an arbitrary monomial $:b_{n_1}\cdots b_{n_a}:$ is defined: one should pull all the annihilation operators to the right, not changing their order. (The last requirement does not matter very much since all the annihilation operators commute.)

We have by definition

$$b_nb_m = :b_nb_m: + [b_nb_m]\,.$$

The following identity generalizes this equality.

5.3. Wick theorem. *Let $c_1,\ldots,c_a$, $c'_1,\ldots,c'_b$ be arbitrary elements of the set $\{b_n,\ n\in\mathbb{Z}\}$. We have*

$$:c_1\cdots c_a::c'_1\cdots c'_b: = :c_1\cdots c'_b: + \sum_{p,q}[c_p c'_q]:c_1\cdots \hat{c}_p\cdots \hat{c}'_q\cdots c'_b:$$
$$+ \sum_{p,p',q,q'}[c_p c'_q][c_{p'} c'_{q'}]:c_1\cdots \hat{c}_p\cdots \hat{c}_{p'}\cdots \hat{c}'_q\cdots \hat{c}'_{q'}\cdots c'_b: + \cdots.\ \square$$

5.4. We define the generating function $\phi'(z)$ by

$$\phi'(z) = -\sum_{n\in\mathbb{Z}} b_n z^{-n-1}.$$

Here z is a formal variable.

The commutation relations 5.1(a) are equivalent to the identity

(a) $$[\phi'(z)\phi'(w)] = \frac{2}{(z-w)^2}.$$

Let us deduce (a) from 5.1(a). We have by definition

$$[\phi'(z)\phi'(w)] = \phi'(z)\phi'(w) - :\phi'(z)\phi'(w): = \sum_{n,m}(b_n b_m - :b_n b_m:)z^{-n-1}w^{-m-1}$$

$$= \sum_{n>0} 2n z^{-n-1}w^{n-1} = 2\partial_w\left(\sum_{n\geq 0} z^{-n-1}w^n\right) = 2z^{-1}\partial_w((1-w/z)^{-1})$$

$$= 2\partial_w\left(\frac{1}{z-w}\right) = \frac{2}{(z-w)^2}.$$

5.5. Given a number $\alpha\in\mathbb{C}$, the *Fock representation* F_α of the Heisenberg algebra $\mathcal{H}$ is the representation defined by one generator v_α and the relations

$$b_n v_\alpha = 0\ \ (n>0),\qquad b_0 v_\alpha = 2\alpha v_\alpha,\qquad \mathbf{1} v_\alpha = v_\alpha.$$

The mapping $x\mapsto x\cdot v_\alpha$ defines the isomorphism of $\mathcal{H}^-$-modules $U\mathcal{H}^- \xrightarrow{\sim} F_\alpha$.

The *shift operator*

$$T_\beta: F_\alpha \to F_{\alpha+\beta}$$

is the unique $\mathcal{H}^-$-linear operator sending v_α to $v_{\alpha+\beta}$.

The *category* $\mathcal{O}$ is the full subcategory of the category of $\mathcal{H}$-modules whose objects are representations M having the following properties:

(a) $\mathbf{1}$ acts as the identity on M.

(b) M is b_0-diagonalizable. All b_0-homogeneous components are finite dimensional.

(c) M is $\mathcal{H}^+$-locally finite. This means that for any $x\in M$, the space $U\mathcal{H}^+\cdot x$ is finite dimensional.

All Fock representations belong to the category $\mathcal{O}$.

5.6. The *Witt algebra* $\mathcal{L}$ is the Lie algebra of algebraic vector fields on $\mathbb{C}^* = \mathbb{C} - \{0\}$. It can be defined by the generators $e_n = -z^{n+1}\, d/dz$ ($n \in \mathbb{Z}$) and the relations
$$[e_n, e_m] = (n-m)e_{n+m} \qquad (n, m \in \mathbb{Z}).$$

The *Virasoro algebra* $\widehat{\mathcal{L}}$ is the Lie algebra defined by the generators L_n ($n \in \mathbb{Z}$) and c, and relations
$$[L_n, L_m] = (n-m)L_{n+m} + \frac{n^3 - n}{12}\delta_{n+m,0}\cdot c, \quad [c, L_n] = 0 \qquad (n, m \in \mathbb{Z}).$$

We have the morphism of Lie algebras $\widehat{\mathcal{L}} \to \mathcal{L}$ sending L_n to e_n, and c to 0. This morphism identifies $\mathcal{L}$ with $\widehat{\mathcal{L}}/\mathbb{C}\cdot c$. It identifies $\widehat{\mathcal{L}}$ with the universal central extension of $\mathcal{L}$.

Let $\widehat{\mathcal{L}}^+$ (resp., $\widehat{\mathcal{L}}^-$, $\widehat{\mathcal{L}}^0$) denote the Lie subalgebras of $\widehat{\mathcal{L}}$ generated by the elements L_n, $n > 0$ (resp., by L_n, $n < 0$, by L_0, c).

The generating function
$$T(z) = \sum_{n \in \mathbb{Z}} L_n z^{-n-2}$$
is called the *stress-energy tensor*.

5.7. Let $\alpha_0 \in \mathbb{C}$. Consider the expressions

(a) $$L_n = \frac{1}{4}\sum_{p+q=n} :b_p b_q: - \alpha_0(n+1)b_n \qquad (n \in \mathbb{Z}).$$

Although they are infinite sums, these expressions are well defined as operators on modules from the category $\mathcal{O}$. We can rewrite them as follows.

(b)
$$L_n = \frac{1}{4}\sum_{p \in \mathbb{Z}} b_{n-p} b_p - \alpha_0(n+1)b_n \quad \text{if } n \neq 0,$$
$$L_0 = \frac{1}{2}\sum_{p \geq 1} b_{-p} b_p + \frac{1}{4}b_0^2 - \alpha_0 b_0.$$

In terms of generating functions, (a) is read as

(c) $$T(z) = \frac{1}{4}:\phi'(z)^2: - \alpha_0 \phi''(z).$$

5.8. Theorem. *The expressions 5.7(a) define the action of the Virasoro algebra on modules from the category $\mathcal{O}$, with the central charge $1 - 24\alpha_0^2$.*

5.9. Lemma. *The operators L_n, 5.7(a), satisfy the following commutation relations:*

(a) $$[b_n, L_m] = n b_{n+m} + 2n(n-1)\alpha_0 \delta_{n,-m}.$$

PROOF. This can be checked easily using the definitions. Let us give an alternative proof, using some simple *chiral calculus*. We claim that (a) is equivalent to

(b) $$\phi'(z)T(w) = -\frac{4\alpha_0}{(z-w)^3} + \frac{\phi'(w)}{(z-w)^2} + \cdots.$$

Here (and below) dots denote the expression regular at $z = w$. Let us prove that (b) implies (a). According to the *Cauchy formula* of the chiral calculus, we have

$$[b_n, T(w)] = -\operatorname*{res}_{z=w} (z^n \phi'(z) T(w))$$

for any n. By the binomial formula,

$$\operatorname*{res}_{z=w} \frac{z^n}{(z-w)^a} = \binom{n}{a-1} w^{n-a+1}.$$

Therefore, (b) implies

(c) $$[b_n, T(w)] = 2n(n-1)\alpha_0 w^{n-2} - n\phi'(w) w^{n-1}$$

which is equivalent to (a).

Let us prove (b). We have, by the Wick theorem,

$$\phi'(z){:}\phi'(w)^2{:} = 2[\phi'(z)\phi'(w)]\phi'(w) + \cdots = \frac{4\phi'(w)}{(z-w)^2} + \cdots,$$

$$\phi'(z)\phi''(w) = \partial_w(\phi'(z)\phi'(w)) = \frac{4}{(z-w)^3} + \cdots.$$

This implies (b) and proves the lemma. □

5.10. Proof of Theorem 5.8. The Virasoro commutation relations can be verified directly, using the previous lemma and 5.7(b), by a tedious computation.

Let us give a proof using the chiral calculus. We must prove the following.

(a) $$T(z)T(w) = \frac{1-24\alpha_0^2}{2(z-w)^4} + \frac{2T(w)}{(z-w)^2} + \frac{T'(w)}{z-w} + \cdots.$$

Let us derive the Virasoro commutation relations from (a). By the Cauchy formula of the chiral calculus, we have

$$[L_n, T(w)] = \operatorname*{res}_{z=w} (z^{n+1} T(z) T(w)).$$

As in the proof of the previous lemma, (a) implies that

(b) $$[L_n, T(w)] = T'(w) w^{n+1} + 2(n+1)T(w) w^n + (1-24\alpha_0^2)\frac{n^3-n}{12} w^{n-2},$$

which is equivalent to the Virasoro commutation relations with the central charge $1 - 24\alpha_0^2$.

Let us prove (a). We have, by the Wick theorem,

$$:\phi'(z)^2::\phi'(w)^2: = 2\left[\phi'(z)\phi'(w)\right]^2 + 4\left[\phi'(z)\phi'(w)\right]:\phi'(z)\phi'(w):$$
$$= \frac{8}{(z-w)^4} + \frac{8}{(z-w)^2}:\phi'(w)^2: + \frac{8}{z-w}:\phi'(w)\phi''(w): + \cdots,$$
$$\phi''(z):\phi'(w)^2: = 2\left[\phi''(z)\phi'(w)\right]\phi'(w) + \cdots$$
$$= 2\partial_z([\phi'(z)\phi'(w)])\phi'(w) = -\frac{8}{(z-w)^3}\phi'(w) + \cdots,$$
$$:\phi'(z)^2:\phi''(w) = 2\left[\phi'(z)\phi''(w)\right]\phi'(z) = 2\partial_w([\phi'(z)\phi'(w)])\phi'(z)$$
$$= \frac{8}{(z-w)^3}\left(\phi'(w) + (z-w)\phi''(w) + \frac{(z-w)^2}{2}\phi'''(w) + \cdots\right)$$
$$= \frac{8}{(z-w)^3}\phi'(w) + \frac{8}{(z-w)^2}\phi''(w) + \frac{4}{z-w}\phi'''(w) + \cdots,$$
$$\phi''(z)\phi''(w) = [\phi''(z)\phi''(w)] + \cdots = \partial_z\partial_w([\phi'(z)\phi'(w)]) + \cdots$$
$$= -\frac{12}{(z-w)^4} + \cdots.$$

Therefore we obtain, after adding,

$$T(z)T(w) = \left(\frac{1}{4}:\phi'(z)^2: - \alpha_0\phi''(z)\right)\left(\frac{1}{4}:\phi'(w)^2: - \alpha_0\phi''(w)\right)$$
$$= \frac{1 - 24\alpha_0^2}{2(z-w)^4} + \frac{(1/2):\phi'(w)^2: - 2\alpha_0\phi''(w)}{(z-w)^2}$$
$$+ \frac{(1/2):\phi'(w)\phi''(w): - \alpha_0\phi'''(w)}{z-w} + \cdots$$
$$= \frac{1-24\alpha_0^2}{2(z-w)^4} + \frac{2T(w)}{(z-w)^2} + \frac{T'(w)}{z-w} + \cdots.$$

This proves (a) and the theorem. □

5.11. We define the representation $F_{\alpha;\alpha_0}$ of the Virasoro algebra as the $\mathcal{H}$-module F_α, regarded as an $\widehat{\mathcal{L}}$-module by means of the formulas in the previous theorem.

The representations F_{α,α_0} will be called *Feigin–Fuchs modules*.

§6. Bosonic vertex operators

The classical works on this subject are [**FF, TK**].

6.1. Let $\alpha, \beta \in \mathbb{C}$. Let us define the operators

$$V_n(\beta)\colon F_\alpha \to F_{\alpha+\beta} \quad (n \in \mathbb{Z})$$

by means of the generating function

$$V(\beta;z) = \sum_{n\in\mathbb{Z}} V_n(\beta)z^{-n}.$$

We set by definition
$$V(\beta; z) = T_\beta \!:\! \exp\left(-\beta \sum_{n \neq 0} \frac{b_n}{n} z^{-n}\right)\!:$$
$$= T_\beta \exp\left(-\beta \sum_{n<0} \frac{b_n}{n} z^{-n}\right) \exp\left(-\beta \sum_{n>0} \frac{b_n}{n} z^{-n}\right).$$

This expression is an operator acting as follows:
$$V(\beta; z): F_\alpha \to F_{\alpha+\beta}((z^{-1})).$$

These operators are called the *bosonic vertex operators*. We also define the operators
$$V_-(\beta; z) = \exp\left(-\beta \sum_{n<0} \frac{b_n}{n} z^{-n}\right): F_\alpha \to F_\alpha[[z]],$$
$$V_+(\beta; z) = \exp\left(-\beta \sum_{n>0} \frac{b_n}{n} z^{-n}\right): F_\alpha \to F_\alpha[z^{-1}].$$

Thus, $V(\beta; z) = T_\beta V_-(\beta; z) V_+(\beta; z)$.

6.2. LEMMA. *We have*
$$[b_n, V_-(\beta; z)] = \begin{cases} 2\beta z^n V_-(\beta; z) & \text{if } n > 0, \\ 0 & \text{otherwise,} \end{cases}$$
$$[b_n, V_+(\beta; z)] = \begin{cases} 2\beta z^n V_+(\beta; z) & \text{if } n < 0, \\ 0 & \text{otherwise.} \end{cases}$$

We leave the easy proof to the reader.

6.3. THEOREM. *For every $n \in \mathbb{Z}$, we have $[b_n, V(\beta; z)] = 2\beta z^n V(\beta; z)$.*

PROOF. Follows at once from the previous lemma. □

6.4. Let us give an alternative proof, which uses the chiral calculus. Let us introduce the expression
$$\widetilde{V}(\beta; z) = z^{2\beta\alpha} V(\beta; z).$$
At this point $z^{\beta\alpha}$ is a formal symbol. It will play its role in the sequel, when we start differentiating. More precisely, $\widetilde{V}(\beta; z)$ is the operator $V(\beta; z)$ considered as a section of a certain de Rham complex with a nontrivial connection.

Our formula is equivalent to $[b_n, \widetilde{V}(\beta; z)] = 2\beta \widetilde{V}(\beta; z)$. We must prove that

(a) $$\phi'(z) \widetilde{V}(\beta; w) = -\frac{2\beta}{z-w} \widetilde{V}(\beta; w) + \cdots.$$

Let us introduce the "*free bosonic field*"
$$\phi(z) = q - b_0 \log(z) + \sum_{n \neq 0} \frac{b_n}{n} z^{-n}.$$

Here q is an operator that commutes with the generators b_n by the formulas
$$[q, b_n] = 2\delta_{n,0} \mathbf{1}.$$

It follows that
$$[b_0, e^{-\beta q}] = 2\beta e^{-\beta q}.$$

We identify the operator $e^{-\beta q}$ with T_β. Thus,
$$\widetilde{V}(\beta;z) = {:}\exp(-\beta\phi(z)){:},$$
where the only nontrivial normal ordering with the operator q is defined as ${:}b_0 q{:} = qb_0$.

We have

(b) $$\phi(z)\phi(w) = 2\log(z-w) + {:}\phi(z)\phi(w){:}\,.$$

It follows that
$$\phi'(z)\phi(w) = \frac{2}{z-w} + \cdots,$$
hence, from the Wick theorem
$$\phi'(z){:}\phi(w)^n{:} = \frac{2n}{z-w}{:}\phi(w)^{n-1}{:} + \cdots$$
($n \geq 0$), so if $f(\phi)$ is any power series in ϕ,
$$\phi'(z){:}f(\phi(w)){:} = \frac{2}{z-w}{:}f'(\phi(w)){:} + \cdots.$$

In particular, we have
$$\phi'(z){:}\exp(-\beta\phi(w)){:} = -\frac{2\beta}{z-w}{:}\exp(-\beta\phi(w)){:} + \cdots,$$
which proves (a) and the theorem. $\square$

6.5. Let us regard the operators $V_n(\beta)$ as operators acting on Feigin–Fuchs modules
$$V_n(\beta)\colon F_{\alpha;\alpha_0} \to F_{\alpha+\beta;\alpha_0}.$$
Let us understand their generating function $V(\beta;z)$ in the same sense.

6.6. THEOREM. *For any $n \in \mathbb{Z}$, we have*
$$[L_n, V(\beta;z)] = \left(z^{n+1}\frac{d}{dz} + ((\beta^2 - 2\alpha_0\beta)(n+1) + 2\alpha\beta)z^n\right)V(\beta;z).$$

In other words,
$$[L_n, \widetilde{V}(\beta;z)] = \left(z^{n+1}\frac{d}{dz} + (\beta^2 - 2\alpha_0\beta)(n+1)z^n\right)\widetilde{V}(\beta;z).$$

PROOF. We must prove that

(a) $$T(z)\widetilde{V}(\beta;w) = \frac{\beta^2 - 2\alpha_0\beta}{(z-w)^2}\widetilde{V}(\beta;w) + \frac{1}{z-w}\widetilde{V}'(\beta;w) + \cdots.$$

Let us prove (a). We have, by the Wick theorem,
$${:}\phi'(z)^2{:}{:}\phi(w)^n{:} = \frac{4n(n-1)}{(z-w)^2}{:}\phi(w)^{n-2}{:} + \frac{4n}{z-w}{:}\phi'(w)\phi(w)^{n-1}{:} + \cdots$$
for any $n \geq 0$, hence
$${:}\phi'(z)^2{:}\widetilde{V}(\beta;w) = \frac{4\beta^2}{(z-w)^2}\widetilde{V}(\beta;w) + \frac{4}{z-w}\widetilde{V}'(\beta;w) + \cdots.$$

It follows from 6.4 (a) that

$$\phi''(z)\widetilde{V}(\beta;w) = \frac{2\beta}{(z-w)^2}\widetilde{V}(\beta;w) + \cdots.$$

Summing, we get (a). This proves the theorem. □

6.7. LEMMA. *We have*

$$V_+(\beta_1;z_1)V_-(\beta_2;z_2) = (1 - z_1^{-1}z_2)^{2\beta_1\beta_2}V_-(\beta_2;z_2)V_+(\beta_1;z_1)$$

(equality in $\mathrm{Hom}(F_\alpha, F_\alpha[[z_1^{-1}, z_2]])$*).*

Here we understand $(1 - z_1^{-1}z_2)^{2\beta_1\beta_2}$ as the formal power series

$$\exp\left(-2\beta_1\beta_2 \sum_{n>0} z_1^{-n}\frac{z_2^n}{n}\right).$$

REMARK. Note that the operator $V_-(\beta_2;z_2)V_+(\beta_1;z_1)$ belongs to the space $\mathrm{Hom}(F_\alpha, F_\alpha[z_1^{-1}, z_2])$.

PROOF. By Lemma 6.2, we have

$$V_+(\beta_1;z_1)b_{-n} = (b_{-n} - 2\beta_1 z_1^{-n})V_+(\beta_1;z_1)$$

for $n > 0$; hence

$$V_+(\beta_1;z_1)\exp\left(\beta_2 \frac{b_{-n}}{n} z_2^n\right) = \exp\left(\beta_2 \frac{b_{-n} - 2\beta_1 z_1^{-n}}{n} z_2^n\right)V^+(\beta_1;z_1)$$

$$= \exp\left(-2\beta_1\beta_2 \frac{1}{n} z_1^{-n} z_2^n\right)\exp\left(\beta_2 \frac{b_{-n}}{n} z_2^n\right)V_+(\beta_1;z_1).$$

Therefore,

$$V_+(\beta_1;z_1)V_-(\beta_2;z_2) = \exp\left(-2\beta_1\beta_2 \sum_{n>0} \frac{1}{n} z_1^{-n} z_2^n\right)V_-(\beta_2;z_2)V_+(\beta_1;z_1)$$

$$= \exp(2\beta_1\beta_2 \log(1 - z_1^{-1}z_2))V_-(\beta_2;z_2)V_+(\beta_1;z_1)$$

$$= (1 - z_1^{-1}z_2)^{2\beta_1\beta_2}V_-(\beta_2;z_2)V_+(\beta_1;z_1). □$$

6.8. Choose complex numbers $\alpha, \beta_1, \ldots, \beta_p$. Let us consider the generating function of all the compositions

$$F_\alpha \xrightarrow{V_{n_p}(\beta_p)} F_{\alpha+\beta_p} \to \cdots \xrightarrow{V_{n_1}(\beta_1)} F_{\alpha+\beta_p+\cdots+\beta_1};$$

we have obviously

$$\sum_{(n_1,\ldots,n_p)\in\mathbb{Z}^p} V_{n_1}(\beta_1)\cdots V_{n_p}(\beta_p) z_1^{-n_1}\cdots z_p^{-n_p} = V(\beta_1;z_1)\cdots V(\beta_p;z_p)$$

as a formal power series. The previous lemma shows that

(a) $$V(\beta_1;z_1)\cdots V(\beta_p;z_p) = \prod_{1\leq i<j\leq p}(1 - z_i^{-1}z_j)^{2\beta_i\beta_j} :V(\beta_1;z_1)\cdots V(\beta_p;z_p): .$$

6.9. Let us look more attentively at the operator $:V(\beta_1;z_1)\cdots V(\beta_p;z_p):$. We have

$$:V(\beta_1;z_1)\cdots V(\beta_p;z_p): = T_{\beta_1+\cdots+\beta_p}\exp\left(-\sum_{n<0}\frac{\beta_1 z_1^{-n}+\cdots+\beta_p z_p^{-n}}{n}b_n\right)$$
$$\times \exp\left(-\sum_{n>0}\frac{\beta_1 z_1^{-n}+\cdots+\beta_p z_p^{-n}}{n}b_n\right).$$

It follows that

(a) the operator $:V(\beta_1;z_1)\cdots V(\beta_p;z_p):$ belongs to the space

$$\operatorname{Hom}(F_\alpha, F_{\alpha+\sum\beta_i}[z_1,z_1^{-1},\ldots,z_p,z_p^{-1}]).$$

In particular, this operator is a holomorphic operator-valued function on the space $\mathbb{C}^p - \bigcup_{i=1}^p\{z_i=0\}$. Also, the above formula shows that

(b) for any bijection $\sigma\colon\{1,\ldots,p\}\xrightarrow{\sim}\{1,\ldots,p\}$, we have

$$:V(\beta_1;z_1)\cdots V(\beta_p;z_p): = :V(\beta_{\sigma(1)};z_{\sigma(1)})\cdots V(\beta_{\sigma(p)}z_{\sigma(p)}):.$$

6.10. Formula 6.8 (a) shows that the formal power series $V(\beta_1;z_1)\cdots V(\beta_p;z_p)$ defines the germ of a holomorphic multivalued function in the domain $|z_1|>\cdots>|z_p|>0$, where $(1-z_i^{-1}z_j)^{2\beta_i\beta_j}$ is understood as $\exp(-2\beta_i\beta_j\sum_{n>0}z_i^{-n}z_j^n/n)$.

6.11. Let us introduce the tilded operators. We set

$$\widetilde{V}(\beta_i;z_i) = :\exp(-\beta_i\phi(z_i)): = T_{\beta_i}z_i^{\beta_i b_0}:\exp\left(-\beta_i\sum_{n\neq 0}\frac{b_n}{n}z_i^{-n}\right):.$$

Since $z_i^{\beta_i b_0}T_{\beta_j} = z^{2\beta_i\beta_j}T_{\beta_j}z_i^{\beta_i b_0}$, we have

$$\widetilde{V}(\beta_1;z_1)\cdots\widetilde{V}(\beta_p;z_p) = T_{\sum\beta_i}\prod_i z_i^{\beta_i b_0}\prod_{i<j}z_i^{2\beta_j\beta_j}V(\beta_1;z_1)\cdots V(\beta_p;z_p)$$
$$= \prod_i z_i^{2\beta_i\alpha}\prod_{i<j}(z_i-z_j)^{2\beta_i\beta_j}:V(\beta_1;z_1)\cdots V(\beta_p;z_p):.$$

Recall that the operator $:V(\beta_1;z_1)\cdots V(\beta_p;z_p):$ belongs to the space

$$\operatorname{Hom}(F_\alpha, F_{\alpha+\sum\beta_i}[z_1,z_1^{-1},\ldots,z_p,z_p^{-1}]).$$

This defines the product $\widetilde{V}(\beta_1;z_1)\cdots\widetilde{V}(\beta_p;z_p)$ as the germ of the multivalued holomorphic function in the domain

$$\{(z_1,\ldots,z_p)\in\mathbb{C}^p\mid z_i\notin\mathbb{R}_{\leq 0}\text{ for all }i; |z_1|>\cdots>|z_p|\}.$$

Here z^γ is understood as $\exp(\gamma\log(z))$, where $\log(z)$ is the branch of the logarithm taking real values for $z\in\mathbb{R}_{>0}$.

Let us regard the previous compositions as operators acting on the Feigin–Fuchs modules

$$\widetilde{V}(\beta_1;z_1)\cdots\widetilde{V}_p(\beta_p;z_p)\colon F_{\alpha;\alpha_0}\to F_{\alpha+\sum\beta_i;\alpha_0}.$$

6.12. THEOREM. *For any $n \in \mathbb{Z}$,*

$$[L_n, \widetilde{V}(\beta_1; z_1) \cdots \widetilde{V}(\beta_p; z_p)]$$
$$= \left(\sum_{i=1}^{p} z_i^{n+1} \partial_{z_i} + (\beta_i^2 - 2\alpha_0 \beta_i)(n+1) z_i^n \right) \widetilde{V}(\beta_1; z_1) \cdots \widetilde{V}(\beta_p; z_p).$$

In other words,

$$[L_n, :V(\beta_1, z_1) \cdots V(\beta_p; z_p):]$$
$$= \left(\sum_{i=1}^{p} (z_i^{n+1} \partial_{z_i} + ((\beta_i^2 - 2\alpha_0 \beta_i)(n+1) + 2\beta_i \alpha) z_i^n) \right.$$
$$\left. + \sum_{1 \leq i < j \leq p} 2\beta_i \beta_j \frac{z_i^{n+1} - z_j^{n+1}}{z_i - z_j} \right) :V(\beta_1; z_1) \cdots V(\beta_p; z_p): .$$

PROOF. By the Cauchy formula,

$$[L_n, \widetilde{V}(\beta_1; z_1) \cdots \widetilde{V}(\beta_p; z_p)] = \sum_{i=1}^{p} \underset{z=z_i}{\mathrm{res}} \left(z^{n+1} T(z) \widetilde{V}(\beta_1; z_1) \cdots \widetilde{V}_p(\beta_p; z_p) \right).$$

Note that

$$\widetilde{V}(\beta_1; z_1) \cdots \widetilde{V}(\beta_p; z_p) = \prod_{i<j} (z_i - z_j)^{2\beta_i \beta_j} :\widetilde{V}(\beta_1; z_1) \cdots \widetilde{V}(\beta_p; z_p): .$$

We claim that

(a) $T(z) :\widetilde{V}(\beta_1; z_1) \cdots \widetilde{V}(\beta_p; z_p):$
$$= \left(\sum_{i=1}^{p} \frac{\beta_i^2 - 2\alpha_0 \beta_i}{(z - z_i)^2} + \sum_{i<j} \frac{2\beta_i \beta_j}{(z - z_i)(z - z_j)} \right) :\widetilde{V}(\beta_1; z_1) \cdots \widetilde{V}(\beta_p; z_p):$$
$$+ \sum_{i=1}^{p} \frac{1}{z - z_i} :\widetilde{V}(\beta_1; z_1) \cdots \widetilde{V}'(\beta_i; z_i) \cdots \widetilde{V}(\beta_p; z_p): + \cdots .$$

Let us prove (a). To shorten the formulas, assume that $p = 2$; the general case is completely similar. By the Wick theorem, we have

$$:\phi'(z)^2: :\phi(z_1)^{n_1} \phi(z_2)^{n_2}:$$
$$= \frac{4n_1(n_1 - 1)}{(z - z_1)^2} :\phi(z_1)^{n_1 - 2} \phi(z_2)^{n_2}: + \frac{4n_2(n_2 - 1)}{(z - z_2)^2} :\phi(z_1)^{n_1} \phi(z_2)^{n_2 - 2}:$$
$$+ \frac{8n_1 n_2}{(z - z_1)(z - z_2)} :\phi(z_1)^{n_1 - 1} \phi(z_2)^{n_2 - 1}: + \frac{4n_1}{z - z_1} :\phi'(z_1) \phi(z_1)^{n_1 - 1} \phi(z_2)^{n_2}:$$
$$+ \frac{4n_2}{z - z_2} :\phi(z_1)^{n_1} \phi'(z_2) \phi(z_2)^{n_2 - 1}: + \cdots$$

for any $n_1, n_2 \geq 0$; hence

$$:\phi'(z)^2: :\widetilde{V}(\beta_1; z_1) \widetilde{V}(\beta_2; z_2): = 4 \left(\frac{\beta_1}{z - z_1} + \frac{\beta_2}{z - z_2} \right)^2 :\widetilde{V}(\beta_1; z_1) \widetilde{V}(\beta_2; z_2):$$
$$+ 4 \left(\frac{\partial_{z_1}}{z - z_1} + \frac{\partial_{z_2}}{z - z_2} \right) :\widetilde{V}(\beta_1; z_1) \widetilde{V}(\beta_2; z_2): + \cdots .$$

Similarly, the application of the Wick theorem gives

$$\phi'(z){:}\widetilde{V}(\beta_1;z_1)\widetilde{V}(\beta_2;z_2){:} = \left(-\frac{2\beta_1}{z-z_1} - \frac{2\beta_2}{z-z_2}\right){:}\widetilde{V}(\beta_1;z_1)\widetilde{V}(\beta_2;z_2){:} + \cdots,$$

whence

$$\phi''(z){:}\widetilde{V}(\beta_1;z_1)\widetilde{V}(\beta_2;z_2){:} = \left(\frac{2\beta_1}{(z-z_1)^2} + \frac{2\beta_2}{(z-z_2)^2}\right){:}\widetilde{V}(\beta_1;z_1)\widetilde{V}(\beta_2;z_2){:} + \cdots.$$

Summing, we get formula (a).

The theorem follows from formula (a) by the above mentioned Cauchy residue formula. One should take into account that

$$\left(\operatorname*{res}_{z=z_i} + \operatorname*{res}_{z=z_j}\right) \frac{z^{n+1}}{(z-z_i)(z-z_j)} = \frac{z_i^{n+1} - z_j^{n+1}}{z_i - z_j}$$

and

$$(z_i^{n+1}\partial_{z_i} + z_j^{n+1}\partial_{z_j})(z_i - z_j)^{2\beta_i\beta_j} = 2\beta_i\beta_j \frac{z_i^{n+1} - z_j^{n+1}}{z_i - z_j}(z_i - z_j)^{2\beta_i\beta_j}.$$

This completes the proof. $\square$

§7. Screening charges

7.1. We fix the parameters $\alpha, \alpha_0 \in \mathbb{C}$ as in the previous section. Let β_+, β_- be the complex numbers defined by

$$\beta_\pm = \alpha_0 \pm \sqrt{\alpha_0 + 1}.$$

Thus, $\beta_\pm$ are the two roots of the equation $\beta^2 - 2\alpha_0\beta = 1$.

The vertex operators $V(\beta_+;z)$, $V(\beta_-;z)$ are called *screening charges*. By theorem 6.6, for any $n \in \mathbb{Z}$,

(a) $$[L_n, V(\beta_\pm;z)] = \left(\frac{d}{dz} + \frac{2\alpha\beta_\pm}{z}\right)(z^{n+1}V(\beta_\pm;z)).$$

In other words,

(b) $$T(z)\widetilde{V}(\beta_\pm;z) = \frac{\widetilde{V}(\beta_\pm;w)}{(z-w)^2} + \frac{\widetilde{V}'(\beta_\pm;w)}{z-w} + \cdots = \partial_w\left(\frac{\widetilde{V}(\beta_\pm;w)}{z-w}\right) + \cdots.$$

7.2. Let us introduce the operators

$$T(L_n;z) = z^{n+1}T(z), \quad V_\pm(e_n;z) = z^n\widetilde{V}(\beta_\pm;z) \quad (L_n \in \widehat{\mathcal{L}}, \ e_n \in \mathcal{L}).$$

It follows from 7.1(b) that these operators satisfy the following cocycle property

(a) $$T(L_n;z)V_\pm(e_m;w) - T(L_m;z)V_\pm(e_n;w) = \frac{V_\pm([e_n,e_m];w)}{z-w} + \cdots.$$

7.3. More generally, suppose that $\beta_1, \ldots, \beta_p$ is a sequence of complex numbers, each β_i being equal to β_- or β_+. Let us call such a sequence a *screening sequence*.

Let us consider the operator $:V(\beta_1;z_1)\cdots V(\beta_p;z_p):$. Note that by the symmetry property 6.9(b), we may actually assume that $\beta_1 = \cdots = \beta_{p'} = \beta_-$ and $\beta_{p'+1} = \cdots = \beta_p = \beta_+$.

It follows from Theorem 6.12, that

$$[L_n, :V(\beta_1;z_1)\cdots V(\beta_p;z_p):]$$
$$= \sum_{i=1}^{p}\left(\partial_{z_i} + \frac{2\alpha\beta_i}{z_i} + \sum_{j\neq i}\frac{2\beta_i\beta_j}{z_i-z_j}\right)(z_i^{n+1}:V(\beta_1;z_1)\cdots V(\beta_p;z_p):)$$

for any screening sequence $\beta_1,\ldots,\beta_p$, and for all $n\in\mathbb{Z}$.

In the sequel we fix the screening sequence $\beta_1 = \cdots = \beta_p = \beta_+$. However, all the constructions below are valid, with the obvious modifications, for an arbitrary screening sequence.

7.4. Let us consider the ring

$$A_p = \mathbb{C}[[z_1,\ldots,z_p]]\left[\prod_{i=1}^{p}z_i^{-1}, \prod_{1\leqslant i<j\leqslant p}(z_i-z_j)^{-1}\right]$$

(thus A_p is the ring of functions on the formal variety X_p, the pth power of the formal punctured disk without diagonals).

For $1\leqslant a\leqslant p$, let Ω^a denote the space of algebraic differential a-forms on X_p. Thus, $\Omega^0 = A_p$, and the elements of Ω^a have the form

$$\sum f(z_1,\ldots,z_p)\,dz_{i_1}\wedge\cdots\wedge dz_{i_a} \qquad (f(z_1,\ldots,z_p)\in A_p).$$

Let us consider the complex

$$\Omega^\cdot : 0 \to \Omega^0 \xrightarrow{d} \Omega^1 \xrightarrow{d} \cdots \xrightarrow{d} \Omega^p \to 0,$$

where the differential is defined by

$$d\eta = d_{DR}(\eta) + \left(\sum_{i=1}^{p}2\alpha\beta_+\frac{dz_i}{z_i} + \sum_{1\leqslant i<j\leqslant p}2\beta_+^2\frac{dz_i-dz_j}{z_i-z_j}\right)\wedge\eta$$

($\eta\in\Omega^a$), d_{DR} being the usual de Rham differential.

Each vector field $\tau = \mu(z)\partial_z \in \mathcal{L}$, $\mu(z)\in\mathbb{C}[z,z^{-1}]$, defines a series of morphisms $i_\tau : \Omega^a \to \Omega^{a-1}$, by

$$i_\tau(f(z_1,\ldots,z_p)\,dz_{i_1}\wedge\cdots\wedge dz_{i_a})$$
$$= \sum_{b=1}^{a}(-1)^{b-1}\mu(z_{i_b})f(z_1,\ldots,z_p)\,dz_{i_1}\wedge\cdots\wedge\hat{dz}_{i_b}\wedge\cdots\wedge dz_{i_a}.$$

Let us define the morphisms of the *Lie derivative* $\mathrm{Lie}_\tau : \Omega^a \to \Omega^a$ by

$$\mathrm{Lie}_\tau(\eta) = di_\tau(\eta) + i_\tau(d\eta).$$

The morphisms $i_\tau, \mathrm{Lie}_\tau$ define the action of the dg Lie algebra $\mathcal{L}^\cdot$ associated with the Witt algebra $\mathcal{L}$ (cf. 4.2), on the complex $\Omega^\cdot$ (cf. 4.2).

7.5. Let us consider the complex $\mathrm{Hom}(F_{\alpha;\alpha_0}, F_{\alpha+p\beta_+;\alpha_0}\otimes\Omega^\cdot)$.

Let us note that the action of the Virasoro algebra $\widehat{\mathcal{L}}$ on the modules $F_{\alpha;\alpha_0}$, $F_{\alpha+p\beta_+;\alpha}$ induces the action of $\widehat{\mathcal{L}}$ on the space $\mathrm{Hom}(F_{\alpha;\alpha_0}, F_{\alpha+p\beta_+;\alpha_0})$ by the usual commutator formula, which factors through $\mathcal{L}$, since the Feigin–Fuchs modules have the same central charge. This in turn induces the action of $\mathcal{L}$ on the complex $\mathrm{Hom}(F_{\alpha;\alpha_0}, F_{\alpha+p\beta_+;\alpha_0}\otimes\Omega^\cdot)$, which we shall call the *first action*. We have also the *second action* of $\mathcal{L}$ on $\mathrm{Hom}(F_{\alpha;\alpha_0}, F_{\alpha+p\beta_+;\alpha_0} \otimes \Omega^{\cdot})$ which is induced by the action of $\mathcal{L}$ on $\Omega^{\cdot}$ through the Lie derivative. The first and second actions commute.

We also have the operators

$$i_\tau \colon \mathrm{Hom}(F_{\alpha;\alpha_0}, F_{\alpha+p\beta_+;\alpha_0} \otimes \Omega^{\cdot}) \to \mathrm{Hom}(F_{\alpha;\alpha_0}, F_{\alpha+p\beta_+;\alpha_0} \otimes \Omega^{\cdot})[-1] \qquad (\tau \in \mathcal{L})$$

induced by the operators of the same name on $\Omega^{\cdot}$. These operators, together with the second action, define the action of the dg Lie algebra $\mathcal{L}^{\cdot}$ on the complex $\mathrm{Hom}(F_{\alpha;\alpha_0}, F_{\alpha+p\beta_+;\alpha_0} \otimes \Omega^{\cdot})$.

Let us define the element $V^{0p} = V^{0p}(z_1, \ldots, z_p) \in \mathrm{Hom}(F_{\alpha;\alpha_0}, F_{\alpha+p\beta_+;\alpha_0} \otimes \Omega^p)$ by

$$V^{0p}(z_1, \ldots, z_p) = {:}V(\beta_+; z_1) \cdots V(\beta_+; z_p){:}\, dz_1 \wedge \cdots \wedge dz_p\,.$$

The formula in 7.3 may be reformulated as the following theorem.

7.6. THEOREM. *The element V^{0p} lies in the subspace of $\mathcal{L}$-invariants*

$$\mathrm{Hom}(F_{\alpha;\alpha_0}, F_{\alpha+p\beta_+;\alpha_0} \otimes \Omega^p)^{\mathcal{L}}.$$

Here the action of the Lie algebra $\mathcal{L}$ on the space $\mathrm{Hom}(F_{\alpha;\alpha_0}, F_{\alpha+p\beta_+;\alpha_0} \otimes \Omega^p)$ is the sum of the first and the second actions. $\square$

7.7. Now let us apply the construction of §4. Consider the double complex

$$C^{\cdot}(\mathcal{L}; \mathrm{Hom}(F_{\alpha;\alpha_0}, F_{\alpha+p\beta_+;\alpha_0} \otimes \Omega^{\cdot}))\,.$$

Here the first differential is the Koszul differential of the cochain complex of the Lie algebra $\mathcal{L}$ with coefficients in the module $\mathrm{Hom}(F_{\alpha;\alpha_0}, F_{\alpha+p\beta_+;\alpha_0} \otimes \Omega^{\cdot})$, with the first $\mathcal{L}$-action. The second differential is induced by the differential in $\Omega^{\cdot}$.

Define the elements

$$V^{a,p-a} \in C^a(\mathcal{L}; \mathrm{Hom}(F_{\alpha;\alpha_0}, F_{\alpha+p\beta_+;\alpha_0} \otimes \Omega^{p-a})) \qquad (0 \leqslant a \leqslant p)$$

by

$$V^{a,p-a}(\tau_1 \wedge \cdots \wedge \tau_a) = i_{\tau_1} \cdots i_{\tau_a}(V^{0p})\,.$$

7.8. THEOREM. *The element $V = (V^{0p}, \ldots, V^{p0})$ is a p-cocycle in the total complex associated with the double complex $C^{\cdot}(\mathcal{L}; \mathrm{Hom}(F_{\alpha;\alpha_0}, F_{\alpha+p\beta_+;\alpha_0} \otimes \Omega^{\cdot}))$.*

The proof is the same as that of Lemma 4.8.

7.9. COROLLARY. *The cocycle V induces the maps*

$$f_i \colon H^i(\Omega^{\cdot})^* \to \mathrm{Ext}^{p-i}_{\widehat{\mathcal{L}}}(F_{\alpha;\alpha_0}, F_{\alpha+p\beta_+;\alpha_0}), \qquad i = 0, \ldots, p\,. \square$$

The map

$$f_p \colon H^p(\mathcal{L})^* \to \mathrm{Hom}_{\widehat{\mathcal{L}}}(F_{\alpha;\alpha_0}, F_{\alpha+p\beta_+;\alpha_0})$$

is the Feigin–Fuchs intertwiner, [**FF**, Chapter 4].

Chapter 3. $\widehat{\mathfrak{sl}}(2)$

The main construction of this and the following chapters was inspired by [**BMP1, BMP2**].

§8. Wakimoto realization

8.1. Let $\mathfrak{g}$ be a Lie algebra and $B(\,\cdot\,,\,\cdot\,)$ an invariant bilinear form on $\mathfrak{g}$. The corresponding *affine Lie algebra* $\widehat{\mathfrak{g}}$ is defined by the generators X_n ($X \in \mathfrak{g}$, $n \in \mathbb{Z}$) and $\mathbf{1}$, and the relations

(a) $\quad [X_n, X_m] = [X, Y]_{m+n} + nB(X,Y)\delta_{m+n,0} \cdot \mathbf{1} \qquad (X, Y \in \mathfrak{g},\ m, n \in \mathbb{Z})$.

The element $\mathbf{1}$ will act as the identity on all our representations.

Let us introduce the generating functions (*currents*) $X(z) = \sum_{n \in \mathbb{Z}} X_n z^{-n-1}$ ($X \in \mathfrak{g}$). Formula (a) is equivalent to

(b) $\qquad X(z)Y(w) = \dfrac{B(X,Y)}{(z-w)^2} + \dfrac{[X,Y](w)}{z-w} + \cdots$.

Here as usual the dots stands for the part regular at $z = w$. One deduces (a) from (b) at once, using the chiral Cauchy formula

$$[X_n, Y(w)] = \operatorname*{res}_{z=w} (z^n X(z) Y(w)).$$

8.2. In this chapter we assume that $\mathfrak{g} = \mathfrak{sl}(2)$ with the standard generators E, F, H. For $X, Y \in \mathfrak{g}$, we set $(X, Y) = \operatorname{tr}(XY)$. Thus, $(E, F) = (F, E) = 1$, $(H, H) = 2$. We fix a complex number k, and set $B(X, Y) = k(X, Y)$.

The bosonization formulas for the algebra $\widehat{\mathfrak{g}}$ presented below were discovered by Wakimoto, [**W**].

8.3. Let $\mathbf{a}$ denote the Lie algebra defined by the generators b_n, a_n, a_n^* ($n \in \mathbb{Z}$), $\mathbf{1}$ and the relations

(a) $[b_n, b_m] = 2n\delta_{n+m,0} \cdot \mathbf{1}$;

(b) $[a_n, a_m^*] = \delta_{n+m,0} \cdot \mathbf{1}$, $[a_n, a_m] = [a_n^*, a_m^*] = 0$;

(c) $[b_n, a_m] = [b_n, a_m^*] = 0$, $\mathbf{1}$ commutes with everything.

Let $\mathbf{a}_+$ denote the Lie subalgebra of $\mathbf{a}$ generated by the elements b_n, a_n^* ($n > 0$), a_n ($n \geqslant 0$). These generators are called *annihilation operators*. One introduces the *normal ordering* of a monomial in $U\mathbf{a}$ in the usual way: all annihilation operators should be pulled to the right.

All $\mathbf{a}$-modules M that we consider be $\mathbf{a}_+$-*locally finite*, i.e., will have the property that for every $x \in M$, the space $U\mathbf{a}_+ x$ is finite-dimensional. The operator $\mathbf{1}$ will act as the identity.

For computational purposes, we shall also use one more operator q, with the only nontrivial commutation relation

$$[q, b_0] = \mathbf{1}.$$

8.4. For $\lambda \in \mathbb{C}$, let F_λ denote the $\mathbf{a}$-module defined by one generator v_λ and relations $\mathbf{a}_+ v_\lambda = 0$, $b_0 v_\lambda = 2\lambda v_\lambda$, $\mathbf{1} v_\lambda = v_\lambda$.

8.5. Let us introduce the generating functions

$$\phi(z) = q - b_0 \log(z) + \sum_{n \neq 0} \frac{b_n}{n} z^{-n}, \qquad p(z) = \phi'(z) = -\sum_n b_n z^{-n-1},$$

$$\beta(z) = \sum_n a_n z^{-n-1}, \qquad \gamma(z) = \sum_n a_n^* z^{-n}.$$

We have

$$\phi(z)\phi(w) = 2\log(z-w) + \cdots, \qquad p(z)\phi(w) = \frac{2}{z-w} + \cdots,$$

$$p(z)p(w) = \frac{2}{(z-w)^2} + \cdots, \qquad \gamma(z)\beta(w) = \frac{1}{z-w} + \cdots,$$

all other products being trivial (not having a singular part).

8.6. Let us define the currents

(a) $\qquad E(z) = \beta(z),$

(b) $\qquad H(z) = 2{:}\gamma(z)\beta(z){:} + \nu p(z),$

(c) $\qquad F(z) = -{:}\gamma(z)^2\beta(z){:} - \nu{:}\gamma(z)p(z){:} - k\gamma'(z),$

where $\nu^2 = k+2$.

8.7. THEOREM. *The previous formulas define the bosonization for $\hat{\mathfrak{g}}$, i.e., the Fourier components of the currents $E(z)$, $H(z)$, $F(z)$ satisfy the commutation relations of $\hat{\mathfrak{g}}$.*

PROOF. We must check the relations (a)–(f) below.

(a) $\qquad H(z)H(w) = \dfrac{2k}{(z-w)^2} + \cdots.$

Indeed, we have (using the Wick theorem)

$$H(z)H(w) = (2{:}\gamma(z)\beta(z){:} + \nu p(z))(2{:}\gamma(w)\beta(w){:} + \nu p(w))$$
$$= 4{:}\gamma(z)\beta(z){:}{:}\gamma(w)\beta(w){:} + \nu^2 p(z)p(w) + \cdots$$

(the terms of the first order cancel out)

$$= 4[\gamma(z)\beta(w)][\beta(z)\gamma(w)] + \nu^2 p(z)p(w) + \cdots$$
$$= \frac{-4+2\nu^2}{(z-w)^2} + \cdots = \frac{2k}{(z-w)^2} + \cdots.$$

(b) $\qquad H(z)E(w) = \dfrac{2E(w)}{z-w} + \cdots.$

Indeed,

$$H(z)E(w) = (2{:}\gamma(z)\beta(z){:} + \nu p(z))\beta(w) = 2[\gamma(z)\beta(w)]\beta(w) + \cdots = \frac{2\beta(w)}{z-w} + \cdots.$$

(c) $\qquad H(z)F(w) = -\dfrac{2F(w)}{(z-w)} + \cdots.$

Indeed,

$$H(z)F(w) = -(2\!:\!\gamma(z)\beta(z)\!:\! + \nu p(z))(:\!\gamma(w)^2\beta(w)\!:\! + \nu\!:\!\gamma(w)p(w)\!:\! + k\gamma'(w)).$$

Let us compute all the nontrivial products. We have

$$:\!\gamma(z)\beta(z)\!::\!\gamma(w)^2\beta(w)\!: = -\frac{2}{(z-w)^2}\,\gamma(w) - \frac{1}{z-w}:\!\beta(w)\gamma(w)^2\!:\! + \cdots,$$

$$:\!\gamma(z)\beta(z)\!::\!\gamma(w)p(w)\!: = -\frac{1}{z-w}:\!\gamma(w)p(w)\!:\! + \cdots,$$

$$:\!\gamma(z)\beta(z)\!:\!\gamma'(z) = -\frac{1}{(z-w)^2}\,\gamma(z) + \cdots$$

$$= -\frac{1}{(z-w)^2}\,\gamma(w) - \frac{1}{z-w}\,\gamma'(w) + \cdots,$$

$$p(z)\!:\!\gamma(w)p(w)\!: = \frac{2}{(z-w)^2}\,\gamma(w) + \cdots.$$

Summing up, we get (c).

(d) $$E(z)E(w) = 0 + \cdots.$$

This is obvious.

(e) $$E(z)F(w) = \frac{k}{(z-w)^2} + \frac{H(w)}{z-w} + \cdots.$$

Indeed,

$$E(z)F(w) = -\beta(z)(:\!\gamma(w)^2\beta(w)\!:\! + \nu\!:\!\gamma(w)p(w)\!:\! + k\gamma'(w))$$

$$= \frac{2}{z-w}:\!\gamma(w)\beta(w)\!:\! + \frac{\nu}{z-w}p(w) + \frac{k}{(z-w)^2} + \cdots$$

$$= \frac{k}{(z-w)^2} + \frac{H(w)}{z-w} + \cdots.$$

(f) $$F(z)F(w) = 0 + \cdots.$$

Indeed,

$$F(z)F(w) = (:\!\gamma(z)^2\beta(z)\!:\! + \nu\!:\!\gamma(z)p(z)\!:\! + k\gamma'(z))$$
$$\times (:\!\gamma(w)^2\beta(w)\!:\! + \nu\!:\!\gamma(w)p(w)\!:\! + k\gamma'(w)).$$

Let us compute the nontrivial products. We have

$$:\!\gamma(z)^2\beta(z)\!::\!\gamma(w)^2\beta(w)\!: = -\frac{4}{(z-w)^2}:\!\gamma(z)\gamma(w)\!:\! + \frac{2}{z-w}:\!\gamma(z)\beta(z)\gamma(w)^2\!:$$

$$- \frac{2}{z-w}:\!\gamma(z)^2\gamma(w)\beta(w)\!:\! + \cdots$$

$$= -\frac{4}{(z-w)^2}:\!\gamma(w)^2\!:\! - \frac{4}{z-w}:\!\gamma(w)\gamma'(w)\!:\! + \cdots,$$

$$:\!\gamma(z)p(z)\!::\!\gamma(w)p(w)\!: = \frac{2}{(z-w)^2}:\!\gamma(z)\gamma(w)\!:\! + \cdots$$

$$= \frac{2}{(z-w)^2}:\!\gamma(w)^2\!:\! + \frac{2}{z-w}:\!\gamma(w)\gamma'(w)\!:\! + \cdots,$$

$$:\gamma(z)^2\beta(z)::\gamma(w)p(w): = -\frac{1}{z-w} :\gamma(w)^2 p(w): + \cdots ,$$

$$:\gamma(z)p(z)::\gamma(w)^2\beta(w): = \frac{1}{z-w} :p(w)\gamma(w)^2: + \cdots ,$$

$$:\gamma(z)^2\beta(z):\gamma'(w) = -\frac{1}{(z-w)^2} :\gamma(z)^2: + \cdots$$

$$= -\frac{1}{(z-w)^2} :\gamma(w)^2: - \frac{2}{z-w} :\gamma'(w)\gamma(w): + \cdots ,$$

$$\gamma'(z):\gamma(w)^2\beta(w): = -\frac{1}{(z-w)^2} :\gamma(w)^2: + \cdots .$$

After summing, we get a zero singular part. This proves (f) and completes the proof of the theorem. $\square$

§9. Screening current

9.1. We keep the assumptions of the previous section. In this section we assume that $\nu \neq 0$ (the level is noncritical).

For $\chi \in \mathbb{C}$, we shall denote by $W_{\chi;\nu}$ the $\hat{\mathfrak{g}}$-module $F_{-\chi/2\nu}$, the $\hat{\mathfrak{g}}$-module structure being defined by formulas 8.6(a)–(c). Such $\hat{\mathfrak{g}}$-modules are called *Wakimoto modules*.

9.2. For $\alpha \in \mathbb{C}$, define the operator

$$V(\alpha; z) = :\exp(-\alpha\phi(z)): = T_\alpha z^{\alpha b_0} \exp\left(-\alpha \sum_{n<0} \frac{b_n}{n} z^{-n}\right) \exp\left(-\alpha \sum_{n>0} \frac{b_n}{n} z^{-n}\right)$$

acting from F_λ to $F_{\lambda+\alpha} \otimes z^{2\alpha\lambda}\mathbb{C}((z^{-1}))$.

9.3. LEMMA. *We have*

$$F(z)(-\beta(w)V(\alpha;w)) = -\frac{2\alpha\nu + k}{(z-w)^2} V(\alpha;w) + \frac{2-2\alpha\nu}{z-w} :\gamma(w)\beta(w):V(\alpha;w):$$
$$+ \frac{\nu}{z-w} :p(w)V(\alpha;w): + \cdots .$$

PROOF. The right-hand side is equal to

$$(:\gamma(z)^2\beta(z): + \nu :\gamma(z)p(z): + k\gamma'(z))(\beta(w)V(\alpha;w)) .$$

Let us compute all the products using the Wick theorem. We have

$$:\gamma(z)^2\beta(z):\beta(w)V(\alpha;w) = \frac{2}{z-w} :\gamma(w)\beta(w):V(\alpha;w) + \cdots ;$$

recalling that by 6.4(a)

$$p(z)V(\alpha;w) = -\frac{2\alpha}{z-w} V(\alpha;w) + \cdots ,$$

we have

$$:\gamma(z)p(z):\beta(w)V(\alpha;w) = -\frac{2\alpha}{(z-w)^2} V(\alpha;w) + \frac{1}{z-w} :p(w)V(\alpha;w):$$
$$- \frac{2\alpha}{z-w} :\gamma(w)\beta(w):V(\alpha;w) + \cdots .$$

Finally,
$$\gamma'(z)\beta(w)V(\alpha;w) = -\frac{1}{(z-w)^2}V(\alpha;w) + \cdots.$$
By summing all up, we get the statement of the lemma. $\square$

9.4. Let us introduce the operators called *screening currents* by
$$S(z) = -\beta(z)V(\nu^{-1};z)\colon W_{\chi;\nu} \to W_{\chi-2;\nu} \otimes z^{-\chi/\nu^2}\mathbb{C}((z^{-1})) \qquad (\chi \in \mathbb{C}).$$
Set
$$S(F;z) = -\nu^2 V(\nu^{-1};z).$$

9.5. THEOREM. *We have*
$$F(z)S(w) = \partial_w\left(\frac{S(F;w)}{z-w}\right) + \cdots, \quad E(z)S(w) = 0 + \cdots, \quad H(z)S(w) = 0 + \cdots.$$

PROOF. The first formula follows from the previous lemma after the substitution $\alpha = \nu^{-1}$, taking into account that
$$\partial_w V(\alpha;w) = -\alpha{:}p(w)V(\alpha;w){:}\,.$$
The second formula is obvious. Let us prove the third formula. We have
$$H(z)S(w) = -(2{:}\gamma(z)\beta(z){:} + \nu p(z))\beta(w)V(\nu^{-1};w).$$
Now,
$$\gamma(z)\beta(z)\beta(w)V(\nu^{-1};w) = \frac{1}{z-w}\beta(w)V(\nu^{-1};w) + \cdots,$$
$$p(z)\beta(w)V(\nu^{-1};w) = -\frac{2\nu^{-1}}{z-w}\beta(w)V(\nu^{-1};w) + \cdots.$$
By adding the two terms, we get the third formula. $\square$

9.6. COROLLARY. *For every* $n \in \mathbb{Z}$,
$$[F_n, S(w)] = \partial_w(w^n S(F;w)), \qquad [E_n, S(w)] = [H_n, S(w)] = 0.\,\square$$

9.7. Let us define the operators $S(X;z)$ ($X \in \mathfrak{g}$) as follows. First, $S(X;z)$ linearly depends on $X \in \mathfrak{g}$. Second, $S(F;z)$ is as above, and $S(E;z) = S(H;z) = 0$.

Now, let us define the operators $S(x;z)$ ($x \in \hat{\mathfrak{g}}$) as follows. First, they depend linearly on $x \in \hat{\mathfrak{g}}$. Second, we set

(a) $\qquad S(X_n;z) = z^n S(X;z) \quad (X \in \mathfrak{g},\, n \in \mathbb{Z}), \qquad S(\mathbf{1};z) = 0.$

In this notation, we can rewrite the previous theorem and corollary as follows.

9.8. THEOREM. (a) *For all* $X \in \mathfrak{g}$,
$$X(z)S(w) = \partial_w\left(\frac{S(X;w)}{z-w}\right) + \cdots.$$

(b) *For all* $x \in \hat{\mathfrak{g}}$,
$$[x, S(w)] = \partial_w S(x;w).\,\square$$

9.9. LEMMA. *We have*

$$E(z)\,S(F;w) = 0 + \cdots,$$

$$H(z)\,S(F;w) = -2\frac{S(F;w)}{z-w} + \cdots, \qquad F(z)\,S(F;w) = \frac{\gamma(w)\,S(F;w)}{z-w} + \cdots.$$

This is proved by a simple direct computation.

9.10. THEOREM. (a) *For any $X, Y \in \mathfrak{g}$, we have*

$$X(z)\,S(Y;w) - Y(z)\,S(X;w) = \frac{S([X,Y];w)}{z-w} + \cdots.$$

(b) *For any $x, y \in \hat{\mathfrak{g}}$, we have*

$$[x, S(y;w)] - [y, S(x;w)] = S([x,y];w).$$

PROOF. (a) follows from the previous lemma. Alternatively, it follows from 9.8(a) and the facts (c), (d) below, using the associativity of the operator products.

(b) follows from (a) and the fact that

(c) in the products $X(z)\,S(Y;w)$ at most first order poles are present.

(This claim is a weaker version of the previous lemma.)

Alternatively, (b) follows from 9.8(b), if we notice that

(d) for generic χ the operator ∂_w is an isomorphism.

This proves the theorem. □

9.11. Set $A = \mathbb{C}((z^{-1}))$. Let us consider the twisted de Rham complex

$$\Omega^{\cdot} : 0 \to \Omega^0 \to \Omega^1 \to 0,$$

where $\Omega^0 = A$, $\Omega^1 = A\,dz$, the differential being equal to

$$d_{DR} - \frac{\chi}{\nu^2} \cdot \frac{dz}{z}.$$

As before, we form the double complex

$$C^{\cdot}(\hat{\mathfrak{g}}; \mathrm{Hom}\,(W_{\chi;\nu}, W_{\chi-2;\nu} \otimes \Omega^{\mathrm{dot}})).$$

Let us define a one-cochain $V = (V^{01}, V^{10})$ in the associated total complex as follows. We set

$$V^{01} = S(z)\,z^{\chi/\nu^2}\,dz, \qquad V^{10}(x) = S(x;z)\,z^{\chi/\nu^2} \qquad (x \in \hat{\mathfrak{g}}).$$

The next theorem is the reformulation of Theorems 9.8 (b) and 9.10 (b).

9.12. THEOREM. *The cochain V is a 1-cocycle.* □

9.13. For an integer $p \geqslant 1$, consider the normally ordered product

$$:S(z_1)\cdots S(z_p):.$$

To simplify the notations, we regard it as an element of

$$\text{Hom}(W_{\chi;\nu}, W_{\chi-2p;\nu} \otimes A_p)$$

(we ignore the powers of z_i). The same concerns the operators $S(x;z_i)$. Recall that

$$A_p = \mathbb{C}[[z_1,\ldots,z_p]]\left[\prod z_i^{-1}, \prod_{i\neq j}(z_i-z_j)^{-1}\right] = \Omega^0(X_p)$$

(cf. 7.4).

Let us consider the twisted de Rham complex

$$\Omega^{\cdot}: 0 \to \Omega^0 \to \cdots \to \Omega^p \to 0,$$

where $\Omega^i = \Omega^i(X_p)$. The differential is

$$d_{DR} - \sum_i \frac{\chi}{\nu^2} \cdot \frac{dz_i}{z_i} + \sum_{i<j} \frac{2}{\nu^2} \cdot \frac{dz_i - dz_j}{z_i - z_j}.$$

9.14. For each $a = 0,\ldots,p$, let us define the operators

$$V^{a,p-a} \in \text{Hom}(\Lambda^a \hat{\mathfrak{g}}; \text{Hom}(W_{\chi;\nu}, W_{\chi-2p;\nu} \otimes \Omega^{p-a}))$$

as follows. By definition,

$$V^{a,p-a}(x_1 \wedge \cdots \wedge x_a)$$
$$= \sum_{1 \leqslant i_1 < \cdots < i_a \leqslant p} (-1)^{\text{sgn}(i_1,\ldots,i_a)}$$
$$\times \left(\sum_{\sigma \in \Sigma_m} (-1)^{\text{sgn}(\sigma)} {:}S(z_1)\cdots S(x_{\sigma(1)};z_{i_1})\cdots S(x_{\sigma(a)};z_{i_a})\cdots S(z_p){:}\right)$$
$$\times dz_1 \wedge \cdots \wedge \hat{dz}_{i_1} \wedge \cdots \wedge \hat{dz}_{i_a} \wedge \cdots \wedge dz_p.$$

Here the sign $\text{sgn}(i_1,\ldots,i_a)$ is defined by induction on a as follows.

$$\text{sgn}(\ ) = 0, \qquad \text{sgn}(i_1,\ldots,i_a) = \text{sgn}(i_1,\ldots,i_{a-1}) + i_a + a.$$

For example,

$$V^{0p} = {:}S(z_1)\cdots S(z_p){:}\, dz_1 \wedge \cdots \wedge dz_p,$$
$$V^{p0}(x_1 \wedge \cdots \wedge x_p) = \sum_{\sigma \in \Sigma_p} (-1)^{\text{sgn}(\sigma)} {:}S(x_{\sigma(1)};z_1)\cdots S(x_{\sigma(p)};z_p){:}\,.$$

9.15. Let us consider the double Koszul complex $C^{\cdot}(\hat{\mathfrak{g}}; \text{Hom}(W_{\chi;\nu}, W_{\chi-2p;\nu} \otimes \Omega^{\cdot}))$. We have a p-cochain $V = (V^{0p},\ldots,V^{p0})$ in the associated total complex.

9.16. THEOREM. *The cochain V is a p-cocycle.* □

Chapter 4. Affine Lie algebras (general case)

The most important and general results on bosonization for arbitrary affine Lie algebras are due to Feigin and Frenkel, see [**Fr**], [**FFR**] and references therein.

§10. Bosonization

For more details, see [**BMP2**].

10.1. Let $\mathfrak{g}$ be the finite-dimensional total complex Lie algebra[1] corresponding to a Cartan matrix $A = (a_{ij})_{i,j=1}^r$ and given by the Chevalley generators E_i, H_i, F_i ($i = 1, \ldots, r$). Let $\mathfrak{h} \subset \mathfrak{g}$ be the Cartan subalgebra generated by $H_1, \ldots, H_r$. Let $\alpha_1, \ldots, \alpha_r \in \mathfrak{h}^*$ be the simple roots; let $(\cdot, \cdot)$ be the symmetric nondegenerate bilinear form on $\mathfrak{h}^*$ such that $a_{ij} = 2(\alpha_i, \alpha_j)/(\alpha_i, \alpha_i)$. This bilinear form defines the isomorphism $\mathfrak{h}^* \xrightarrow{\sim} \mathfrak{h}$. Using this isomorphism, we carry over the bilinear form to $\mathfrak{h}$; this last bilinear form will also be denoted by $(\cdot, \cdot)$. Finally, Δ_+ will denote the set of positive roots.

Let g be the dual Coxeter number of the root system of $\mathfrak{g}$. We fix a complex parameter $\nu \neq 0$ and set $k = \nu^2 - g$.

10.2. Let $\mathbf{a}$ be the Lie algebra defined by the generators b_n^i ($i = 1, \ldots, r$, $n \in \mathbb{Z}$), a_n^α, $a_n^{\alpha*}$ ($\alpha \in \Delta_+$, $n \in \mathbb{Z}$), $\mathbf{1}$ and the relations

(a) $[b_n^i, b_m^j] = (H_i, H_j) n \delta_{ij} \delta_{n+m,0} \cdot \mathbf{1}$;

(b) $[a_n^\alpha, a_m^{\beta*}] = \delta_{\alpha\beta} \delta_{n+m,0} \cdot \mathbf{1}$;

(c) all other commutators between the generators vanish.

Let $\mathbf{a}_+$ denote the Lie subalgebra of $\mathbf{a}$ generated by the elements b_n^i, a_n^α ($n > 0$, $\alpha \in \Delta_+$), $a_n^{\alpha*}$ ($n \geqslant 0$, $\alpha \in \Delta_+$). These generators are called *annihilation operators*. One introduces the *normal ordering* of a monomial in $U\mathbf{a}$ in the usual way: all annihilation operators should be pulled to the right.

All the $\mathbf{a}$-modules M that we shall consider will be $\mathbf{a}_+$-*locally finite*, i.e., have the property that for every $x \in M$, the space $U\mathbf{a}_+ x$ is finite-dimensional. The operator $\mathbf{1}$ will act as the identity.

For computational purposes, we shall also use the operators q^i ($i = 1, \ldots, r$) with the only nontrivial commutation relations

$$[q^i, b_0^j] = (H_i, H_j) \cdot \mathbf{1}.$$

10.3. For $\lambda \in \mathfrak{h}^*$, let $\mathcal{F}_\lambda$ denote the $\mathbf{a}$-module defined by one generator v_λ and the relations $\mathbf{a}_+ v_\lambda = 0$, $b_0^i v_\lambda = \langle H_i, \lambda \rangle \lambda v_\lambda$, $\mathbf{1} v_\lambda = v_\lambda$.

10.4. Let us introduce the generating functions

$$\phi^i(z) = q^i - b_0^i \log(z) + \sum_{n \neq 0} \frac{b_n^i}{n} z^{-n},$$

$$p^i(z) = \phi^{i\prime}(z) = -\sum_n b_n^i z^{-n-1} \quad (i = 1, \ldots, r),$$

$$\beta^\alpha(z) = \sum_n a_n^\alpha z^{-n-1}, \quad \gamma^\alpha(z) = \sum_n a_n^{\alpha*} z^{-n} \quad (\alpha \in \Delta_+).$$

---

[1] We suspect that most of what appears below is true for the Kac–Moody algebra corresponding to an arbitrary symmetrizable generalized Cartan matrix.

We have
$$\phi^i(z)\phi^j(w) = (H_i, H_j)\log(z-w) + \cdots, \qquad \gamma^\alpha(z)\beta^{\alpha'}(w) = \frac{\delta_{\alpha\alpha'}}{z-w} + \cdots,$$
all other products being trivial.

To shorten the notation, we shall write $\beta^i(z)$, $\gamma^i(z)$ instead of $\beta^{\alpha_i}(z)$, $\gamma^{\alpha_i}(z)$.

10.5. The bosonization formulas for the affine Lie algebra $\hat{\mathfrak{g}}$ with central charge k have the following form.

(a) $\quad E_i(z) = {:}\mathcal{E}_i(\gamma^\alpha(z), \beta^{\alpha'}(z)){:},$

(b) $\quad H_i(z) = \displaystyle\sum_{\alpha \in \Delta_+} \langle H_i, \alpha\rangle {:}\gamma^\alpha(z)\beta^\alpha(z){:} + \nu p^i(z),$

(c) $\quad F_i(z) = {:}\mathcal{F}_i(\gamma^\alpha(z), \beta^{\alpha'}(z)){:} - \nu\dfrac{(\alpha_i, \alpha_i)}{2}{:}\gamma^i(z)p^i(z){:} - c_i(\nu)\dfrac{d\gamma^i(z)}{dz}.$

Here $\mathcal{E}_i(\gamma^\alpha, \beta^{\alpha'})$, $\mathcal{F}_i(\gamma^\alpha, \beta^{\alpha'})$ are certain polynomials, linear in the β's. They depend on the Poincaré–Birkhoff–Witt isomorphism for the enveloping algebra $\mathcal{U}\mathfrak{n}_-$, which identifies this algebra with the symmetric algebra on root vectors F_α ($\alpha \in \Delta_+$). For their definition, see [**BMP2**]. The coefficients $c_i(\nu)$ are certain numbers.

§11. Screening currents

11.1. For $\chi \in \mathfrak{h}^*$, we define the Wakimoto module $W_{\chi;\nu}$ as the $\hat{\mathfrak{g}}$-module $\mathcal{F}_{-\chi/\nu}$, the $\hat{\mathfrak{g}}$-module structure being defined by the bosonization formulas 10.5(a)–(c).

11.2. For $\mu = \sum_i \mu_i \alpha_i \in \mathfrak{h}^*$, define the bosonic vertex operator
$$V(\mu; z) = {:}\exp\left(-\sum_i \mu_i \frac{(\alpha_i, \alpha_i)}{2}\phi^i(z)\right){:}.$$

This operator acts from $\mathcal{F}_\lambda$ to $\mathcal{F}_{\lambda+\mu} \otimes z^{(\lambda,\mu)}\mathbb{C}((z^{-1}))$.

11.3. We have the *product formula* (cf. 6.11)
$$V(\mu_1; z_1)V(\mu_2; z_2) = (z_1 - z_2)^{(\mu_1,\mu_2)}{:}V(\mu_1; z_1)V(\mu_2; z_2){:}.$$

11.4. By definition, the *screening currents* are defined by
$$S_i(z) = \mathcal{S}_i(\gamma^\alpha(z), \beta^{\alpha'}(z))V(\nu^{-1}\alpha_i; z) \colon W_{\chi;\nu} \to W_{\chi-\alpha_i;\nu} \otimes z^{-(\chi,\alpha_i)/\nu^2}\mathbb{C}((z^{-1})).$$

Here $\mathcal{S}_i(\gamma^\alpha, \beta^{\alpha'})$ are certain polynomials depending on a PBW decomposition for $\mathcal{U}\mathfrak{n}_-$, see [**BMP2**].

We set
$$S_i(F_j; z) = -\delta_{ij}\nu^2 V(\nu^{-1}\alpha_i; z).$$

11.5. In all proven cases (including the case $\mathfrak{g} = \mathfrak{sl}(n)$ as well as some other simple algebras) we have the following basic formulas
$$F_i(z)S_j(w) = \partial_w\left(\frac{S_j(F_i; w)}{z-w}\right) + \cdots,$$
$$E_i(z)S_j(w) = 0 + \cdots, \qquad H_i(z)S_j(w) = 0 + \cdots.$$

11.6. Let us define the operators $S_i(X;z)$ ($X \in \mathfrak{g}$, $i = 1, \ldots, r$) as follows. First, $S_i(X;z)$ linearly depends on $X \in \mathfrak{g}$. Second, $S_i(F_j;z)$ is as above, and $S_i(E_j;z) = S_i(H_j;z) = 0$.

Now, let us define the operators $S(x;z)$ ($x \in \hat{\mathfrak{g}}$) as follows. First, they depend linearly on $x \in \hat{\mathfrak{g}}$. Second, we set

(a) $\qquad S(X_n;z) = z^n S(X;z) \quad (X \in \mathfrak{g}, n \in \mathbb{Z}), \qquad S(1;z) = 0$.

In this notation, the formulas in 11.5 maybe rewritten as follows.

11.7. (a) *For all $X \in \mathfrak{g}$, $i = 1, \ldots, r$, we have*

$$X(z)\, S_i(w) = \partial_w \left(\frac{S_i(X;w)}{z - w} \right) + \cdots.$$

(b) *For all $x \in \hat{\mathfrak{g}}$, $i = 1, \ldots, r$, we have*

$$[x, S_i(w)] = \partial_w S_i(x;w).$$

11.8. THEOREM. *For all $x, y \in \hat{\mathfrak{g}}$, $i = 1, \ldots, r$, we have*

$$[x, S_i(y;w)] - [y, S_i(x;w)] = S_i([x,y];w).$$

PROOF. Let us differentiate both sides of the equality in w. The resulting equality is true by (b) above. Since for generic χ the operator ∂_w is an isomorphism, the initial equality is also true for generic χ, hence for all χ. $\square$

11.9. We suspect that for any $X, Y \in \mathfrak{g}$, i, one has

$$X(z)\, S_i(Y;w) - Y(z)\, S_i(X;w) = \frac{S_i([X,Y];w)}{z - w} + \cdots.$$

11.10. Let us fix a sequence $i_1, \ldots, i_p$, where $1 \leqslant i_j \leqslant r$ for all j.

Let us consider the twisted de Rham complex

$$\Omega^{\cdot} : 0 \to \Omega^0 \to \cdots \to \Omega^p \to 0,$$

where $\Omega^i = \Omega^i(X_p)$ are the same spaces as in 9.13. The differential is by definition equal to

$$d_{DR} - \sum_{j=1}^p \frac{(\chi, \alpha_{i_j})}{\nu^2} \cdot \frac{dz_j}{z_j} + \sum_{1 \leqslant j' < j'' \leqslant p} \frac{(\alpha_{i_{j'}}, \alpha_{i_{j''}})}{\nu^2} \cdot \frac{dz_{j'} - dz_{j''}}{z_{j'} - z_{j''}}.$$

11.11. Set $\alpha = \sum_{j=1}^p \alpha_{i_j}$. For each $a = 0, \ldots, p$, let us define the operators

$$V^{a,p-a} \in \operatorname{Hom}(\Lambda^a \hat{\mathfrak{g}}; \operatorname{Hom}(W_{\chi;\nu}, W_{\chi-\alpha;\nu} \otimes \Omega^{p-a}))$$

as follows. By definition,

$V^{a,p-a}(x_1 \wedge \cdots \wedge x_a)$

$= \sum_{1 \leqslant j_1 < \cdots < j_a \leqslant p} (-1)^{\operatorname{sgn}(j_1,\ldots,j_a)}$

$\times \left(\sum_{\sigma \in \Sigma_m} (-1)^{\operatorname{sgn}(\sigma)} {:} S_{i_1}(z_1) \cdots S_{i_{j_1}}(x_{\sigma(1)}; z_{j_1}) \cdots S_{i_{j_a}}(x_{\sigma(a)}; z_{j_a}) \cdots S_{i_p}(z_p) {:} \right)$

$\times dz_1 \wedge \cdots \wedge \hat{dz}_{j_1} \wedge \cdots \wedge \hat{dz}_{j_a} \wedge \cdots \wedge dz_p$

(cf. 9.14). The sign $\operatorname{sgn}(j_1, \ldots, j_a)$ is defined in 9.14.

For example,
$$V^{0p} = {:}S_{i_1}(z_1)\cdots S_{i_p}(z_p){:}\,dz_1\wedge\cdots\wedge dz_p,$$
$$V^{p0}(x_1\wedge\cdots\wedge x_p) = \sum_{\sigma\in\Sigma_p}(-1)^{\mathrm{sgn}(\sigma)}{:}S_{i_1}(x_{\sigma(1)};z_1)\cdots S_{i_p}(x_{\sigma(p)};z_p){:}\,.$$

11.12. Let us consider the double Chevalley–Eilenberg complex
$$C^{\cdot}(\hat{\mathfrak{g}};\mathrm{Hom}(W_{\chi;\nu},W_{\chi-\alpha;\nu}\otimes\Omega^{\cdot})).$$
We have a p-cochain $V=(V^{0p},\ldots,V^{p0})$ in the associated total complex.

11.13. THEOREM. *The cochain V is a p-cocycle.* □

References

[BD] A. Beilinson and V. Drinfeld, *Chiral algebras*, Preprint (1996).

[BMP1] P. Bowknegt, J. McCarthy, and K. Pilch, *Quantum group structure in the Fock space resolutions of the $\widehat{sl}(n)$ representations*, Comm. Math. Phys. **131** (1990), 125–155.

[BMP2] ———, *Free field approach to two-dimensional conformal field theory*, Prog. Theor. Phys. Suppl. **102** (1990), 67–135.

[Br] J.-L. Brylinski, *Non-commutative Ruelle-Sullivan type currents*, The Grothendieck Festschrift (P. Cartier et al., eds.), vol. I, Birkhäuser, Boston, 1990, pp. 477–498.

[FFR] B. Feigin, E. Frenkel, and N. Reshetikhin, *Gaudin model, Bethe Ansatz and critical level*, Comm. Math. Phys. **166** (1994), 27–62.

[FF] B. Feigin and D. Fuchs, *Representations of the Virasoro algebra*, Representations of Lie groups and related topics (A. M. Vershik and D. P. Zhelobenko, eds.), Gordon and Breach, New York, 1990, pp. 465–554.

[F] G. Felder, *BRST approach to minimal models*, Nuclear Phys. B **317** (1989), 215–236.

[FS] M. Finkelberg and V. Schechtman, *Localization of modules over small quantum groups*, J. of Math. Sciences **82** (1996), 3127–3164.

[Fr] E. Frenkel, *Free field realizations in representation theory and conformal field theory*, Address to the ICM, Zürich, August 3–11 (1994).

[K] V. Kac, *Infinite dimensional Lie algebras*, 2nd Edition, Cambridge University Press, Cambridge, 1985.

[L] G. Lusztig, *Introduction to quantum groups*, Birkhäuser, Boston, 1993.

[S] A. Sebbar, *Quantum screening operators and q-de Rham cohomology*, PhD Thesis, Stony Brook (1997).

[TK] A. Tsuchiya and Y. Kanie, *Fock space representations of the Virasoro algebra*, Publ. Res. Inst. Math. Sci. **22** (1986), 259–327.

[W] M. Wakimoto, *Fock representations of the affine Lie algebra $A_1^{(1)}$*, Comm. Math. Phys. **104** (1986), 605–609.

DEPARTMENT OF MATHEMATICS, UNIVERSITY OF CHICAGO, CHICAGO ILLINOIS 60637, USA
E-mail address: `ginzburg@math.uchicago.edu`

MAX-PLANCK-INSTITUT FÜR MATHEMATIK, GOTTFRIED-CLAREN-STRASSE 26, 53225 BONN, DEUTSCHLAND
E-mail address: `vadik@mpim-bonn.mpg.de`

Interlacing Measures

Sergei Kerov

To A. A. Kirillov on his 60th birthday

ABSTRACT. We introduce the notion of interlacing measures generalizing the notion of interlacing sequences. Let $\tau = \tau_+ - \tau_-$ be the Jordan decomposition of a signed measure of bounded variation on the real line. Then the measures τ_+ and τ_- interlace if and only if there exists a probability distribution μ such that the ordinary Cauchy–Stieltjes transform of μ equals the multiplicative version of this transform for τ:

$$\int \frac{\mu(du)}{z-u} = \exp \int \ln \frac{1}{z-u} \, \tau(du), \qquad \operatorname{Im} z \neq 0.$$

We survey various applications of this identity, including (1) distributions of mean values of Poisson–Dirichlet random measures; (2) Markov moment problem; (3) exponential representations of functions of negative imaginary type; (4) spectral shift function of a self-adjoint operator; (5) random growth of Young diagrams with respect to Plancherel measures of symmetric groups.

Assuming that the measures μ, τ have finite moments h_n, p_n for all $n = 0, 1, 2, \ldots$, the above identity implies that $\sum_{n=0}^{\infty} h_n z^n = \exp \sum_{n=1}^{\infty} p_n z^n / n$. We emphasize the combinatorial approach based on the study of moments of the measures μ, τ.

Contents

Introduction
1. Limits of interlacing sequences
2. Functions of negative imaginary type
3. The moments of interlacing measures
4. Poisson–Dirichlet random measures
5. Combinatorics of continued fractions
6. Interlacing measures in operator theory
7. Interlacing measures as generalized Young diagrams
8. Exponential representations related to Gaussian measure

References

1991 *Mathematics Subject Classification.* Primary 05E05, 05E35, 60G57; Secondary 30E05.

Key words and phrases. Interlacing sequences, Young diagrams, Markov moment problem, functions of negative imaginary type, Poisson–Dirichlet random measures, continued fractions, spectral shift functions, partition structures.

Introduction

The goal of this paper is to study the relationship between a probability distribution μ and a bounded signed measure τ on the real line $\mathbb{R}$, satisfying the identity

$$(1) \qquad \int_{-\infty}^{\infty} \frac{\mu(du)}{z-u} = \exp \int_{-\infty}^{\infty} \ln \frac{1}{z-u} \tau(du), \qquad z \in \mathbb{C} \setminus \mathbb{R}.$$

In other words, the ordinary Cauchy–Stieltjes transform of μ equals the multiplicative version of this transform for τ. By virtue of the Stieltjes inversion formula, the measures μ, τ determine each other uniquely.

The identity (1) arises in quite a number of apparently unrelated contexts, such as:

1. distributions of mean values of Poisson–Dirichlet random measures, see [**10**];
2. the Markov moment problem, see [**28**];
3. exponential representations of functions of negative imaginary type, see [**3, 20**];
4. the spectral shift function of a self-adjoint operator, see [**7, 29**];
5. the Plancherel growth of Young diagrams, see [**23, 24**].

As a basic introductory example motivating the study of equation (1), consider a rational function $R(z)$ written in two different ways:

$$(2) \qquad \sum_{k=1}^{n} \frac{\mu_k}{z-x_k} = \frac{(z-y_1)\cdots(z-y_{n-1})}{(z-x_1)(z-x_2)\cdots(z-x_n)}.$$

It is well known that the following properties of the fraction (2) are equivalent:

$$\mu_1,\ldots,\mu_n > 0, \quad \sum_{k=1}^{n} \mu_k = 1 \quad \Longleftrightarrow \quad x_1 < y_1 < x_2 < \cdots < x_{n-1} < y_{n-1} < x_n.$$

Assuming that these conditions hold, denote by μ the discrete distribution with the weights μ_k at the points x_k. Let τ be the signed measure with the weights $+1$ at the points x_k, and the weights -1 at the points y_k. Then the identity (1) specializes to (2).

In order to generalize this example, we introduce the notion of interlacing measures. By definition, two finite positive measures τ_+, τ_- on the real line *interlace*, if

$$\tau_+\{(-\infty,u)\} \geqslant \tau_-\{(-\infty,u)\}, \qquad \tau_+\{(u,\infty)\} \geqslant \tau_-\{(u,\infty)\}$$

for all $u \in \mathbb{R}$, and $\tau_+\{\mathbb{R}\} - \tau_-\{\mathbb{R}\} = 1$. As an obvious example of interlacing measures, one can take the uniform discrete measures τ_+, τ_- with unit weights at the points of two interlacing sequences. Another simple example is provided by any probability distribution $\tau_+ = \tau$, and a zero measure $\tau_- \equiv 0$.

Consider a bounded signed measure τ with the Jordan decomposition $\tau = \tau_+ - \tau_-$. Then a probability distribution μ satisfying (1) exists if and only if the Jordan components of τ interlace.

Assume that the measures μ, τ in equation (1) have finite moments

$$h_n = \int_{-\infty}^{\infty} u^n \mu(du), \qquad p_n = \int_{-\infty}^{\infty} u^n \tau(du)$$

of all orders $n = 1, 2, \ldots$, and that they are uniquely determined by the moments. Then the identity (1) can be written in the form

$$\sum_{n=0}^{\infty} h_n z^n = \exp \sum_{n=1}^{\infty} \frac{p_n}{n} z^n. \tag{3}$$

This equality of formal power series is equivalent to a series of polynomial equations

$$h_n = \frac{1}{n!} \sum_{\pi \in \mathfrak{S}_n} \prod_{k \geq 1} p_k^{r_k(\pi)}, \qquad n = 0, 1, 2, \ldots,$$

where $r_k(\pi)$ denotes the number of cycles of length k in a permutation π of n objects.

A solid foundation for the study of identity (1) is provided by the theory of functions of negative imaginary type (i.e., the functions $w = R(z)$ analytic in the half-plane $\operatorname{Im} z > 0$ and having there a nonpositive imaginary part, $\operatorname{Im} R(z) \leq 0$), see [**3, 20, 28**]. The Cauchy–Stieltjes transforms of probability distributions,

$$R(z) = \int_{-\infty}^{\infty} \frac{\mu(du)}{z - u}, \tag{4}$$

form a part $\mathcal{N}$ in the space of functions of negative imaginary type, characterized by the condition $\lim_{y \to +\infty} iy \, R(iy) = 1$. Every function $R \in \mathcal{N}$ admits an *exponential representation* of the form

$$R(z) = \frac{1}{z} \exp\left(-\int_{-\infty}^{0} \frac{F(u)}{z - u} du + \int_{0}^{\infty} \frac{1 - F(u)}{z - u} du \right), \tag{5}$$

where the integrable function F satisfies the conditions $0 \leq F(u) \leq 1$ and

$$\int_{-\infty}^{-1} F(u) \frac{du}{|u|} < \infty, \qquad \int_{1}^{\infty} (1 - F(u)) \frac{du}{u} < \infty.$$

In the special case of a function F of finite variation, equation (5) can be written as

$$R(z) = \exp \int_{-\infty}^{\infty} \ln \frac{1}{z - u} \, dF(u). \tag{6}$$

The distribution μ and the function F in the canonical representations (4), (5) of a function $R \in \mathcal{N}$ determine each other uniquely. We refer to the nonlinear transformation $F \mapsto \mu$ as to the *Markov transform*.

It was probably Markov who initiated the study of equation (1). In his paper [**34**], published a century ago, he proved (see equation (8) in [**34**]) that for every bounded density $0 \leq f(u) \leq 1$, the coefficients $c_1, c_2, \ldots$ of the continued fraction expansion

$$\frac{1}{z} \exp \int_0^1 \frac{f(u)}{z - u} du = \cfrac{1}{z - \cfrac{c_1}{1 - \cfrac{c_2}{z - \cfrac{c_3}{1 - \cdots}}}}$$

are all positive. It follows from the celebrated memoir [37] by Stieltjes that such a fraction converges to the integral $\int (z-u)^{-1}\mu(du)$, with a positive measure μ.

A number of fundamental ideas were introduced in the field by M. G. Krein and his colleagues (see [20, 28, 29]). In particular, the Markov moment problem on a finite interval was reduced to the well-studied Hausdorff moment problem. Krein also developed to a much greater generality the idea of a spectral shift function suggested by Lifshits in [30]. As a powerful tool in these investigations, Krein used Nevanlinna's theory of integral representations of functions of negative imaginary type, which he extended and generalized considerably. Important general results on the exponential representations of functions of negative imaginary type, and their relationship to the corresponding additive representations, were obtained by Aronszajn and Donoghue in [3]. See equation (2.3.5) below for an example of fascinating properties of exponential representations found in [3].

Both Markov and Krein dealt mostly with compactly supported measures, which can be adequately studied in terms of their moments. On the contrary, in the approach to equation (1) based on the study of distributions of mean values of Poisson–Dirichlet random measures (see [10, 43]), one works with almost general probability distributions on the real line. We specialize and simplify the Aronszajn–Donoghue theory in such a way that it generalizes and unifies the two approaches above.

An important new idea concerning the representation (5), (6) is to treat the function F as a generalized Young diagram. More precisely, the function F in the exponential representation (5) can be written, under mild assumptions, in the form

$$F(x) = \frac{1}{2}(1 + \omega'(x)), \qquad 1 - F(x) = \frac{1}{2}(1 - \omega'(x)),$$
$$\omega(x) = \int_{-\infty}^{x} F(u)\, du + \int_{x}^{\infty} (1 - F(u))\, du.$$

If a measure μ is the Markov transform of F, we say that the function ω is the *diagram* of μ. For the particular case in which ω is the graph of a true Young diagram, the measure μ describes the transition probabilities of the Plancherel growth process, see [23, 24]. This analogy was used in [24] to obtain a new probabilistic procedure providing the Markov transform μ of a general pair of interlacing measures, the *interval shrinkage process*. In [23], the same idea led to a convergence theorem for interlacing roots of orthogonal polynomials. See also [25] for the equivalence between the Plancherel growth of generalized Young diagrams, and the Burgers–Hopf equation

$$\frac{\partial}{\partial t} R(z,t) + R(z,t) \frac{\partial}{\partial z} R(z,t) = 0.$$

The present paper is, to a great extent, expository and intended to exhibit a variety of applications of the identity (1). Still, the author believes the following results may be new.
1. Characterization (3.5.2) of moments of interlacing sequences and measures in terms of characters of symmetric groups.
2. Formula (3.7.4) for the difference derivatives of the polynomials (3.7.3) and the related generalization of the basic identity (3) to the general *partition structures* in the sense of Kingman [26].

3. The combinatorial description (5.2.6) for the asymptotic series of the logarithmic derivative of a real J-type continued fraction.
4. A direct combinatorial description for the moments of the spectral shift function of a bounded self-adjoint operator, see §6.
5. Explicit exponential representation (8.1.4) for the Cauchy–Stieltjes transform of the standard Gaussian measure.

The contents of the paper is as follows. In §1 we introduce the basic definitions of interlacing measures and Rayleigh functions, and define topology in the corresponding spaces. In §2 we establish a homeomorphism between the space of Rayleigh functions and that of probability distributions, based on the theory of functions of negative imaginary type. The moments of interlacing measures are defined in §3. We prove an identity for the difference derivatives of the moments of the measures μ, τ related by the identity (1) and present a characterization of the moment sequences of interlacing measures. The connection with Poisson–Dirichlet random measures is sketched in §4. In §5 we present a combinatorial description of moments of interlacing measures similar to the Flajolet theorem. A version of this result in the context of operator theory is outlined in §6. In §7 we briefly mention the important connection between interlacing measures and Young diagrams. In the final §8 we provide an explicit exponential representation for the Cauchy–Stieltjes transform of the Gaussian measure and for its deformation to the semi-circle distribution related to the "associated Hermite polynomials" of Askey and Wimp.

Acknowledgments. At various stages of the work on this paper the author had fruitful discussions with many colleagues. My interest in equation (1) originates in a joint result [41, 42] with A. Vershik, and I have certainly learned a lot from my teacher. M. Gordin explained the Markov moment problem to me, and provided a number of important references. X. Viennot introduced me to his combinatorial theory of orthogonal polynomials. At Harvard, P. Diaconis gave me a chance to learn the approach to equation (1) via Poisson–Dirichlet random measures, which was entirely new for me at the time. V. Vasyunin drew my attention to the connection between equation (1) and the idea of the spectral shift function. I appreciate very much the inspiring interest in this work and encouragement I received from G.-C. Rota and from my friend G. Olshanski.

I would also like to thank the Ministère de l'Enseignement Supérieur et de la Recherche for financial support and my colleagues J. Bétréma, M. Bousquet-Melou, N. K. Nikolski, and especially A. Zvonkin for their hospitality at the University Bordeaux-I.

§1. Limits of interlacing sequences

1.1. Interlacing sequences. Let $x = \{x_k\}_{k=1}^n$, $y = \{y_k\}_{k=1}^m$ be two finite sets[1] of real numbers. They are said to be *interlacing* if there is a point of x in between any two points of y, and a point of y in between any two points of x. As a matter of fact, there is exactly one point of one sequence in between any two *neighboring* points of the other.

The numbers of elements of interlacing sequences can differ at most by one, $|n - m| \leqslant 1$. In this paper we focus on the particular case when $m = n - 1$, so that

---

[1] We regard a finite set of reals as a sequence by enumerating its elements in increasing order.

the sequences x, y interlace if and only if

(1.1.1) $$x_1 < y_1 < x_2 < \cdots < x_{n-1} < y_{n-1} < x_n.$$

The interlacing sequences of this type arise in quite a variety of contexts. We mention a few examples which are intended to motivate further consideration.

(1.1.2) EXAMPLE. *Rayleigh theorem.* Let $x_1 < x_2 < \cdots < x_n$ be the lengths of semi-axes of an ellipsoid X in $\mathbb{R}^n$. Denote by $Y = X \cap H$ the intersection of X with some hyperplane $H \subset \mathbb{R}^n$, and let $y_1 < \cdots < y_{n-1}$ be the lengths of semi-axes of the ellipsoid Y. Then the sequences x, y interlace.

(1.1.3) EXAMPLE. *Orthogonal polynomials.* Denote by $\{P_n(x)\}_{n=0}^{\infty}$ a sequence of polynomials orthonormal with respect to a probability measure μ. Then the roots of two consecutive polynomials $P_n(x)$, $P_{n-1}(x)$ interlace.

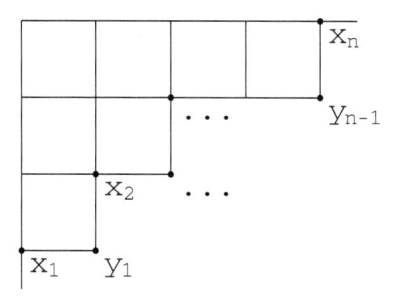

FIGURE 1. Interlacing sequences representing a Young diagram.

(1.1.4) EXAMPLE. *Young diagrams.* Let λ be a Young diagram with n boxes considered in the usual way as a subset of the plane (see Figure 1). Call the number $c(b) = j - i$ the *content* of a point $b = (i,j)$ at the crossing of row i and column j. Then the contents of all corner points at the border of the diagram λ form a pair of interlacing sequences. For instance, Figure 1 represents the Young diagram of an integer partition $7 = 4+2+1$, and the corresponding sequences are $x = (-3,-1,1,4)$ and $y = (-2,0,3)$. Interlacing sequences which arise from Young diagrams have two specific features: their elements are *integers*, and $\sum_{k=1}^{n} x_k = \sum_{k=1}^{n-1} y_k$.

1.2. **Convergence of interlacing sequences.** In the examples above it is quite natural to look for the *limits* of interlacing sequences as $n \to \infty$. In order to define a topology on the set $\mathcal{F}_0$ of interlacing sequences (1.1.1), we shall embed this set in a space of measures.

Note that the function

(1.2.1) $$F_{x,y}(u) = \#\{k : x_k < u\} - \#\{k : y_k < u\}, \quad u \in \mathbb{R},$$

determines a pair of interlacing sequences uniquely. We say that $F_{x,y}(u)$ is the *distribution function* of $(x,y) \in \mathcal{F}_0$ (see Figure 2). This function is bounded, $0 \leq F_{x,y}(u) \leq 1$ for all $u \in \mathbb{R}$, hence integrable on every finite interval in $\mathbb{R}$. We consider $F_{x,y}(u)$ as the *density* of an infinite absolutely continuous measure

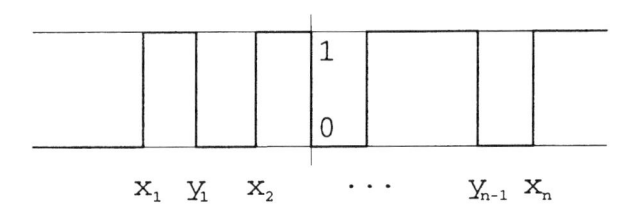

FIGURE 2. A distribution function of interlacing sequences.

$F_{x,y}(u)\,du$ in $\mathbb{R}$, and endow the set $\mathcal{F}_0$ with the topology of proper weak convergence of such measures.

(1.2.2) DEFINITION. Given a sequence $\{(x^{(n)}, y^{(n)})\}_{n=1}^{\infty}$ in $\mathcal{F}_0$, denote by F_n the distribution function of the pair $(x^{(n)}, y^{(n)})$. The sequence is said to *converge* if the following two conditions hold:
1. there exists an integrable function F such that

$$\lim_{n \to \infty} \int_a^b F_n(u)\,du = \int_a^b F(u)\,du$$

for every finite interval $(a, b) \subset \mathbb{R}$;

2. for every $\varepsilon > 0$ there exists an $A > 0$ such that

$$\int_{-\infty}^{-A} \frac{F_n(u)}{1+|u|}\,du < \varepsilon, \qquad \int_A^\infty \frac{1-F_n(u)}{1+u}\,du < \varepsilon$$

for all $n = 1, 2, \ldots$.

The second condition is equivalent to the assumption

$$\int_{-\infty}^0 \frac{F(u)}{1+|u|}\,du < \infty, \qquad \int_0^\infty \frac{1-F(u)}{1+u}\,du < \infty.$$

The purpose of imposing this condition in the definition of convergence will be clarified in subsection 2.3. In a few words, our definition of convergence corresponds to the *proper* weak convergence of probability distributions to be associated to interlacing sequences.

(1.2.3) EXAMPLE. For $n = 1, 2, \ldots$ define a pair $(x^{(n)}, y^{(n)})$ of interlacing sequences as follows:

$$x_k^{(n)} = 2(k-n), \quad k = 0, 1, \ldots, 2n, \qquad y_k^{(n)} = 2(k-n) - 1, \quad k = 1, \ldots, 2n.$$

Then the functions $F_n(u)$ converge pointwise, $\lim_{n\to\infty} F_n(u) = F(u)$, and condition (1) holds. Still, the family does not converge since condition (2) fails.

If there exists an interval $[\alpha, \beta]$ such that $\alpha \leqslant x_1^{(n)}$ and $x_n^{(n)} \leqslant \beta$ for all $n = 1, 2, \ldots$, then condition (2) is redundant.

(1.2.4) EXAMPLE. Consider a family of interlacing sequences

$$x_k^{(n)} = \frac{2k}{2^n} - 1, \quad k = 0, 1, \ldots, 2^n, \qquad y_k^{(n)} = \frac{2k-1}{2^n} - 1, \quad k = 1, \ldots, 2^n.$$

This family converges, the limiting function F being

$$F(u) = \begin{cases} 0, & \text{if } u < -1, \\ 1/2, & \text{if } -1 < u < 1, \\ 1, & \text{if } 1 < u. \end{cases}$$

Note that F is *not* a pointwise limit of F_n (see Figure 3).

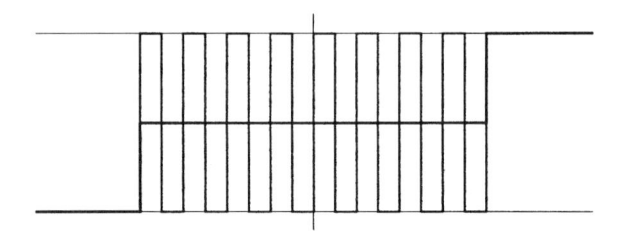

FIGURE 3. Interlacing sequences approximating a distribution function.

The function F in Example 1.2.3 is the distribution function of a discrete probability measure τ with two equal point masses $1/2$ at the points $x = -1$ and $x = 1$. As this example suggests, distribution functions of interlacing sequences may converge to a distribution function of a probability measure. We introduce the notion of interlacing measures which includes both probability measures and interlacing sequences.

1.3. **Interlacing measures.** Given a sequence $x = \{x_k\}_{k=1}^n$ and a real number $u \in \mathbb{R}$, denote by $L_x(u) = \#\{k : x_k < u\}$, $R_x(u) = \#\{k : x_k > u\}$ the numbers of points in x to the left and to the right of u. Note that the sequences $x = \{x_k\}_{k=1}^n$ and $y = \{y_k\}_{k=1}^{n-1}$ interlace if and only if $L_x(u) \geqslant L_y(u)$ and $R_x(u) \geqslant R_y(u)$ for all $u \in \mathbb{R}$. In this form the definition can be generalized to measures in $\mathbb{R}$.

Let τ be a finite positive measure in $\mathbb{R}$. We denote by $L_\tau(u) = \tau\{(-\infty, u)\}$ the distribution function of the measure τ, and by $R_\tau(u) = \tau\{(u, \infty)\}$ its hazard function.

(1.3.1) DEFINITION. We say that two finite positive measures τ_+, τ_- *interlace* if the following four conditions hold:

(1.3.2) $\qquad L_{\tau_+}(u) \geqslant L_{\tau_-}(u), \quad R_{\tau_+}(u) \geqslant R_{\tau_-}(u), \qquad u \in \mathbb{R};$

(1.3.3) $\qquad \tau_+\{(-\infty, \infty)\} - \tau_-\{(-\infty, \infty)\} = 1;$

(1.3.4) $\qquad \int_{-\infty}^{\infty} \ln(1 + |u|) \tau_+(du) < \infty, \quad \int_{-\infty}^{\infty} \ln(1 + |u|) \tau_-(du) < \infty;$

(1.3.5) $\qquad$ the measures $\tau_\pm$ are mutually singular.

(1.3.6) EXAMPLE. Let τ_+, τ_- be uniform discrete measures with unit weights at the points of interlacing sequences x, y correspondingly. Then the measures $\tau_\pm$ interlace.

(1.3.7) EXAMPLE. Let $\tau_+ = \tau$ be any probability distribution for which the first integral in (1.3.4) converges, and let τ_- be the zero measure. Then τ_+ and τ_- interlace.

Denote by $\tau = \tau_+ - \tau_-$ the difference of the interlacing measures $\tau_\pm$. By condition (1.3.5), $\tau = \tau_+ - \tau_-$ is the Jordan decomposition of the signed measure τ, hence τ determines τ_+ and τ_- uniquely. One can restate the definition of interlacing measures in terms of their difference.

(1.3.8) DEFINITION. Let τ be a bounded measure in $\mathbb{R}$, not necessarily positive. We say that τ is a *Rayleigh measure*, if the following three conditions hold:

(1.3.9) $$0 \leqslant \tau\{(-\infty, u)\} \leqslant 1, \quad u \in \mathbb{R};$$

(1.3.10) $$\tau\{(-\infty, \infty)\} = 1;$$

(1.3.11) $$\int_{-\infty}^{\infty} \ln(1+|u|)\,|\tau(du)| < \infty.$$

The distribution function of a sample Rayleigh measure is shown in Figure 4. Two positive measures interlace if and only if their difference is a Rayleigh measure. For each Rayleigh measure τ its Jordan components τ_+, τ_- interlace.

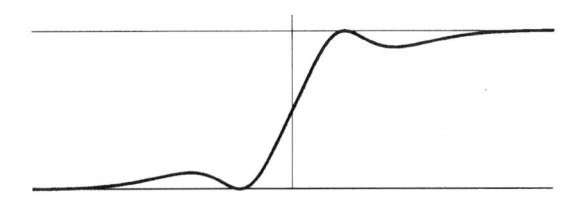

FIGURE 4. A distribution function of interlacing measures.

1.4. Rayleigh functions. The set of Rayleigh measures is only part of a larger space $\mathcal{F}$ which arises as a completion of the space $\mathcal{F}_0$ of interlacing sequences. We proceed with the description of $\mathcal{F}$.

(1.4.1) DEFINITION. Any measurable function $F\colon \mathbb{R} \to \mathbb{R}$ such that

(1.4.2) $$0 \leqslant F(u) \leqslant 1, \quad u \in \mathbb{R},$$

(1.4.3) $$\int_{-\infty}^{0} \frac{F(u)}{1+|u|}\,du < \infty, \quad \int_{0}^{\infty} \frac{1-F(u)}{1+u}\,du < \infty,$$

will be referred to as a *Rayleigh function*. If the difference of two Rayleigh functions vanishes a.e., we consider such functions to be identical.

(1.4.4) EXAMPLE. Choose two *infinite* sequences x, y such that $x_1 < y_1 < x_2 < y_2 < \cdots$ and $\lim y_n = \infty$, and define their distribution function $F(u) = F_{x,y}(u)$ by equation (1.2.1). Then condition (1.4.3) is equivalent to the convergence of the series

(1.4.5) $$\sum_{k=1}^{\infty} \frac{x_{k+1} - y_k}{y_k} < \infty.$$

This implies that the infinite product

$$(1.4.6) \qquad R(z) = \prod_{k=1}^{\infty} \frac{z - y_k}{z - x_k}$$

converges for $z \neq x_k$, $k = 1, 2, \ldots$, and that the sum of the coefficients of the partial fractions in the decomposition

$$(1.4.7) \qquad R(z) = \sum_{k=1}^{\infty} \frac{\mu_k}{z - x_k}$$

equals one, $\sum \mu_k = 1$. In turn, the latter condition yields (1.4.5).

By definition, the limit F of a converging family of interlacing sequences is always a Rayleigh function. Every Rayleigh function can be approximated by a family of interlacing sequences.

A Rayleigh function $F(u)$ determines two finite absolutely continuous positive measures on left and right semi-axes, with the densities

$$F^-(u) = \frac{F(u)}{1 + |u|}, \quad u < 0, \qquad F^+(u) = \frac{1 - F(u)}{1 + u}, \quad u > 0.$$

We define the convergence of Rayleigh functions as the proper weak convergence of the measures $F^-(u)\,du$, $F^+(u)\,du$.

(1.4.8) DEFINITION. A sequence $\{F_n\}_{n=1}^{\infty}$ of Rayleigh functions *converges* to a function F if

$$(1.4.9) \qquad \begin{aligned} \lim_{n \to \infty} \int_{-\infty}^{x} \frac{F_n(u)}{1 + |u|}\,du &= \int_{-\infty}^{x} \frac{F(u)}{1 + |u|}\,du, \\ \lim_{n \to \infty} \int_{x}^{\infty} \frac{1 - F_n(u)}{1 + |u|}\,du &= \int_{x}^{\infty} \frac{1 - F(u)}{1 + |u|}\,du \end{aligned}$$

for all $x \in \mathbb{R}$.

The definition conforms to that of convergence of interlacing sequences in subsection 1.2. The limiting function F is a Rayleigh function also.

We denote the space of Rayleigh functions by $\mathcal{F}$. The convergence $F_n \to F$ to a function $F \in \mathcal{F}$ simply means that

$$\lim_{n \to \infty} \int_a^b F_n(u)\,du = \int_a^b F(u)\,du$$

for all finite intervals (a, b). The subset $\mathcal{F}_0$ of interlacing sequences is dense in $\mathcal{F}$.

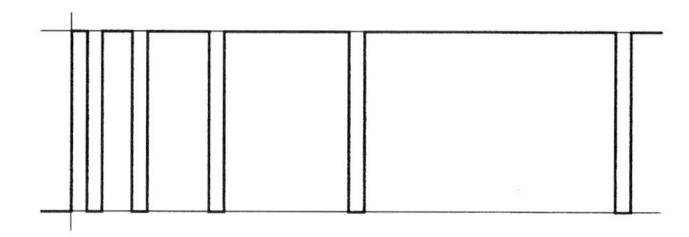

FIGURE 5. A Rayleigh function of infinite variation.

If τ is a Rayleigh measure, its distribution function $F(u) = L_\tau(u)$ (see Figure 4) is of bounded variation, hence it is Lebesgue integrable in every finite interval, and the condition (1.4.2) holds. The condition (1.4.3) is equivalent to the assumption (1.3.11). Thus, we have identified the Rayleigh measures with Rayleigh functions of bounded variation. We denote by $\mathcal{F}_b$ the subspace of such functions. Figure 5 represents a Rayleigh function of infinite variation.

§2. Functions of negative imaginary type

2.1. **Integral representations.** Let $w = R(z)$ denote a function of a complex variable z, analytic in the upper half-plane $\operatorname{Im} z > 0$. Recall that R is said to be of *negative imaginary type* (see [6]) if

$$(2.1.1) \qquad \operatorname{Im} z > 0 \implies \operatorname{Im} R(z) \leqslant 0.$$

We denote the space of such functions by $\widetilde{\mathcal{N}}$, and we endow it with the topology of uniform convergence on compact subsets in the open upper half-plane.

Assume that a function $R \in \widetilde{\mathcal{N}}$ is rational, and that

$$(2.1.2) \qquad \lim_{z \to \infty} z R(z) = 1.$$

Then R is the ratio of unimodal polynomials with interlacing real simple roots. Conversely, for each pair $(x, y) \in \mathcal{F}_0$ of interlacing sequences (1.1.1), the function

$$(2.1.3) \qquad R(z) = \frac{(z - y_1) \cdots (z - y_{n-1})}{(z - x_1)(z - x_2) \cdots (z - x_n)}$$

is of negative imaginary type and condition (2.1.2) holds.

A proper rational fraction can also be represented by its partial fraction decomposition,

$$(2.1.4) \qquad R(z) = \sum_{k=1}^{n} \frac{\mu_k}{z - x_k}.$$

Note that $R \in \widetilde{\mathcal{N}}$ if and only if $\mu_k > 0$ for all $k = 1, \ldots, n$. The condition (2.1.2) is equivalent to the identity $\sum \mu_k = 1$.

Formula (2.1.4) is a particular case of the classical integral representation valid for all functions $R \in \widetilde{\mathcal{N}}$.

(2.1.5) THEOREM (R. Nevanlinna, see [28, Theorem A.2]). *Every function R of negative imaginary type can be uniquely represented in the form*

$$(2.1.6) \qquad R(z) = -\alpha z + \beta + \int_{-\infty}^{\infty} \left(\frac{1}{z - t} + \frac{t}{1 + t^2} \right) \nu(dt),$$

where $\alpha > 0$, $\beta \in \mathbb{R}$, and ν is a positive measure such that

$$\int_{-\infty}^{\infty} \frac{\nu(dt)}{1 + t^2} < \infty.$$

(2.1.7) COROLLARY (see [**36**, Lemma II.2.2]). *A function $w = R(z)$ can be represented as a Cauchy–Stieltjes transform*

$$(2.1.8) \qquad R(z) = \int_{-\infty}^{\infty} \frac{\mu(dt)}{z-t}$$

of a probability distribution μ if and only if it has a negative imaginary type, and

$$(2.1.9) \qquad \lim_{y \to +\infty} iy\, R(iy) = 1.$$

The measure μ in the representation (2.1.8) is determined by the function R uniquely, and can be recovered by the Stieltjes–Perron inversion formula

$$(2.1.10) \qquad \int_a^b \mu(du) = -\frac{1}{\pi} \lim_{y \to +\infty} \int_a^b \operatorname{Im} R(x+iy)\, dx.$$

(2.1.11) EXAMPLE. Consider the rational fraction (2.1.3) whose poles and zeros are given by the sequences $x = (-3, -1, 1, 4)$, $y = (-2, 0, 3)$ of Example 1.1.4 (these sequences correspond to the Young diagram in Figure 1). Figure 6 represents the curve $w = R(u + 0.1i)$ in the lower half-plane, $-\infty < u < \infty$. In Figure 7 one can see the density $-\pi^{-1} \operatorname{Im} R(u + 0.1\, i)$ approximating the discrete measure with the weights $9/28$, $1/5$, $1/4$, $8/35$ at the points $-3, -1, 1, 4$, correspondingly. The latter is a transition distribution of the Markov chain on the Young lattice, called the *Plancherel growth process* (see equation (7.1.1) below). The graph of the function $1 + \pi^{-1} \arg R(u + 0.1\, i)$ is shown in Figure 8; it approaches the Rayleigh function of Figure 2.

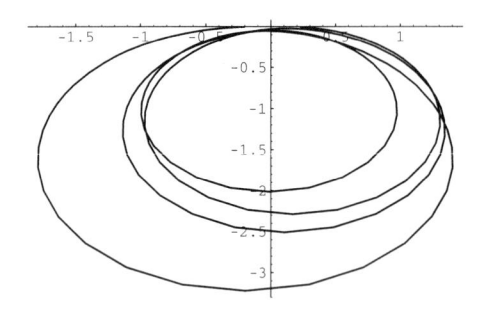

FIGURE 6. The curve $w = R(u + 0.1i)$ for the Young diagram of Example 2.1.11.

We denote by $\mathcal{N}$ the set of all functions $R \in \widetilde{\mathcal{N}}$ satisfying condition (2.1.9). A sequence of functions $R_n \in \mathcal{N}$ is said to converge *properly* if it converges in $\widetilde{\mathcal{N}}$, and for any $\varepsilon > 0$ there exists $A > 0$ such that $|iy\, R_n(iy) - 1| < \varepsilon$ for all $y > A$ and all $n = 1, 2, \dots$. The limit of such a sequence belongs to $\mathcal{N}$. If a sequence R_n converges in $\widetilde{\mathcal{N}}$ to a function $R \in \mathcal{N}$, the convergence is the proper one.

Let $\mathcal{M}$ be the space of all probability distributions in $\mathbb{R}$ endowed with the topology of proper weak convergence (see [**13**]). Then the representation (2.1.8) establishes a homeomorphism between the spaces $\mathcal{M}$ and $\mathcal{N}$. It is clear from (2.1.4) that the subset $\mathcal{M}_0$ of discrete distributions $\mu \in \mathcal{M}$ with finite support corresponds to the subset $\mathcal{N}_0 \subset \mathcal{N}$ of rational functions.

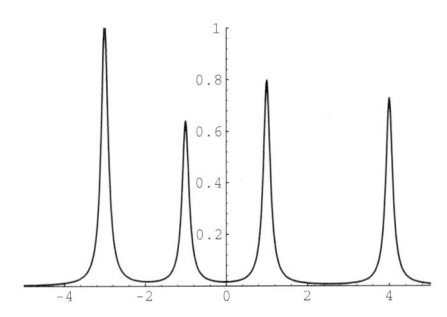

FIGURE 7. The density $-\frac{1}{\pi}\operatorname{Im} R(u+0.1\,i)$ for the curve in Figure 6.

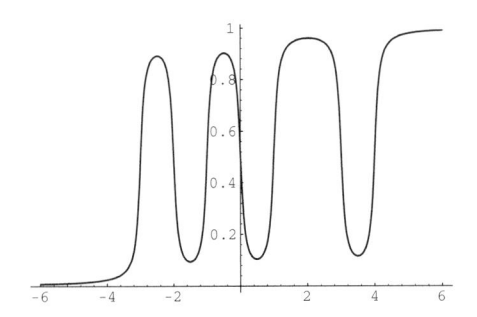

FIGURE 8. The distribution function $1+\frac{1}{\pi}\arg R(u+0.1\,i)$ for the curve in Figure 6.

2.2. **Exponential representations.** For every function $R \in \widetilde{\mathcal{N}}$, its natural logarithm $\ln R$ also belongs to $\widetilde{\mathcal{N}}$, hence admits an integral representation of type (2.1.6). Since the imaginary part of the logarithm is bounded, $-\pi \leqslant \operatorname{Im}\ln R(z) = \arg R(z) \leqslant 0$, its integral representation has specific features.

(2.2.1) THEOREM (see [**28**, Theorem A.3] or [**3**, 1.10]). *Every function $R \in \widetilde{\mathcal{N}}$ admits a unique representation*

$$(2.2.2) \qquad R(z) = -C \exp \int_{-\infty}^{\infty} \left(\frac{1}{u-z} - \frac{u}{1+u^2} \right) F(u)\,du,$$

where $C > 0$ and $0 \leqslant F(u) \leqslant 1$ for all $u \in \mathbb{R}$.

When specialized to the subclass $\mathcal{N} \subset \widetilde{\mathcal{N}}$, the *exponential representation* (2.2.2) can be written as follows.

(2.2.3) THEOREM. *Every function $R \in \mathcal{N}$ admits a unique representation of the form*

$$(2.2.4) \qquad R(z) = \frac{1}{z} \exp\left(-\int_{-\infty}^{0} \frac{F(u)}{z-u}\,du + \int_{0}^{\infty} \frac{1-F(u)}{z-u}\,du \right),$$

where $F \in \mathcal{F}$ is a Rayleigh function. Given a Rayleigh function F, the integrals in the right-hand side of (2.2.4) *converge for all nonreal z, and $R \in \mathcal{N}$.*

PROOF. Assume that $\operatorname{Im} z > 0$. Since $\operatorname{Im} R(z) \leqslant 0$, we have $\ln R(z) = -i\pi + \ln(-R(z))$, and it follows from (2.2.2) that

$$\ln R(z) = -i\pi + \ln C + \int_{-\infty}^{\infty} \left(\frac{1}{u-z} - \frac{u}{1+u^2}\right) F(u)\,du.$$

One can easily check (see [**3**, equation (1.9i)]) that

$$\ln\left(-\frac{1}{z}\right) = \int_0^{\infty} \left(\frac{1}{u-z} - \frac{u}{1+u^2}\right) du$$

and hence

$$0 = -i\pi - \ln\frac{1}{z} + \int_0^{\infty} \left(\frac{1}{u-z} - \frac{u}{1+u^2}\right) du.$$

It follows that

$$\ln R(z) = \ln\frac{C}{z} + \int_{-\infty}^{0} \left(\frac{1}{u-z} - \frac{u}{1+u^2}\right) F(u)\,du$$
$$- \int_0^{\infty} \left(\frac{1}{u-z} - \frac{u}{1+u^2}\right) (1 - F(u))\,du.$$

Since F is a Rayleigh function, the convergence of the integrals (1.4.3) implies that the integrals

$$\int_{-\infty}^{0} \frac{uF(u)\,du}{1+u^2} < \infty, \qquad \int_0^{\infty} \frac{u(1-F(u))\,du}{1+u^2} < \infty$$

converge, too. Set $C_1 = C_2 C$, where

$$C_2 = \exp\left(-\int_{-\infty}^{0} \frac{u}{1+u^2} F(u)\,du + \int_0^{\infty} \frac{u}{1+u^2}(1-F(u))\,du\right).$$

Then $R(z)$ may be written in the form

$$R(z) = \frac{C_1}{z} \exp\left(-\int_{-\infty}^{0} \frac{F(u)}{z-u}\,du + \int_0^{\infty} \frac{1-F(u)}{z-u}\,du\right),$$

and it is easy to see that condition (2.1.9) is equivalent to $C_1 = 1$. □

(2.2.5) EXAMPLE. Let $F(u) = F_{x,y}(u)$ be the distribution function of a pair of interlacing sequences defined by equation (1.2.1). Then the exponential representation (2.2.4) simplifies to (2.1.3).

The topology of the space $\mathcal{F}$ of Rayleigh functions in 1.4 was designed in such a way that the representation (2.2.4) establishes a homeomorphism of the spaces $\mathcal{F}$ and $\mathcal{N}$. The subset $\mathcal{F}_0$ of interlacing sequences corresponds to the subset $\mathcal{N}_0$ of rational functions in $\mathcal{N}$.

The exponential representation (2.2.4) takes a particularly simple form if F is a Rayleigh function of *bounded variation*, i.e., if it is the distribution function $F(u) = \tau\{(-\infty, u)\}$ of a Rayleigh measure τ (see Theorem 2.4.1 below).

2.3. The Markov transform. An immediate consequence of the existence of the canonical representations (2.2.4) and (2.1.8) is the following fact which is central for the present paper.

(2.3.1) THEOREM. *For every Rayleigh function $F \in \mathcal{F}$ there exists a unique probability distribution $\mu \in \mathcal{M}$ such that*

$$(2.3.2) \qquad \int_{-\infty}^{\infty} \frac{\mu(dt)}{z-t} = \frac{1}{z} \exp\left(-\int_{-\infty}^{0} \frac{F(u)}{z-u}\,du + \int_{0}^{\infty} \frac{1-F(u)}{z-u}\,du\right)$$

for all $z \in \mathbb{C} \setminus \mathbb{R}$. The transformation $F \mapsto \mu$ establishes a homeomorphism between the space $\mathcal{F}$ of Rayleigh functions and the space $\mathcal{M}$ of probability distributions.

We refer to (2.3.2) as the *master identity*, and we call μ the *Markov transform* of F. The relationship between the two basic integral representations (2.1.6), (2.2.2) of a generic function R of negative imaginary type had been studied in great detail (see [**20, 28**] and the very informative papers [**3, 4**]). We derive a number of basic properties of the Markov transform just as simplified versions of the more general results of [**3**].

We start with a simple remark concerning the supports of F and μ. If the Rayleigh function F in the master identity is identically zero in some interval, $F(u) \equiv 0$ for $a < u < b$, then $\mu\{(a,b)\} = 0$, and the same conclusion holds if $F(u) \equiv 1$ for $a < u < b$. We say that a Rayleigh function F is *supported* by an interval $[a,b]$ if $F(u) \equiv 0$ for $u < a$ and $F(u) \equiv 1$ for $u > b$. The smallest interval supporting a Rayleigh function F coincides with the smallest interval supporting its Markov transform μ.

Another nice feature of the transform $F \mapsto \mu$ is that it commutes with the affine transformations of the real line: if F and μ are related by (2.3.2), then the measure $\widetilde{\mu}$ with the distribution function $\widetilde{\mu}\{(-\infty, u)\} = \mu\{(-\infty, au+b)\}$ is the Markov transform of the Rayleigh function $\widetilde{F}(u) = F(au+b)$, for any $a > 0, b \in \mathbb{R}$.

We turn now to a characterization of the discrete and singular parts of the measure μ in terms of the corresponding Rayleigh function F. Recall that a probability measure μ is said to be *singular*, if $\mu(A) = 1$ for a measurable set $A \subset \mathbb{R}$ of zero Lebesgue measure.

(2.3.3) THEOREM (Aronszajn and Donoghue [**3**, Theorem IX]). *Let $\mu \in \mathcal{M}$ and $F \in \mathcal{F}$ be related by the master identity (2.3.2). Then μ is singular if and only if $F = \chi_A$ is (mod 0 equivalent to) a characteristic function of a measurable set $A \subset \mathbb{R}$. In particular, if A is a union of a finite number of intervals in $\mathbb{R}$, then the measure μ is discrete and finitely supported.*

This beautiful result is worth specializing to the still simpler case of compactly supported measures.

(2.3.4) COROLLARY. *The identity*

$$(2.3.5) \qquad \int_{0}^{1} \frac{\mu(du)}{z-u} = \frac{1}{z} \exp \int_{B} \frac{du}{z-u}, \qquad z \in \mathbb{C} \setminus [0,1],$$

establishes a bijective correspondence between the singular probability distributions μ in the interval $[0,1]$, and the (classes mod 0 of) measurable subsets $B \in [0,1]$.

One can easily describe the *discrete* part of a measure μ in terms of the corresponding Rayleigh function F.

(2.3.6) THEOREM (Aronszajn and Donoghue [**3**, Theorem IV]). *Let μ be the Markov transform of a Rayleigh function F (i.e., the master identity (2.3.2) holds).*

Then the weight $\mu\{c\}$ of μ at a point $c \in \mathbb{R}$ equals

$$(2.3.7) \qquad \mu\{c\} = \exp\left(-\int_{-\infty}^{c} \frac{F(u)}{|u-c|} du - \int_{c}^{\infty} \frac{1-F(u)}{|u-c|} du\right).$$

In particular, $\mu\{c\} = 0$ unless both integrals in (2.3.7) converge.

2.4. The master identity for interlacing measures. If a Rayleigh function F is of bounded variation, $F \in \mathcal{F}_b$, it can be regarded as the distribution function of some Rayleigh measure τ. In this case the master identity can be written in a more attractive form.

(2.4.1) COROLLARY. *Assume that $F(u) = L_\tau(u)$ is the distribution function of a Rayleigh measure τ. Then*

$$\frac{1}{z} \exp\left(-\int_{-\infty}^{0} \frac{F(u)}{z-u} du + \int_{0}^{\infty} \frac{1-F(u)}{z-u} du\right) = \exp\left(-\int_{-\infty}^{\infty} \ln(z-u)\,\tau(du)\right),$$

and the master identity (2.3.2) takes the form

$$(2.4.2) \qquad \int_{-\infty}^{\infty} \frac{\mu(du)}{z-u} = \exp \int_{-\infty}^{\infty} \ln\frac{1}{z-u}\,\tau(du), \qquad \operatorname{Im} z > 0.$$

PROOF. Direct integration by parts. $\square$

Note that the right-hand side of (2.4.2) is nothing else than the multiplicative version of the Cauchy–Stieltjes transform. Since

$$\frac{d}{dz} \int_{-\infty}^{\infty} \ln(z-u)\,d\tau(u) = \int_{-\infty}^{\infty} \frac{d\tau(u)}{z-u},$$

the master identity (2.4.2) can be written in the equivalent form

$$(2.4.3) \qquad -\frac{d}{dz} \ln \int_{-\infty}^{\infty} \frac{\mu(du)}{z-u} = \int_{-\infty}^{\infty} \frac{\tau(du)}{z-u}.$$

Assume that τ is a bounded signed measure on the real line such that the integral $\int_{-\infty}^{\infty} \ln(1+|u|)\,|\tau(du)|$ converges. Then the right-hand side of (2.4.2) determines a function $R(z)$ of negative imaginary type if and only if τ is a Rayleigh measure.

(2.4.4) EXAMPLE. Take for τ the Cauchy distribution

$$\tau(du) = \frac{1}{\pi} \frac{du}{1+u^2}.$$

One can easily check that

$$\frac{1}{\pi} \int_{-\infty}^{\infty} \frac{du}{(z-u)(1+u^2)} = \frac{1}{z+i}, \qquad \operatorname{Im} z > 0,$$

$$\exp\left(-\frac{1}{\pi} \int_{-\infty}^{\infty} \frac{\ln(z-u)}{(1+u^2)} du\right) = \frac{1}{z+i}, \qquad \operatorname{Im} z > 0,$$

so that the additive and the multiplicative Cauchy–Stieltjes transforms of τ coincide. It follows that in this case the Markov transform μ is identical with τ. As it was shown in [44], the Cauchy distributions are the only fixed points of the Markov transform.

Let μ denote the Markov transform of a Rayleigh measure τ. It follows from a result of Aronszajn and Donoghue ([**3**, Theorem VIII]) that μ is absolutely continuous, except possibly for a finite set of point masses. Cifarelly and Regazzini have found in [**8**] the following explicit formula for the density of μ in the particular case of a *positive* measure τ (the same formula was independently conjectured in [**24**] for *all* Rayleigh measures).

(2.4.5) THEOREM (Cifarelly and Regazzini [**8**]). *Let μ be the Markov transform of a probability distribution τ in a finite interval $[a,b]$. Then the measure μ is absolutely continuous, and its density equals*

$$(2.4.6) \qquad \frac{\mu(dx)}{dx} = \frac{\sin \pi L_\tau(x)}{\pi} \exp \int_a^b \ln \frac{1}{|x-u|} \tau(du),$$

where $L_\tau(x) = \tau\{(-\infty, x)\}$ is the distribution function of τ.

Recall that the point masses of μ were described by equation (2.3.7).

§3. The moments of interlacing measures

3.1. The definition of moments. It follows from a result by Aronszajn and Donoghue ([**3**, Theorem A-d]) that a Rayleigh measure τ has the moment

$$(3.1.1) \qquad p_n = \int_{-\infty}^{\infty} u^n \tau(du)$$

of some order n if and only if its Markov transform μ has the moment

$$(3.1.2) \qquad h_n = \int_{-\infty}^{\infty} u^n \mu(du)$$

of the same order. Because of the close connection between interlacing measures $\tau_\pm$ and the Rayleigh measure $\tau = \tau_+ - \tau_-$, we shall sometimes call p_n the *nth moment of interlacing measures* τ_+, τ_-.

(3.1.3) EXAMPLE. The mth moment of interlacing sequences (1.1.1) equals

$$p_m = \sum_{k=1}^{n} x_k^m - \sum_{k=1}^{n-1} y_k^m, \qquad m = 0, 1, 2, \ldots.$$

The mth moment of the distribution μ determined by the partial fraction decomposition (2.1.4) can be written as

$$h_m = \sum_{k=0}^{m} (-1)^k \sum y_{i_1} \cdots y_{i_k} x_{j_1} \cdots x_{j_{m-k}},$$

where the indices $i_1 < \cdots < i_k$ in the second sum increase strictly, while the indices $j_1 \leqslant \cdots \leqslant j_{m-k}$ increase weakly.

3.2. The Markov transform in terms of moments. Assume that a probability distribution $\mu \in \mathcal{M}$ has finite moments of all orders. Then its Cauchy–Stieltjes transform can be considered as the moment generating function,

$$(3.2.1) \qquad \int_{-\infty}^{\infty} \frac{\mu(du)}{z-u} = \sum_{n=0}^{\infty} \frac{h_n}{z^{n+1}}.$$

The series in the right-hand side may very well diverge for all $z \in \mathbb{C}$, but it is always the asymptotic series of the function in the left-hand side, as $z \to \infty$, with respect to every domain $\operatorname{Im} z \geqslant \varepsilon > 0$. This means that

$$R(z) = \sum_{n=0}^{N-1} \frac{h_n}{z^{n+1}} + o\left(\frac{1}{z^N}\right), \qquad N = 1, 2, \ldots,$$

if $|z| \to \infty$, $|\operatorname{Im} z| > \varepsilon > 0$ (see [**36**, Theorem II.2.1] or [**18**, Chapter 9]). This implies that

$$h_n = \lim_{z \to \infty} z^{n+1} \left(R(z) - \sum_{k=0}^{n-1} \frac{h_k}{z^{k+1}} \right),$$

so that the moments are truly determined by the function $R(z)$. For a compactly supported measure μ, the series in (3.2.1) converges for sufficiently large $|z|$, and its sum equals $R(z)$.

In a similar way,

$$(3.2.2) \qquad \ln z - \int_{-\infty}^{\infty} \ln(z - u) \tau(du) = \sum_{n=1}^{\infty} \frac{p_n}{nz^n}$$

is the asymptotic series of the function in the left-hand side. One can readily derive the relations between the moments of the measures τ and μ.

(3.2.3) THEOREM. *Assume that a Rayleigh measure τ and its Markov transform μ have moments of all orders. Then the master identity (2.4.2) implies the identity*

$$(3.2.4) \qquad \sum_{n=0}^{\infty} h_n z^n = \exp \sum_{n=1}^{\infty} \frac{p_n}{n} z^n$$

of moment generating functions (to be understood as the equality of formal power series).

In many situations the moments (3.1.1) determine the measure μ uniquely, though this is by no means true in the general case. One says (see, e.g., [**36**]) that the moment problem for the measure μ is *determined* or *indeterminate* accordingly. Note that the moment problem is determined for every compactly supported measure.

(3.2.5) COROLLARY. *The moment problem for a Rayleigh measure τ is determined if and only if the moment problem for its Markov transform μ is determined. If the measures τ, μ are determined by their moments uniquely, the master identity (2.4.2) is equivalent to the identity (3.2.4).*

(3.2.6) REMARK. Let μ be a probability distribution with finite moments $\{h_n\}_{n=0}^{\infty}$. Recall that *cumulants* (or *semi-invariants*) $\{s_n\}_{n=1}^{\infty}$ of the measure μ can be defined by their exponential generating series

$$(3.2.7) \qquad \sum_{n=1}^{\infty} \frac{s_n}{n!} z^n = \ln \sum_{n=0}^{\infty} \frac{h_n}{n!} z^n.$$

The series in the right-hand side of (3.2.7) is nothing but the Laplace transform $\int e^{zu} \mu(du)$ of the measure μ. It follows that $s_n(\mu * \nu) = s_n(\mu) + s_n(\nu)$ for all

$n = 1, 2, \ldots,$ where $\mu * \nu$ is the convolution of the measures μ, ν. The same property holds for the numbers $\beta_n = s_n/(n-1)!$. Equation (3.2.7) takes the form

$$\ln \sum_{n=0}^{\infty} \frac{h_n}{n!} z^n = \sum_{n=1}^{\infty} \frac{\beta_n}{n} z^n,$$

similar to equation (3.2.4). Nevertheless, there seem to be no simple relations between the numbers β_n and p_n.

3.3. **Explicit formulas.** The identity (3.2.4) of formal power series can be written as a series of polynomial identities between their coefficients. For instance, one can easily check that

$$(3.3.1) \qquad h_n = \sum_{\rho \vdash n} \frac{p_1^{r_1} p_2^{r_2} \cdots}{1^{r_1} r_1! \, 2^{r_2} r_2! \cdots}, \qquad n = 1, 2, \ldots,$$

where the sum runs over all partitions $\rho = (1^{r_1}, 2^{r_2}, \ldots)$ of $n = r_1 + 2r_2 + \cdots$. Since $n!/(1^{r_1} r_1! \, 2^{r_2} r_2! \cdots)$ equals the number of permutations $\pi \in \mathfrak{S}_n$ with exactly $r_k = r_k(\pi)$ cycles of length k, one can rewrite (3.3.1) in the form

$$(3.3.2) \qquad h_n = \frac{1}{n!} \sum_{\pi \in \mathfrak{S}_n} \prod_{k \geq 1} p_k^{r_k(\pi)}.$$

Taking logarithmic derivatives of both sides of equation (3.2.4), one readily derives a useful recurrence relation (see [**32**, I.2.11])

$$(3.3.3) \qquad n h_n = \sum_{k=1}^{n} h_{n-k} p_k .$$

The identities relating the coefficients h_n, p_n of the series in (3.2.4) may also be written as the following determinantal identities (see [**32**, I.2, Ex. 8]):

$$(3.3.4) \qquad h_n = \frac{1}{n!} \begin{vmatrix} p_1 & -1 & 0 & \cdots & 0 \\ p_2 & p_1 & -2 & \cdots & 0 \\ \vdots & \vdots & \vdots & \ddots & \vdots \\ p_{n-1} & p_{n-2} & p_{n-3} & \cdots & -n+1 \\ p_n & p_{n-1} & p_{n-2} & \cdots & p_1 \end{vmatrix}.$$

The first few equations are as follows:

$$(3.3.5) \qquad \begin{aligned} h_1 &= p_1, \\ 2!\, h_2 &= p_1^2 + p_2, \\ 3!\, h_3 &= p_1^3 + 3 p_1 p_2 + 2 p_3, \\ 4!\, h_4 &= p_1^4 + 6 p_1^2 p_2 + 3 p_2^2 + 8 p_1 p_3 + 6 p_4 . \end{aligned}$$

Conversely,

$$(3.3.6) \qquad p_n = (-1)^{n-1} \begin{vmatrix} h_1 & 1 & 0 & \cdots & 0 \\ 2 h_2 & h_1 & 1 & \cdots & 0 \\ \vdots & \vdots & \vdots & \ddots & \vdots \\ n h_n & h_{n-1} & h_{n-2} & \cdots & h_1 \end{vmatrix}.$$

In particular,

(3.3.7)
$$\begin{aligned}p_1 &= h_1, \\ p_2 &= 2h_2 - h_1^2, \\ p_3 &= 3h_3 - 3h_1 h_2 + h_1^3, \\ p_4 &= 4h_4 - 4h_1 h_3 - 2h_2^2 + 4h_1^2 h_2 - h_1^4.\end{aligned}$$

3.4. The moments of Rayleigh functions. Using integration by parts, the definition of the moment p_n in (3.1.1) can be stated in the form

$$(3.4.1) \qquad p_n = -\int_{-\infty}^0 n u^{n-1} F(u)\, du + \int_0^\infty n u^{n-1}(1 - F(u))\, du,$$

where $F(u) = \tau\{(-\infty, u)\}$ is the distribution function of τ. The right-hand side makes sense even if the function F has infinite variation.

(3.4.2) DEFINITION. Given a Rayleigh function $F \in \mathcal{F}$ and an integer $n = 1, 2, \ldots$, assume that the integrals in the right-hand side of (3.4.1) converge absolutely. Then the real number p_n determined by this formula will be called the nth *moment of the Rayleigh function* F. We also set $p_0 \equiv 1$ for all $F \in \mathcal{F}$. The linear functional

$$(3.4.3) \qquad \mathcal{M}_F(Q) = Q(0) - \int_{-\infty}^0 Q'(u) F(u)\, du + \int_0^\infty Q'(u)(1 - F(u))\, du,$$

$Q \in \mathbb{C}[u]$, will be referred to as the *moment functional* of F.

As a matter of fact, one can also write the moment functional as

$$(3.4.4) \qquad \mathcal{M}_F(Q) = Q(c) - \int_{-\infty}^c Q'(u) F(u)\, du + \int_c^\infty Q'(u)(1 - F(u))\, du,$$

since the right-hand side does not depend on the choice of c.

The definition is consistent with that of the moments of a Rayleigh measure, and equations (3.2.4), (3.3.2), (3.3.4) hold for all $F \in \mathcal{F}$.

(3.4.5) EXAMPLE. Let $F = F_{x,y}$ be the distribution function (1.2.1) of a pair of interlacing sequences $x_1 < y_1 < x_2 < \cdots < y_{n-1} < x_n$. Then the moment functional of F equals

$$(3.4.6) \qquad \mathcal{M}_F(Q) = \sum_{k=1}^n Q(x_k) - \sum_{k=1}^{n-1} Q(y_k).$$

3.5. Interlacing measures and symmetric group characters. According to a well-known theorem due to Hamburger, a sequence $\{h_n\}_0^\infty$ is a moment sequence of a positive measure if and only if the inequalities $\det(h_{i+j})_{i,j=0}^n \geq 0$ hold for all $n = 0, 1, \ldots$. Using (3.3.2), this result can be restated as a characterization of moment sequences of Rayleigh functions. Remarkably enough, the irreducible characters of symmetric groups are involved in this characterization.

(3.5.1) THEOREM. *For $n = 1, 2, \ldots$ let χ_n denote the irreducible character of the symmetric group $\mathfrak{S}_N$ of order $N = n(n+1)$ corresponding to the Young diagram $\lambda^{(n)} = (n^{n+1})$ with $n+1$ rows of equal length n. A sequence $\{p_n\}_{n=1}^{\infty}$ of real numbers is a moment sequence of some Rayleigh function $F \in \mathcal{F}$ if and only if*

$$(3.5.2) \quad (-1)^{n(n+1)/2} \sum_{\pi \in \mathfrak{S}_{n(n+1)}} \chi_n(\pi) \prod_{k \geq 1} \frac{1}{r_k(\pi)!} \left(\frac{p_k}{k}\right)^{r_k(\pi)} \geq 0, \quad n = 1, 2, \ldots.$$

PROOF. Consider the Schur symmetric function s_λ corresponding to the Young diagram $\lambda = (n^{n+1})$. By the Jacobi–Trudi identity (see [**32**, I.3, Ex. 8]),

$$s_\lambda = \begin{vmatrix} h_n & h_{n+1} & \cdots & h_{2n} \\ h_{n-1} & h_n & \cdots & 0 \\ \vdots & \vdots & \ddots & \vdots \\ h_0 & h_1 & \cdots & h_n \end{vmatrix} = (-1)^{n(n+1)/2} \begin{vmatrix} h_0 & h_1 & \cdots & h_n \\ h_1 & h_2 & \cdots & h_{n+1} \\ \vdots & \vdots & \ddots & \vdots \\ h_n & h_{n+1} & \cdots & h_{2n} \end{vmatrix}.$$

It is well known that a sequence $\{h_n\}_{n=0}^{\infty}$ is a moment sequence of a positive measure μ iff the last determinant is positive for each $n = 0, 1, 2, \ldots$. It remains to recall the formula expressing the Schur function s_λ in terms of the power sum symmetric polynomials p_n,

$$s_\lambda = \sum_{\pi \in \mathfrak{S}_{n(n+1)}} \chi_n(\pi) \prod_{k \geq 1} \frac{1}{r_k(\pi)!} \left(\frac{p_k}{k}\right)^{r_k(\pi)}$$

(see [**32**, I.7.10]). $\square$

As follows from the character tables in [**17**], the first inequalities (3.5.2) are $p_2 - p_1^2 \geq 0$, $18p_2p_4 - 18p_1^2p_4 - 16p_3^2 + 24p_1p_2p_3 + 8p_1^3p_3 - 9p_2^3 - 9p_1^2p_2^2 + 3p_1^4p_2 - p_1^6 \geq 0$.

3.6. **An identity for the difference derivatives.** Let τ be a probability distribution in the interval $[0, 1]$, and denote by p_n its moment (3.1.1) of order n. Define a sequence h_n, $n = 0, 1, \ldots$, by equation (3.3.2). It follows from the definition of the Markov transform that $\{h_n\}_{n=0}^{\infty}$ is also the moment sequence of a probability measure. We give a direct proof of this simple fact. An interesting feature of the proof is that it links the Markov transform with Kingman's theory of partition structures (see [**26**]).

By the classical Hausdorff theorem, a sequence $\{h_n\}_{n=0}^{\infty}$ is a moment sequence of some positive measure in the interval $[0, 1]$ if and only if it is *completely monotonic*. By definition, this means that

$$(3.6.1) \quad (\Delta^k h)_n \equiv \sum_{j=0}^{k} (-1)^j \binom{k}{j} h_{n+j} \geq 0$$

for all $k, n = 0, 1, \ldots$. Here Δ is the difference derivative operator $(\Delta h)_n = h_n - h_{n+1}$.

(3.6.2) THEOREM. *Assume that a sequence of real numbers $\{p_n\}_{n=0}^{\infty}$ is completely monotonic, and define a new sequence $\{h_n\}_{n=0}^{\infty}$ by equation (3.3.2). Then $\{h_n\}_{n=0}^{\infty}$ is completely monotonic, too.*

This result is an immediate consequence of the following explicit formula for the difference derivatives of the sequence $\{h_n\}_{n=0}^{\infty}$.

(3.6.3) LEMMA. *Given a sequence* $\{p_0 = 1, p_1, p_2, \ldots\}$, *let* $\{h_0 = 1, h_1, h_2, \ldots\}$ *denote the sequence defined by the equations*

$$(3.6.4) \qquad h_n = \frac{1}{n!} \sum_{w \in \mathfrak{S}_n} \prod_{k \geq 1} p_k^{r_k(w)},$$

$n = 0, 1, 2, \ldots$. *Then*

$$(3.6.5) \qquad (\Delta^k h)_n = \frac{1}{(n+k)!} \sum_{\pi \in \mathfrak{S}_{n+k}} \prod_{j=1}^{l(\pi)} (\Delta^{k_j(\pi)} p)_{m_j(\pi)},$$

where $l(\pi)$ *is the number of cycles in a permutation* $\pi \in \mathfrak{S}_{n+k}$, $m_j(\pi) + k_j(\pi)$ *is the number of elements in the jth cycle, and* $m_j(\pi), k_j(\pi)$ *denote the numbers of elements in the cycle which are* $\leq n$ *or* $> n$, *correspondingly.*

(3.6.6) EXAMPLE. If one takes $n = 1$ and $k = 2$, the identity simplifies to

$$h_1 - 2h_2 + h_3 = \frac{1}{6} \big(p_1(1-p_1)^2 + (1-p_1)(p_1-p_2) + p_1(1-2p_1+p_2)$$
$$+ (1-p_1)(p_1-p_2) + (p_1-2p_2+p_3) + (p_1-2p_2+p_3) \big).$$

The summands in the right-hand side correspond to the permutations $(1)(2)(3)$, $(12)(3)$, $(1)(23)$, $(12)(3)$, (123), $(132) \in \mathfrak{S}_3$.

PROOF. To prove formula (3.6.5), we introduce the appropriate notation. For any $j = 1, \ldots, n$ and a permutation $\pi \in \mathfrak{S}_n$, we denote by $d_j(\pi) \in \mathfrak{S}_{n-1}$ the permutation obtained from π by removing the element j from its cycle. Clearly, the operators d_i, d_j commute. The operator $D_k = (1-d_{m+1})(1-d_{m+2})\cdots(1-d_{m+k})$, when applied to a permutation $\pi \in \mathfrak{S}_{m+k}$, produces a formal sum of permutations

$$D_k \pi = \sum_I (-1)^{k-|I|} d_I \pi,$$

where I runs over all subsets $I \subset \{m+1, \ldots, m+k\}$, and $d_I = \prod_{i \in I} d_i$ denotes the operation of removing all the elements in I from the cycles of π. We say that the elements of the set $\{1, \ldots, m\}$ are *small*, and the elements of $\{m+1, \ldots, m+k\}$ are *big*.

To simplify notations, we shall simply write p_π instead of $p_{\rho(\pi)}$, and $p_g = \sum g_\pi p_\pi$ for an element $g = \sum g_\pi \pi$ of the group algebra $\mathbb{R}[\mathfrak{S}_n]$. The proof of (3.6.5) is a result of three simple steps.

If $\pi \in \mathfrak{S}_{m+k}$ is a permutation with a single cycle of length $n = m+k$, then the permutation $d_J \pi$ is cyclic, too. Obviously, $p_\pi = p_n$ and $p_{d_J \pi} = p_{n-j}$ for $j = |J|$, so that

$$p_{D_k \pi} = \sum_{j=0}^k (-1)^{k-j} \binom{k}{j} p_{m+k-j} = (\Delta^k p)_m.$$

For a generic permutation $\pi \in \mathfrak{S}_{m+k}$ with $l = l(\pi)$ cycles, denote by m_i the number of "small" elements in the ith cycle, and by k_i the number of "big" elements in this cycle. The choice of a set J of big elements is equivalent to independent choices of subsets $J_1, \ldots, J_l$ of big elements in the cycles of permutation π. Hence,

$$p_{D_k \pi} = \sum_{j_1=0}^{k_1} \cdots \sum_{j_l=0}^{k_l} (-1)^{k-j} \binom{k_1}{j_1} \cdots \binom{k_l}{j_l} p_{m_1+k_1-j_1} \cdots p_{m_l+k_l-j_l} = \prod_{i=1}^{l(\pi)} (\Delta^{k_i} p)_{m_i}.$$

Let us fix a subset J of big elements, and let $\mathfrak{S}_{\bar{J}}$ denote the group of permutations of the remaining set $\bar{J} = \{1, \ldots, m+k\} \setminus J$. One can easily see that for each permutation $\sigma \in \mathfrak{S}_{\bar{J}}$ there are exactly $(m+k-j+1)(m+k-j+2)\cdots(m+k)$ permutations $\pi \in \mathfrak{S}_{m+k}$ such that $d_J\pi = \sigma$. It follows that

$$(3.6.7) \qquad \frac{1}{(m+k)!} \sum_\pi p_{d_J\pi} = \frac{1}{(m+k-j)!} \sum_\sigma p_\sigma = h_{m+k-j},$$

hence

$$\frac{1}{(m+k)!} \sum_\pi p_{D_k\pi} = \sum_{j=0}^{k} \sum_{J:|J|=j} \frac{1}{(m+k)!} \sum_\pi p_{d_J\pi} = (\Delta^k h)_m,$$

and identity (3.6.5) follows. □

3.7. General partition structures.
An advantage of Theorem 3.6.2, apart from its simplicity, is that it suggests a generalization of the Markov transform.

Choose a probability distribution M_n on the symmetric group $\mathfrak{S}_n$ for all $n = 1, 2, \ldots$ in such a way that the probability $M_n(\pi)$ only depends on the cycle lengths of a permutation $\pi \in \mathfrak{S}_n$. Recall (see [26]) that the family $\{M_n\}_{n=1}^{\infty}$ is called[2] a *partition structure*, if

$$(3.7.1) \qquad M_n(\pi) = \sum_{\sigma: \sigma' = \pi} M_{n+1}(\sigma), \qquad \pi \in \mathfrak{S}_n, \ n = 1, 2, \ldots.$$

Here σ' denotes the permutation in $\mathfrak{S}_n$ obtained from $\sigma \in \mathfrak{S}_{n+1}$ by deleting the element $n+1$ from its cycle.

Formula (3.6.5) can be generalized as follows.

(3.7.2) THEOREM. *Let $\{M_n\}_{n=1}^{\infty}$ be a partition structure. Given a sequence of real numbers $p_0 = 1, p_1, p_2, \ldots$, define a new sequence $h_0 = 1, h_1, h_2, \ldots$ by the formula*

$$(3.7.3) \qquad h_n = \sum_{\pi \in \mathfrak{S}_n} M_n(\pi) p_1^{r_1(\pi)} p_2^{r_2(\pi)} \cdots, \qquad n = 1, 2, \ldots.$$

Then the difference derivatives satisfy the equation

$$(3.7.4) \qquad (\Delta^k h)_n = \sum_{\pi \in \mathfrak{S}_{n+k}} M_{n+k}(\sigma) \prod_{j=1}^{l(\pi)} (\Delta^{k_j(\pi)} p)_{m_j(\pi)},$$

with the same notations as in formula (3.6.5).

PROOF. One can argue exactly as in the proof of Lemma (3.6.3), except for formula (3.6.7). It should read as follows in the present, more general setup:

$$(3.7.5) \qquad \sum_{\pi \in \mathfrak{S}_{m+k}} p_{d_J\pi} M_{m+k}(\pi) = \sum_{\sigma \in \mathfrak{S}_{m+k-j}} p_\sigma M_{m+k-j}(\sigma) = h_{m+k-j}.$$

(3.7.6) COROLLARY. *For every partition structure $\{M_n\}_{n=1}^{\infty}$, and every probability distribution τ in the interval $[0,1]$, there exists a unique probability distribution μ in $[0,1]$ such that the moments (3.1.1), (3.1.2) are related by (3.7.3).*

---

[2] Our terminology at this point is slightly different from the original one.

(3.7.7) EXAMPLE. A distinguished family of partition structures is provided by the *Ewens sampling formula*

$$(3.7.8) \qquad M_n(\pi) = \frac{\theta^{l(\pi)}}{\theta(\theta+1)\cdots(\theta+n-1)}, \qquad \pi \in \mathfrak{S}_n,$$

where $\theta > 0$ is a parameter. In this case the relation between the measures τ and μ can be stated in the form

$$(3.7.9) \qquad \int_0^1 \frac{\mu(du)}{(z-u)^\theta} = \exp\left(-\theta \int_0^1 \ln(z-u)\,\tau(du)\right).$$

If $\theta = 1$, we return to the master identity (2.4.2).

§4. Poisson–Dirichlet random measures

4.1. Distributions of mean values of random measures.
Assuming that τ is a probability distribution, there exists a simple probabilistic construction providing its Markov transform μ (see [**8, 10**]). Let us start with the elementary case of a finitely supported measure τ.

(4.1.1) EXAMPLE. Assume that a discrete probability distribution τ has positive weights $\tau_0, \tau_1, \ldots, \tau_n$ (where $\sum \tau_k = 1$) at the points $x_0 < x_1 < \cdots < x_n$ of the real line. Denote by Δ_n the simplex of all vectors $p = (p_0, p_1, \ldots, p_n)$ with nonnegative components such that $\sum p_k = 1$. The formula

$$(4.1.2) \qquad \frac{p_0^{\tau_0-1} p_1^{\tau_1-1} \cdots p_n^{\tau_n-1}}{\Gamma(\tau_0)\Gamma(\tau_1)\cdots\Gamma(\tau_n)}\, dp_1 \cdots dp_n$$

determines the so called *Dirichlet distribution* with parameters $\tau_0, \tau_1, \ldots, \tau_n$ on the simplex Δ_n. Denote by M the discrete probability measure with a weight p_k at the point x_k, $k = 0, 1, \ldots, n$, where $(p_0, p_1, \ldots, p_n)$ is a random vector in Δ_n distributed according to (4.1.2). Then the distribution μ of the mean value $X = \sum_{k=0}^n x_k p_k$ of the random measure M is the Markov transform of τ. The measure μ is absolutely continuous, and it follows from the general formula (2.4.6) that its density equals

$$(4.1.3) \qquad \frac{\mu(dx)}{dx} = \frac{1}{\pi} \sin\left(\pi \sum_{j=0}^k \tau_j\right) \prod_{k=0}^n \frac{1}{|x-x_k|^{\tau_k}} \qquad \text{if } x_k < x < x_{k+1}.$$

A more general construction, known as the *Poisson–Dirichlet process* (see [**27**]), provides a random probability distribution M associated to every finite positive measure τ. The above example generalizes as follows.

(4.1.4) THEOREM (see [**10**]). *Assume that τ is a probability distribution and that the integral (1.3.11) converges. Let M denote a random probability measure obtained as a sample realization of the Dirichlet process with the parameter measure τ. Then the random integral*

$$(4.1.5) \qquad X = \int_{-\infty}^{\infty} x M(dx)$$

converges absolutely for almost all M and its distribution μ coincides with the Markov transform of τ:

$$\int_{-\infty}^{\infty} \frac{\mu(du)}{z-u} = \exp \int_{-\infty}^{\infty} \ln \frac{1}{z-u} \tau(du), \qquad \operatorname{Im} z \neq 0.$$

If the parameter measure of the Poisson–Dirichlet process is positive and has a finite total mass θ, we can write it in the form $\theta \tau(du)$, where τ is a probability measure. In this case the distribution μ of the mean value (4.1.5) of the random measure M is related to the parameters θ, $\tau(du)$ by the formula

$$(4.1.6) \qquad \int_{-\infty}^{\infty} \frac{\mu(du)}{(z-u)^\theta} = \exp\left(-\theta \int_{-\infty}^{\infty} \ln(z-u)\,\tau(du)\right), \qquad \operatorname{Im} z \neq 0,$$

generalizing (3.7.9).

4.2. Exchangeable sequences. A useful tool in dealing with random measures is the notion of exchangeability. By definition, a sequence $X_1, X_2, \ldots, X_n, \ldots$ of real random variables is called *exchangeable* if its joint distributions are invariant under every permutation of a finite number of its elements. Obviously, a sequence of independent random variables with a common distribution M is exchangeable. More generally, one can use a two-stage sampling: first choose a random measure M, then a sequence of independent variables with the distribution M. By the classical de Finetti theorem, every exchangeable sequence arises as a result of this procedure from an appropriate random measure M, which is determined by the sequence uniquely.

Every statement concerning a random measure M can be restated, in principle, in terms of the corresponding exchangeable sequence $X_1, X_2, \ldots$. For instance, the nth moment of the mean value $X = \int x M(dx)$ of a random measure M equals the average of the product of n distinct terms in the exchangeable sequence:

$$(4.2.1) \qquad \left\langle \left(\int_{-\infty}^{\infty} x M(dx)\right)^n \right\rangle = \langle X_1 \cdots X_n \rangle.$$

The exchangeable sequence $X_1, X_2, \ldots$ corresponding to the Poisson–Dirichlet random measure M with the parameter distribution τ can be obtained by a simple procedure known as the *Chinese restaurant* construction (see [10]). One should imagine a series of large circular tables indexed by random numbers $x_1, x_2, \ldots$ taken independently from the distribution τ. The first guest takes a place at the first table, and we set $X_1 = x_1$. The nth guest takes a place either at the first empty table or to the immediate right of one of the $n-1$ persons already seated, with equal chances $1/n$. We assign to the nth variable X_n the label x_k of the table where the nth person took a seat, $X_n = x_k$.

There is an obvious generalization of the Chinese restaurant construction for arbitrary partition structures in place of the uniform distributions in the above description. Given a partition structure $\{M_n\}_{n=1}^{\infty}$, we associate an exchangeable sequence to every probability distribution τ in $\mathbb{R}$ as follows.

Let X_1 be a random number chosen from the distribution τ, and denote by $\pi_1 \in \mathfrak{S}_1$ the identity permutation. Arguing by induction, assume that we have already chosen the first n elements $X_1, \ldots, X_n$ of the sequence and a permutation $\pi_n \in \mathfrak{S}_n$. Independently of all previous choices, choose a permutation $\pi_{n+1} \in \mathfrak{S}_{n+1}$ such that $\pi'_{n+1} = \pi_n$ (i.e., π_n arises from π_{n+1} by deleting the element $n+1$ from its cycle), with probability $M_{n+1}(\pi_{n+1})/M_n(\pi_n)$. If the element $n+1$ is a fixed point of π_{n+1}, choose X_{n+1} independently from the distribution τ. Otherwise, if the cycle of $n+1$ contains an element $k \leqslant n$, set $X_{n+1} = X_k$ (by the construction, the variables X_k indexed by the elements of the same cycle are equal).

The generalized Chinese restaurant construction provides a random variable with the moments h_n related to the moments p_n of the parameter measure τ by the formula (3.7.3).

(4.2.2) THEOREM. *Let $X_1, \ldots, X_n, \ldots$ be the exchangeable sequence obtained via the Chinese restaurant construction from a probability distribution τ and a partition structure $\{M_n\}_{n=1}^\infty$. Denote by μ the distribution of the mean value $X = \int x M(dx)$, where the random measure M is associated with the sequence by the de Finetti theorem. Then the moments (3.1.1), (3.1.2) of the measures τ, μ are related by the formula*

$$h_n = \sum_{\pi \in \mathfrak{S}_n} M_n(\pi) \prod_{k \geqslant 1} p_k^{r_k(\pi)}, \qquad n = 1, 2, \ldots.$$

The proof coincides almost literally with that of Theorem 4.1.4 in §2 of [**10**].

§5. Combinatorics of continued fractions

Assume that a positive measure μ has finite moments of all orders, and that μ is determined by its moments uniquely. Then the Cauchy–Stieltjes transform of the measure μ, $R(z) = \int (z-u)^{-1} \mu(du)$, can be represented by a continued fraction

$$R(z) = \cfrac{1}{z - b_0 - \cfrac{\lambda_1}{z - b_1 - \cfrac{\lambda_2}{z - b_2 - \cdots}}}$$

which converges for all nonreal $z \in \mathbb{C}$. The elements $b_0, b_1, \ldots, \lambda_1, \lambda_2, \ldots$ of the fraction are real and $\lambda_n > 0$ for all $n = 1, 2, \ldots$ (such fractions are called *real J-type continued fractions*).

The moments $h_n = \int x^n \mu(dx)$ of the measure μ can be written as polynomials in the elements of the continued fraction. Flajolet [**14**] suggested a simple combinatorial description of these polynomials (see [**43**] for a more detailed account of combinatorial theory of orthogonal polynomials). We provide a similar description for the moments p_n of the Rayleigh measure τ associated to μ by the master identity (2.4.2). By the equivalent form (2.4.3) of this identity, the moments p_n can be interpreted as the coefficients of the asymptotic series at infinity,

$$-\frac{R'(z)}{R(z)} = \sum_{n=0}^\infty \frac{p_n}{z^{n+1}},$$

of the logarithmic derivative of the continued fraction above.

5.1. Motzkin paths and the power series of a continued fraction.

Consider a formal power series decomposition of a real J-type continued fraction,

$$(5.1.1) \quad \cfrac{1}{z - b_0 - \cfrac{\lambda_1}{z - b_1 - \cfrac{\lambda_2}{z - b_2 - \cdots}}} = \frac{1}{z} + \sum_{n=1}^{\infty} \frac{h_n}{z^{n+1}}.$$

The coefficients h_n can be written as polynomials in the elements $b_0, b_1, b_2, \ldots$, $\lambda_1, \lambda_2, \ldots$ of the fraction in the left-hand side. The first few examples are simple enough to be written explicitly:

$$(5.1.2) \quad \begin{aligned} h_1 &= b_0, \\ h_2 &= b_0^2 + \lambda_1, \\ h_3 &= b_0^3 + 2b_0\lambda_1 + b_1\lambda_1, \\ h_4 &= b_0^4 + 3b_0^2\lambda_1 + 2b_0 b_1 \lambda_1 + b_1^2 \lambda_1 + \lambda_1^2 + \lambda_1\lambda_2. \end{aligned}$$

For higher moments the equations are more cumbersome.

Since the coefficients of the polynomials (5.1.2) are positive integers, one can try to interpret these polynomials as generating functions of appropriate combinatorial objects. In order to identify these objects, consider the particular example of the continued fraction

$$(5.1.3) \quad \cfrac{1}{z - 1 - \cfrac{1}{z - 1 - \cfrac{1}{z - 1 - \cdots}}} = \frac{1}{z} + \frac{1}{z^2} + \frac{2}{z^3} + \frac{4}{z^4} + \frac{9}{z^5} + \cdots,$$

with all elements equal to one. Then the coefficients $h_n = M_n$ are the so called Motzkin numbers.

Recall that a *Motzkin path* can be defined as a sequence $s = (i_1, \ldots, i_n)$ of nonnegative integers such that $i_n = 0$ and $|i_k - i_{k-1}| \leqslant 1$ for all $k = 1, \ldots, n$ (we assume that $i_0 = 0$). We denote the set of Motzkin paths of length n by $\mathcal{M}_n$. It is common to visualize Motzkin paths as the lattice paths in the upper half-plane, with the endpoints at zero level (see Figure 9 in subsection 5.2).

Define the *Motzkin number* $M_n = |\mathcal{M}_n|$ as the number of Motzkin paths. It is easy to see that these numbers satisfy the recurrence relations

$$(5.1.4) \quad M_{n+1} = M_n + (M_0 M_{n-1} + M_1 M_{n-2} + \cdots + M_{n-1} M_0)$$

(for the properties and various interpretations of Motzkin numbers see [11]). It follows that the generating function

$$(5.1.5) \quad R(z) = \frac{1}{z} + \sum_{n=1}^{\infty} \frac{M_n}{z^{n+1}}$$

satisfies a quadratic equation

$$(5.1.6) \quad R^2(z) - (z-1)R(z) + 1 = 0,$$

and one readily obtains the explicit formula

(5.1.7) $$R(z) = \frac{z - 1 - \sqrt{z^2 - 2z - 3}}{2}.$$

Since the continued fraction in the left-hand side of (5.1.3) also satisfies equation (5.1.6), the identity (5.1.3) follows.

One can see from (5.1.3) that the coefficients of the polynomial representing h_n sum up to M_n; hence the monomials are likely to be associated with Motzkin paths. To see this, one labels the entries i_k of a Motzkin path $s \in \mathcal{M}_n$ by the appropriate elements m_k of the continued fraction (5.1.1), to obtain the monomial $m_1 \cdots m_n$ associated with the path s.

The labeling rule is remarkably simple. Given a path $s = (i_1, \ldots, i_n) \in \mathcal{M}_n$, define the factors m_k as

$$m_k = \begin{cases} \lambda_{i_k+1}, & \text{if } i_k = i_{k-1} - 1, \\ b_{i_k}, & \text{if } i_k = i_{k-1}, \\ 1, & \text{if } i_k = i_{k-1} + 1 \end{cases}$$

(see Figure 9), and set $m_s = m_1 \cdots m_n$ for the monomial in the elements $b_0, b_1, \ldots, \lambda_1, \lambda_2, \ldots$ of the fraction.

(5.1.8) THEOREM (Flajolet [14]). *The nth coefficient h_n in the power series decomposition of the continued fraction (5.1.1) equals the sum of all monomials associated with Motzkin paths of length n,*

(5.1.9) $$h_n = \sum_{s \in \mathcal{M}_n} m_s.$$

5.2. Cyclic Motzkin paths and the logarithm of a continued fraction.

Let us write the formal power series of the logarithmic derivative of a J-type continued fraction as

(5.2.1) $$-\frac{d}{dz} \ln \frac{1}{z - b_0 - \cfrac{\lambda_1}{z - b_1 - \cfrac{\lambda_2}{z - b_2 - \cdots}}} = \frac{1}{z} + \sum_{n=1}^{\infty} \frac{p_n}{z^{n+1}}.$$

We aim at the combinatorial description of the coefficients p_n as polynomials in the elements of the fraction in the left-hand side, similar to that of the Flajolet theorem.

The first few examples of such polynomials are as follows:

(5.2.2)
$$\begin{aligned} p_1 &= b_0, \\ p_2 &= b_0^2 + 2\lambda_1, \\ p_3 &= b_0^3 + 3b_0\lambda_1 + 3b_1\lambda_1, \\ p_4 &= b_0^4 + 4b_0^2\lambda_1 + 4b_0 b_1 \lambda_1 + 4b_1^2 \lambda_1 + 2\lambda_1^2 + 4\lambda_1 \lambda_2. \end{aligned}$$

Note that if one derives these equations from (3.3.7) and (5.1.2), it is rather surprising that all the coefficients are positive.

The sums of coefficients can be obtained from the particular example of the fraction (5.1.3). To see this, set

$$R(z) \equiv \frac{1}{z} + \sum_{n=1}^{\infty} \frac{M_n}{z^{n+1}}$$

and denote by E_n the coefficients of the power series of the logarithmic derivative,

$$-\frac{R'(z)}{R(z)} = \frac{1}{z} + \sum_{n=1}^{\infty} \frac{E_n}{z^{n+1}}.$$

From equation (5.1.6) one readily derives that $-R(z)/R'(z) = z - 1 - 2R(z)$, and it follows from (5.1.7) that the generating function for the numbers E_n equals

(5.2.3) $$-\frac{R'(z)}{R(z)} = \frac{1}{\sqrt{z^2 - 2z - 3}} = \frac{1}{z} + \frac{1}{z^2} + \frac{3}{z^3} + \frac{7}{z^4} + \frac{19}{z^5} + \cdots.$$

The first coefficients $E_1 = 1$, $E_2 = 3$, $E_3 = 7$, $E_4 = 19$ conform to the formulas (5.2.2). These numbers arise in combinatorics in the so called "king walks problem" studied by Euler himself (see [35]). In Euler's problem, the number E_n represents the constant term in the decomposition of $(z + 1 + z^{-1})^n$.

Instead of the chess board king walks of Euler, we prefer to speak of shifted Motzkin paths.

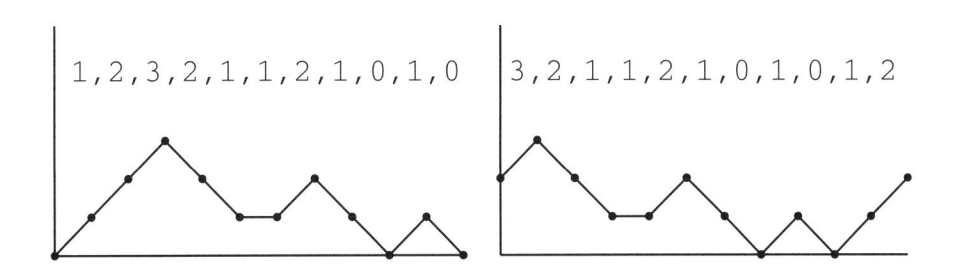

FIGURE 9. Ordinary and shifted Motzkin paths with the monomial $m_s = b_1 \lambda_1^2 \lambda_2^2 \lambda_3$.

(5.2.4) DEFINITION. A sequence $s = (i_1, \ldots, i_n)$ is called a *shifted Motzkin path* if it is conjugate to a true Motzkin path, i.e., it can be obtained as a cyclic shift of an ordinary Motzkin path (see Figure 9). If s is a cyclic shift of some $t \in \mathcal{M}_n$, we set $m_s = m_t$. The set of all shifted Motzkin n-paths is denoted by $\mathcal{S}_n$.

In accordance with the examples (5.2.2), the monomials for nineteen cyclic Motzkin paths of length 4 are distributed as follows: there are four paths for each of the monomials $b_0^2 \lambda_1$, $b_0 b_1 \lambda_1$, $b_1^2 \lambda_1$, $\lambda_1 \lambda_2$, two paths for the monomial λ_1^2, and one for the monomial b_0^4. The latter three are $(1, 0, 1, 0)$, $(0, 1, 0, 1)$, and $(0, 0, 0, 0)$.

By the solution (5.2.3) of Euler's king walks problem, $E_n = |\mathcal{S}_n|$ is the number of shifted Motzkin paths.

(5.2.5) THEOREM. *The nth coefficient p_n in the formal power series expansion of the logarithm of the continued fraction (5.2.1) equals the sum of all monomials associated with shifted Motzkin paths of length n,*

$$(5.2.6) \qquad p_n = \sum_{s \in S_n} m_s.$$

PROOF. The theorem will follow as a direct corollary of a more general result in the next section. □

(5.2.7) EXAMPLE. Assume that $b_0 = b_1 = \cdots = 0$ and $\lambda_1 = \lambda_2 = \cdots = 1$. Then $h_{2n} = \binom{2n}{n}/(n+1)$ is the Catalan number. The number of shifted Dyck paths of length $2n$ (with the increments ± 1 only) equals $p_{2n} = \binom{2n}{n}$. Note that the measures μ and τ in this example are the semi-circle law and the arcsine law, correspondingly,

$$\mu(du) = \frac{1}{2\pi}\sqrt{4-u^2}\,du, \qquad \tau(du) = \frac{1}{\pi}\frac{du}{\sqrt{4-u^2}}.$$

The identity (3.2.4) in this example is equivalent to formula (9.6) in Chapter XII.9 of Feller's book [13], applied to the simple symmetric random walk over integers.

5.3. **An abstract lemma.** Consider an alphabet A and a set P of finite words in A^* called *prime*. Let W denote the set of all words representable as finite concatenations of prime words, and assume that such a representation is unique for all words in W.

(5.3.1) EXAMPLE. Let $A = \mathbb{N}$ be the set of nonnegative integers and P the set of *prime Motzkin paths* (a path $s = (i_1, i_2, \ldots, i_n)$ is said to be prime if $i_n = 0$ is its only zero entry). Every Motzkin path splits canonically into prime Motzkin subpaths.

Let $R = \mathbb{Q}[P]$ be the ring of polynomials in the variables m_p, $p \in P$. With every word $w \in W$ we associate a monomial $m_w = m_{p_1} \cdots m_{p_k} \in R$, where $w = p_1 \cdots p_k$ is the unique representation of w as a concatenation of prime subwords. Write

$$(5.3.2) \qquad h_n = \sum_{w \in W_n} m_w \in R$$

for the sum of monomials corresponding to the words of length n in W.

Denote by S the set of words conjugate to some words in W (i.e., to the words representable as cyclic shifts of words in W). The elements of S are called *shifted* words. If a shifted word $s \in S$ is a cyclic shift of a word $w \in W$, we set $m_s = m_w$ for the monomial associated with s. Denote by

$$(5.3.3) \qquad p_n = \sum_{s \in S_n} m_s \in R$$

the sum of monomials associated with all shifted words of length n.

(5.3.4) LEMMA. *The polynomials (5.3.2), (5.3.3) satisfy the identities*

$$(5.3.5) \qquad nh_n = \sum_{k=1}^n h_{n-k} p_k, \qquad n = 1, 2, \ldots.$$

PROOF. Let $[n] \times W_n$ be the set of pairs (j, w), where $w \in W_n$ and the integer j *marks* the jth character of the word w. Let $\widetilde{W}_n = \bigcup_{k=1}^n W_{n-k} \times S_k$ be the set of pairs (v, s), where $v \in W_{n-k}$ and $s \in S_k$ for some $k = 1, \ldots, n$. In order to prove (5.3.5), we establish a bijection $I \colon [n] \times W_n \to \widetilde{W}_n$ such that $I(j, w) = (v, s)$ implies $m_w = m_v m_s$.

Given $(j, w) \in [n] \times W_n$, denote by p the prime subword of w containing the jth (marked) character. Let v be the concatenation of all prime subwords of w strictly to the left of p, and denote by $\tilde{s}$ the concatenation of p and of the other prime factors of w to the right of p. Let $s \in S$ be the cyclic shift of $\tilde{s}$ such that the marked character comes first. By definition, $I(j, w) = (v, s)$.

The inverse bijection is obvious. Given a pair $(v, s) \in \widetilde{W}_n$, denote by $\tilde{s} \in W$ the only cyclic shift of s such that the first character a of s belongs to the first prime factor of $\tilde{s}$. Set the marker j to the position of a in $\tilde{s}$, and denote by $w = v\tilde{s}$ the concatenation of v and $\tilde{s}$. Then $I(j, w) = (v, s)$, and the Lemma follows. □

(5.3.6) EXAMPLE. Take the Motzkin path $w = (1\,2\,1\,1\,0\,0\,1\,0\,1\,1\,\mathbf{0}\,1\,2\,1\,0) \in \mathcal{M}_{15}$ with the marked entry $i_{11} = 0$. The prime subpaths of w are $(1\,2\,1\,1\,0)$, (0), $(1\,0)$, $(1\,1\,0)$ and $(1\,2\,1\,0)$. Hence, $v = (1\,2\,1\,1\,0\,0\,1\,0)$, $\tilde{s} = (1\,1\,\mathbf{0}\,1\,2\,1\,0)$ and $s = (\mathbf{0}\,1\,2\,1\,0\,1\,1)$.

5.4. A bijective proof of identity (3.3.2). The formula

$$(5.4.1) \qquad n! \, h_n = \sum_{\pi \in \mathfrak{S}_n} \prod_{k \geq 1} p_k^{r_k(\pi)}$$

is equivalent to the identity (5.3.5), hence is true under the assumptions of Lemma 5.3.4. Nevertheless, we present a direct bijective proof of (5.4.1). We use the notations and terminology of 5.3. In particular, A is an alphabet, P a set of *prime* words, W the set of concatenations of words in P, and S the set of words in A^* conjugate to some words in W (shifted words).

Consider a permutation $\pi \in \mathfrak{S}_n$ in the cycle notation,

$$\pi = (j_1^{(1)}, \ldots, j_{k_1}^{(1)})(j_1^{(2)}, \ldots, j_{k_2}^{(2)}) \cdots (j_1^{(l)}, \ldots, j_{k_l}^{(l)}).$$

We denote by l the number of cycles in π, and by $j_2^{(t)} = \pi(j_1^{(t)}), \ldots, j_1^{(t)} = \pi(j_{k_t}^{(t)})$ the elements of the tth cycle. Let $\widetilde{\mathfrak{S}}_n$ be the set of pairs (π, s), where $\pi \in \mathfrak{S}_n$ and

$$s = (i_1^{(1)}, \ldots, i_{k_1}^{(1)})(i_1^{(2)}, \ldots, i_{k_2}^{(2)}) \cdots (i_1^{(l)}, \ldots, i_{k_l}^{(l)})$$

is a family of shifted words $s_t = (i_1^{(t)}, \ldots, i_{k_t}^{(t)})$, $t = 1, \ldots, l$, matching the cycle decomposition of π. Denote by $m(\pi, s) = m_{s_1} \cdots m_{s_l} \in R$ the product of monomials m_{s_t} associated to shifted words s_t in 5.3. It is obvious that

$$\sum_{\pi \in \mathfrak{S}_n} \prod_{k \geq 1} p_k^{r_k(\pi)} = \sum_{(\pi, s) \in \widetilde{\mathfrak{S}}_n} m(\pi, s),$$

and even more obvious that

$$n! \, h_n = \sum_{(\sigma, w) \in \mathfrak{S}_n \times W_n} m_w.$$

In order to prove (5.4.1), it suffices to establish a bijection $G\colon \mathfrak{S}_n \times W_n \to \widetilde{\mathfrak{S}}_n$ such that $G(\sigma, w) = (\pi, s)$ implies $m(\pi, s) = m_w$.

We proceed with the description of the bijection G. Given a pair $(\sigma, w) \in \mathfrak{S}_n \times W_n$, write σ as a sequence of its images,

(5.4.2) $$\sigma = (\sigma(1), \ldots, \sigma(n)).$$

Split the word $w = p_1 \cdots p_d$ into its prime subwords, and let $\sigma = \sigma_1 \cdots \sigma_d$ be the factorization of the sequence (5.4.2) into subsequences $\sigma_1, \ldots, \sigma_d$ of lengths matching the lengths of the prime factors $p_1, \ldots, p_d$ of the word w.

Arrange the sequences $\sigma_1, \ldots, \sigma_d$ in the increasing order of their first elements, and let $\rho \in \mathfrak{S}_d$ be the permutation that restores the original order of the pieces $\sigma_1, \ldots, \sigma_d$. Write the permutation ρ of the sequences $\sigma_1, \ldots, \sigma_d$ in the cycle notation to obtain a permutation $\pi \in \mathfrak{S}_n$. In a similar way, replace each σ_k in the cycle notation of ρ by the corresponding prime word p_k to obtain a family s of shifted words matching the cycles of π. Set $G(\sigma, w) = (\pi, s)$.

(5.4.3) EXAMPLE. We apply here the map G to the pair

$$\binom{\sigma}{w} = \begin{pmatrix} 7 & 16 & 3 & 4 & 13 & 15 & 10 & 8 & 2 & 12 & 9 & 1 & 5 & 11 & 14 & 6 \\ 1 & 1 & 0 & 0 & 1 & 0 & 1 & 1 & 2 & 1 & 0 & 1 & 2 & 1 & 1 & 0 \end{pmatrix}$$

in $\mathfrak{S}_{16} \times \mathcal{M}_{16}$. The path w factors as a product of prime Motzkin paths,

$$w = (1\ 1\ 0)\ (0)\ (1\ 0)\ (1\ 1\ 2\ 1\ 0)\ (1\ 2\ 1\ 1\ 0),$$

and the matching factorization of σ results in a sequence of words

$$(7\ 16\ 3)\ (4)\ (13\ 15)\ (10\ 8\ 2\ 12\ 9)\ (1\ 5\ 11\ 14\ 6).$$

We arrange these words in the increasing order of their first elements,

$$a(1\ 5\ 11\ 14\ 6)\ b(4)\ c(7\ 16\ 3)\ d(10\ 8\ 2\ 12\ 9)\ e(13\ 15).$$

To restore the original order, one must apply the permutation

$$\rho = \begin{pmatrix} a & b & c & d & e \\ c & b & e & d & a \end{pmatrix} = (a\ c\ e)\ (b)\ (d).$$

Replacing each letter a, b, c, d, e by the corresponding word, we derive the permutation

$$\pi = (1\ 5\ 11\ 14\ 6\ 7\ 16\ 3\ 13\ 15)\ (4)\ (2\ 12\ 9\ 10\ 8),$$

and the corresponding family of shifted Motzkin paths:

$$(1\ 2\ 1\ 1\ 0\ 1\ 1\ 0\ 1\ 0)\ (0)\ (2\ 1\ 0\ 1\ 1).$$

Note that the word $d = (2\ 12\ 9\ 10\ 8)$ and the corresponding prime Motzkin path $(2\ 1\ 0\ 1\ 1)$ were shifted cyclically according to the convention that each cycle of a permutation should start with its least element.

To describe the inverse bijection G^{-1}, start with a pair $(\pi, s) \in \widetilde{\mathfrak{S}}_n$. Read off the shifted words $s_1, \ldots, s_l$ corresponding to the cycles of π and split each of these words into prime subwords in P. Split the cycles of π into subsequences $\sigma_1, \ldots, \sigma_d$ matching the factorization of the words $s_1, \ldots, s_l$. Let $\rho \in \mathfrak{S}_d$ be the permutation induced on the set $\{\sigma_1, \ldots, \sigma_d\}$ by the cycles of π. Denote by $\sigma^1, \ldots, \sigma^d$ the sequence of pieces $\sigma_1, \ldots, \sigma_d$ arranged in the increasing order of their first elements, and consider the concatenation of sequences $\rho(\sigma^1), \ldots, \rho(\sigma^d)$. This is a sequence of integers $1, \ldots, n$, all entries being distinct, so that it determines a permutation $\sigma \in \mathfrak{S}_n$. Denote by w the concatenation of the prime words $p^{(1)}, \ldots, p^{(d)}$ matching the pieces $\rho(\sigma^1), \ldots, \rho(\sigma^d)$. It is easy to see that $G(\sigma, w) = (\pi, s)$.

(5.4.4) EXAMPLE. We apply here the map G^{-1} to the pair $(\pi, s) \in \widetilde{\mathfrak{S}}_{16}$, where

$$\binom{\pi}{s} = \binom{(1 \ \ 14 \ \ 12 \ \ 6 \ \ 10 \ \ 13 \ \ 9)}{(1 \ \ \ 0 \ \ \ 1 \ \ 2 \ \ \ 1 \ \ \ 0 \ \ 1)} \ \binom{(2 \ \ 3 \ \ 15 \ \ 16 \ \ 8 \ \ 5 \ \ 11 \ \ 4)}{(0 \ \ 1 \ \ \ 1 \ \ \ 0 \ \ 1 \ \ 1 \ \ \ 0 \ \ 1)} \ \binom{(7)}{(0)}.$$

The system s of shifted Motzkin paths splits into prime Motzkin subpaths (1 2 1 0), (1 1 0), (1 1 0), (1 1 0), (1 0), (0). The corresponding splitting of π results in subsequences (12 6 10 13), (9 1 14), (3 15 16), (8 5 11), (4 2), (7). We arrange these pieces in the order of their first elements, $a = (3\ 15\ 16)$, $b = (4\ 2)$, $c = (7)$, $d = (8\ 5\ 11)$, $e = (9\ 1\ 14)$, $f = (12\ 6\ 10\ 13)$. The cycles of π induce the permutation

$$\rho = (a\,d\,b)\,(c)\,(e\,f) = \begin{pmatrix} a & b & c & d & e & f \\ d & a & c & b & f & e \end{pmatrix}.$$

Hence,

$$\binom{\sigma}{w} = \binom{8 \ \ 5 \ \ 11 \ | \ 3 \ \ 15 \ \ 16 \ | \ 7 \ | \ 4 \ \ 2 \ | \ 12 \ \ 6 \ \ 10 \ \ 13 \ | \ 9 \ \ 1 \ \ 14}{1 \ \ 1 \ \ \ 0 \ | \ 1 \ \ \ 1 \ \ \ 0 \ | \ 0 \ | \ 1 \ \ 0 \ | \ 1 \ \ 2 \ \ \ 1 \ \ \ 0 \ | \ 1 \ \ 1 \ \ \ 0}.$$

§6. Interlacing measures in operator theory

We want now to interpret a measure μ and a function F related by the master identity (2.3.2) in terms of a self-adjoint operator A in a Hilbert space H and a cyclic unit vector $\xi \in H$. The content of this section is closely related to the theory of the *spectral shift function*, see [**14, 15**].

6.1. **Statement of the theorem.** Let A denote a bounded self-adjoint operator in a separable Hilbert space H. Given a unit vector $\xi \in H$, $\|\xi\| = 1$, one can define the *spectral measure of A at the vector ξ* by the identity

$$((zI - A)^{-1}\xi, \xi) = \int_{-\infty}^{\infty} \frac{\mu(du)}{z - u}, \qquad z \in \mathbb{C} \setminus \mathbb{R},$$

or, equivalently, by the equation

(6.1.1) $$(Q(A)\xi, \xi) = \int_{-\infty}^{\infty} Q(u)\,\mu(du)$$

valid for every polynomial $Q \in \mathbb{C}[x]$. Since we assume (for simplicity) that the operator A is bounded, the measure μ is compactly supported and has the moments $h_n = \int u^n \mu(du)$ of all orders. Using (6.1.1), one can also write the moments directly in terms of A and ξ,

(6.1.2) $$h_n = (A^n\xi, \xi), \qquad n = 0, 1, 2, \ldots.$$

Denote by H_ξ the hyperplane in H orthogonal to the vector ξ, and by P_ξ the orthogonal projection $P_\xi \eta = \eta - (\eta, \xi)\xi$, $\eta \in H$, on the hyperplane H_ξ. Given a bounded operator B in H, we write $B_\xi = P_\xi B P_\xi$ for the restriction of B to the hyperplane H_ξ.

For each polynomial $Q \in \mathbb{C}[x]$, the operator $Q(A) - Q(A_\xi)$ is of trace class, and we consider the linear functional $Q \mapsto \operatorname{tr}(Q(A) - Q(A_\xi))$.

(6.1.3) THEOREM. *There exists a Rayleigh function F such that*

(6.1.4) $$\operatorname{tr}(Q(A) - Q(A_\xi)) = \mathcal{M}_F(Q)$$

is the moment functional (3.4.3) of F. The function F and the spectral measure μ of the operator A with respect to a vector ξ determine each other via the master identity

(6.1.5) $$\int_{-\infty}^\infty \frac{\mu(dt)}{z-t} = \frac{1}{z} \exp\left(-\int_{-\infty}^0 \frac{F(u)}{z-u}\,du + \int_0^\infty \frac{1-F(u)}{z-u}\,du\right).$$

The idea of the *trace formula* (6.1.4) belongs to Lifshits [30] and was extensively developed by Krein (see [29]). The function F in this context is called the *spectral shift function* (see [7] for a recent survey of the theory).

A direct combinatorial proof of the Theorem based on the equation (3.2.4) for the moments of F and μ will be given in 6.3. Let us first consider the finite-dimensional case.

6.2. The finite-dimensional case. Let $H = \mathbb{C}^n$ have the finite dimension n. With no loss of generality we assume that the vector ξ is *cyclic* for the operator A, i.e., the vectors $\xi, A\xi, A^2\xi, \ldots$ generate all of H. As a result, the eigenvalues of A are all distinct,

$$x_1 < \cdots < x_n.$$

Choose a basis $\{\eta_k\}_{k=1}^n$ in H in which the matrix of A is diagonal, $A\eta_k = x_k\eta_k$, $k = 1,\ldots,n$. Since the vector $\xi = \sum_{k=1}^n a_k \eta_k$ is cyclic, all of its coordinates a_k are nonzero.

(6.2.1) LEMMA. *The spectral measure μ of the operator A at the vector ξ is discrete; it has positive weights $\mu_k = |a_k|^2$ at the points x_k, $k = 1,\ldots,n$. The eigenvalues $y_1,\ldots,y_{n-1}$ of the restriction A_ξ of A to the hyperplane H_ξ orthogonal to ξ are all distinct and interlace with the eigenvalues of A,*

(6.2.2) $$x_1 < y_1 < x_2 < \cdots < x_{n-1} < y_{n-1} < x_n.$$

The eigenvalues of A_ξ and A are related to the spectral measure μ by the identity

(6.2.3) $$\prod_{k=1}^{n-1}(z-y_k)\prod_{k=1}^n(z-x_k)^{-1} = \sum_{k=1}^n \frac{\mu_k}{z-x_k}.$$

PROOF. It is immediate that

$$(Q(A)\xi,\xi) = \sum_{k=1}^n Q(x_k)|a_k|^2 = \sum_{k=1}^n Q(x_k)\mu_k$$

for any polynomial $Q \in \mathbb{C}[x]$, hence the spectral measure is as indicated in the Lemma. *Define* the numbers $y_1,\ldots,y_{n-1}$ by formula (6.2.3). Since the coefficients of the partial fractions are positive, $\mu_k > 0$, inequalities (6.2.2) hold. It remains to show that the eigenvalues of A_ξ coincide with $y_1,\ldots,y_{n-1}$.

Define the vectors $\zeta_j \in H$, $j = 1,\ldots,n-1$, as

(6.2.4) $$\zeta_j = \sum_{k=1}^n \frac{a_k \eta_k}{y_j - x_k},$$

and note that $\zeta_j \in H_\xi$ since

$$(\zeta_j, \xi) = \sum_{k=1}^{n} \frac{\mu_k}{y_j - x_k} = 0.$$

It is also clear that

$$A\zeta_j = \sum_{k=1}^{n} \frac{a_k x_k \eta_k}{y_j - x_k} = y_j \zeta_j - \xi,$$

hence $A_\xi \zeta_j = y_j \zeta_j$, and we are done. $\square$

Equation (6.1.4) follows immediately, since $\operatorname{tr} Q(A) = \sum_{k=1}^{n} Q(x_k)$, $\operatorname{tr} Q(A_\xi) = \sum_{k=1}^{n-1} Q(y_k)$, and

$$(6.2.5) \qquad \mathcal{M}(Q) = \sum_{k=1}^{n} Q(x_k) - \sum_{k=1}^{n-1} Q(y_k) = \operatorname{tr}(Q(A) - Q(A_\xi))$$

(see Example 3.4.5).

6.3. **Proof of Theorem 6.1.3.** It suffices to prove that $p_n = \operatorname{tr}(A^n - A_\xi^n)$ for all $n = 1, 2, \ldots$, where the moments p_n of the Rayleigh function F are defined by equation (3.4.1).

Choose an orthonormal basis $\{\xi_1, \xi_2, \ldots\}$ in H_ξ and set $\xi_0 \equiv \xi$. Then $\{\xi_0, \xi_1, \xi_2, \ldots\}$ is a basis in H, and we write $a_{ij} = (A\xi_i, \xi_j)$ for the matrix elements of the operator A in this basis, $i, j = 0, 1, 2, \ldots$. The matrix of A_ξ may be obtained from that of A by replacing the entries of the first row and of the first column of A by zeros.

Note that the moment h_n of the spectral measure μ is a polynomial of degree n in the matrix elements,

$$h_n = (A^n \xi, \xi) = \sum_{k_1=0}^{\infty} \cdots \sum_{k_{n-1}=0}^{\infty} a_{0k_1} a_{k_1 k_2} \cdots a_{k_{n-1} 0}.$$

This is true for the moments p_n, too. The trace $\operatorname{tr} A = \sum_{k=0}^{\infty} a_{kk}$ of the operator A may very well diverge, but the matrix

$$A - A_\xi = \begin{pmatrix} a_{00} & a_{01} & a_{02} & \cdots \\ a_{10} & 0 & 0 & \cdots \\ a_{20} & 0 & 0 & \cdots \\ \vdots & \vdots & \vdots & \ddots \end{pmatrix}$$

has finite trace, namely $p_1 = a_{00}$. The matrix elements of the squares are

$$(A^2)_{ij} = \sum_{k=0}^{\infty} a_{ik} a_{kj}, \qquad (A_\xi^2)_{ij} = \sum_{k=1}^{\infty} a_{ik} a_{kj},$$

so that

$$A^2 - A_\xi^2 = \begin{pmatrix} \sum a_{0k} a_{k0} & \sum a_{0k} a_{k1} & \sum a_{0k} a_{k2} & \cdots \\ \sum a_{1k} a_{k0} & a_{10} a_{01} & a_{10} a_{02} & \cdots \\ \sum a_{2k} a_{k0} & a_{20} a_{01} & a_{20} a_{02} & \cdots \\ \vdots & \vdots & \vdots & \ddots \end{pmatrix}$$

and the trace of the matrix $A^2 - A_\xi^2$ equals

$$p_2 = a_{00}^2 + 2\sum_{k=1}^{\infty} a_{0k} a_{k0}.$$

It is finite, since $h_2 = a_{00}^2 + \sum_{k=1}^{\infty} a_{0k} a_{k0}$ is a matrix element of A^2. One derives that $h_2 = (p_1^2 + p_2)/2$, which is the second of the desired equations (3.3.5).

In order to settle the general case we introduce more notation. A finite sequence $w = (i_1, \ldots, i_n)$ of nonnegative integers terminating with $i_n = 0$ will be referred to as a *chain*, or *n-chain*. Let $\mathcal{H}_n$ denote the set of n-chains. With every such chain w we associate a monomial

(6.3.1) $$a_w = a_{i_n i_1} a_{i_1 i_2} \cdots a_{i_{n-1} i_n}$$

in the matrix elements a_{ij} of the operator A. A chain w is *prime* if $i_n = 0$ is its only zero entry. Every chain factors canonically as a concatenation of prime subchains.

Any sequence $s = (i_1, \ldots, i_n)$ of nonnegative integers will be called a *shifted chain* if it has at least one zero entry, not necessarily the last one. Let $\mathcal{P}_n$ be the set of all shifted n-chains. Define the monomial m_s of a shifted chain by equation (6.3.1). The monomial m_s does not change if one shifts the chain cyclically.

(6.3.2) LEMMA. *The polynomials h_n, p_n coincide with the generating functions of the sets $\mathcal{H}_n$, $\mathcal{P}_n$,*

(6.3.3) $$h_n \equiv (A^n \xi, \xi) = \sum_{s \in \mathcal{H}_n} a_s, \qquad p_n \equiv \operatorname{tr}(A^n - A_\xi^n) = \sum_{s \in \mathcal{P}_n} a_s.$$

They satisfy equations (3.3.2) and (3.3.3).

PROOF. The ith diagonal matrix element of the matrix A^n equals

$$(A^n)_{ii} = \sum_{i_1=0}^{\infty} \cdots \sum_{i_{n-1}=0}^{\infty} a_{i i_1} a_{i_1 i_2} \cdots a_{i_{n-1} i}.$$

Since the operator A^n is bounded, the series converges absolutely. In particular, the moment $h_n = (A^n)_{00}$ is finite and coincides with the first generating function in (6.3.3). A similar equation holds for the matrix elements of the operator A_ξ^n,

$$(A_\xi^n)_{ii} = \sum_{i_1=1}^{\infty} \cdots \sum_{i_{n-1}=1}^{\infty} a_{i i_1} a_{i_1 i_2} \cdots a_{i_{n-1} i},$$

but the indices now run over positive integers only. Hence p_n is the generating function of shifted chains in $\mathcal{P}_n$. The series converge absolutely, since

(6.3.4) $$\sum_{s \in \mathcal{P}_n} |a_s| \leqslant n \sum_{s \in \mathcal{H}_n} |a_s| < \infty.$$

The last claim of the Lemma follows at once from the abstract Lemma 5.3.4. Indeed, for the alphabet $\mathcal{A} = \mathbb{N}^2$ take the set of all pairs (i, j) of nonnegative integers. A chain $(i_1, \ldots, i_n)$ determines a word $w = ((0, i_1), (i_1, i_2), \ldots, (i_{n-1}, 0))$ in $\mathcal{A}^*$ and every such word factors uniquely as a product of *prime* words (corresponding to prime chains). The valuation $v: (i, j) \mapsto a_{ij}$ determines a ring homomorphism $v: R \to \mathbb{C}$, and equations (3.3.2), (3.3.3) follow from (5.3.5) and (5.4.1), respectively. $\square$

(6.3.5) REMARK. Theorem 6.1.3 is actually equivalent to our Theorem 5.2.5. Indeed, one can always choose the basis $\{\xi_n\}_{n=0}^{\infty}$ in such a way that the matrix of A is a three-diagonal Jacobi matrix. Then the set $\mathcal{M}_n$ of Motzkin paths is a subset in $\mathcal{H}_n$, and $a_w = 0$ for $w \in \mathcal{H}_n \setminus \mathcal{M}_n$. In a similar fashion, the set $\mathcal{S}_n$ of shifted Motzkin paths is a subset of $\mathcal{P}_n$. The generating functions (6.3.3) coincide with those of equations (5.1.9), (5.2.6), and equation (6.1.4) is equivalent to (5.2.6).

§7. Interlacing measures as generalized Young diagrams

7.1. **The Plancherel growth of Young diagrams.** Recall that to every Young diagram λ there corresponds a positive integer $d(\lambda)$, which admits several equivalent definitions:[3]

$d(\lambda)$ is the number of standard Young tableaux of shape λ;

$d(\lambda)$ is the dimension of the irreducible representation of the symmetric group $\mathfrak{S}_n$ associated with λ;

$d(\lambda)$ is given by the *hook formula* $d(\lambda) = n! / \prod h_{ij}$;

$d(\lambda)$ is the only solution of the recurrence relation $d(\nu) = \sum_{\lambda: \lambda \nearrow \nu} d(\lambda)$ satisfying the initial condition $d(\varnothing) = 1$ for the empty diagram $\lambda = \varnothing$. (We write $\lambda \nearrow \nu$ to show that a Young diagram ν differs from μ by exactly one extra box.)

The numbers $d(\lambda)$ also satisfy the identity $\sum_{\nu: \lambda \nearrow \nu} d(\nu) = (N+1) d(\lambda)$, where $N = |\lambda|$ denotes the number of boxes in λ. By virtue of this formula we associate the transition distribution

$$(7.1.1) \qquad p(\lambda, \nu) = \frac{d(\nu)}{(N+1) d(\lambda)}, \qquad \lambda \nearrow \nu,$$

to each Young diagram λ. The corresponding Markov chain on the set $\mathbb{Y}$ of all Young diagrams (starting with the empty diagram) is called the *Plancherel growth process*. One can easily check that the probability of obtaining a Young diagram λ in the process of Plancherel growth equals $d^2(\lambda)/|\lambda|!$, the probability of λ with respect to the Plancherel measure of the symmetric group $\mathfrak{S}_n$.

Recall that a Young diagram λ can be represented by a pair (1.1.1) of interlacing sequences (see Example 1.1.4). If there is a legal transition from λ to a Young diagram ν, $\lambda \nearrow \nu$, then the content $c(b)$ of the new box $b = \nu \setminus \lambda$ coincides with a point in the first sequence $x_1, \ldots, x_n$. Conversely, to every point x_k there corresponds a unique Young diagram ν_k such that $\lambda \nearrow \nu_k$ and $c(\nu_k \setminus \lambda) = x_k$. Using this remark, we consider the Plancherel transition distribution (7.1.1) as a discrete probability measure μ on the real line. The following observation was first made in [22].

(7.1.2) THEOREM. *Let $x_1 < y_1 < x_2 < \cdots < y_{n-1} < x_n$ be the interlacing sequences of contents of consecutive corner points at the border of a Young diagram λ (see Figure 1). Denote by $\mu_k = d(\nu_k)/|\nu_k| d(\lambda)$ the Plancherel transition probability of attaching to λ a new box with the content $c(\nu_k \setminus \lambda) = x_k$, $k = 1, \ldots, n$. Then*

$$(7.1.3) \qquad \sum_{k=1}^{n} \frac{\mu_k}{z - x_k} = \frac{(z - y_1) \cdots (z - y_{n-1})}{(z - x_1)(z - x_2) \cdots (z - x_n)}$$

for all $z \in \mathbb{C} \setminus \{x_1, \ldots, x_n\}$.

---

[3] See [32] for the terminology and basic facts related to Young diagrams and tableaux.

Since (7.1.3) is a particular case of the master identity (2.4.2), one can restate the theorem as follows.

(7.1.4) COROLLARY. *The Plancherel transition distribution of a Young diagram λ coincides with the Markov transform of the interlacing sequences (1.1.1) associated with λ.*

Note that the set of Rayleigh functions corresponding to Young diagrams (see Figure 2) is dense in the space $\mathcal{F}$ of all Rayleigh functions. The Markov transform of a Rayleigh function generalizes the correspondence between a Young diagram and its Plancherel transition distribution. More precisely, the Markov transform is simply the unique continuous extension of the above correspondence.

7.2. **Representing Rayleigh functions by diagrams.** Given a pair of interlacing sequences

(7.2.1) $$x_1 < y_1 < x_2 < \cdots < x_{n-1} < y_{n-1} < x_n,$$

one can visualize it using a shape similar to that of a Young diagram.

(7.2.2) DEFINITION. A continuous piecewise linear function $\omega\colon \mathbb{R} \to \mathbb{R}$ is called a *rectangular diagram* if $\omega'(u) = \pm 1$ and there exists a constant c such that $\omega(u) = |u - c|$ for sufficiently large $|u|$ (see Figure 10).

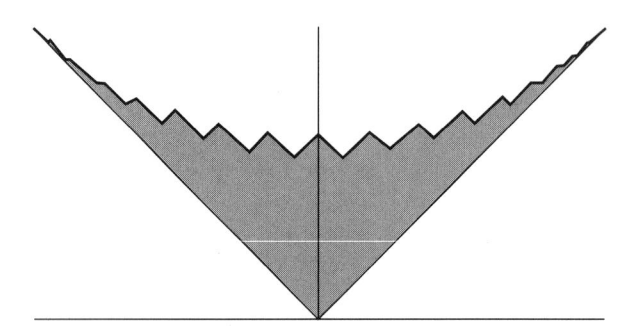

FIGURE 10. A rectangular diagram.

With a rectangular diagram we associate a pair (7.2.1) of its interlacing maxima and minima points. Conversely, every pair (7.2.1) determines a rectangular diagram such that $c = \sum x_k - \sum y_k$ and $\omega'(u) = 1$ for $u \in (x_k, y_k)$; $\omega'(u) = -1$ for $u \in (y_k, x_{k+1})$, $k = 1, \ldots, n-1$. If two polynomials $Q(u) = \prod(x - y_k)$, $P(u) = \prod(x - x_k)$ have interlacing roots (7.2.1), then $\omega'(u) = \operatorname{sign}(Q(u)/P(u))$. For instance, Figure 10 represents the rectangular diagram of interlacing roots of Chebyshev polynomials of the second kind of degrees 15 and 16.

This correspondence between interlacing sequences and rectangular diagrams can be extended to general Rayleigh functions.

Assume that a function $\omega\colon \mathbb{R} \to \mathbb{R}$ satisfies the Lipshitz condition

(7.2.3) $$|\omega(u_2) - \omega(u_1)| \leqslant |u_2 - u_1|, \qquad u_1, u_2 \in \mathbb{R}.$$

Then its derivative $\omega'(u)$ exists for a.a. $u \in \mathbb{R}$, and $|\omega'(u)| \leqslant 1$.

(7.2.4) DEFINITION. We say that a function $\omega\colon \mathbb{R} \to \mathbb{R}$ *represents a diagram* if it satisfies condition (7.2.3) and the following integrals converge:

(7.2.5) $$\int_{-\infty}^{-1} (1+\omega'(u)) \frac{du}{|u|} < \infty, \qquad \int_{1}^{\infty} (1-\omega'(u)) \frac{du}{u} < \infty.$$

Two such functions are *equivalent* if they differ by a constant. A *diagram* is a class of equivalent functions satisfying the conditions (7.2.3), (7.2.5) (see Figure 11).

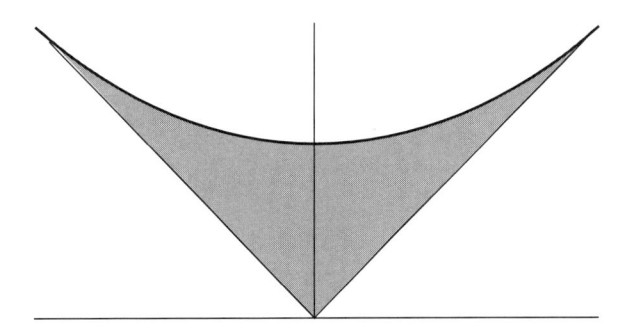

FIGURE 11. A diagram corresponding to a Rayleigh function.

We endow the space $\mathcal{D}$ of all diagrams with the topology of uniform convergence on compact subsets in $\mathbb{R}$. More precisely, a family ω_n of diagrams *converges* to the diagram $\omega \in \mathcal{D}$ if one can choose the representing functions to be uniformly converging on finite intervals.

Obviously, the equations
(7.2.6)
$$F(x) = \frac{1}{2}(1+\omega'(x)), \quad 1 - F(x) = \frac{1}{2}(1-\omega'(x)), \quad \omega(x) = \int_0^x (2F(u)-1)\,du$$

establish a homeomorphism between the space of diagrams $\mathcal{D}$ and the space of Rayleigh functions $\mathcal{F}$.

(7.2.7) EXAMPLE. If $F(u) = 1/2 + (1/\pi)\arctan u$ is the Cauchy distribution function, then the corresponding diagram is represented by the function $\omega(u) = (2/\pi)(u \arctan u - \ln\sqrt{1+u^2})$, so that $\omega'(u) = (2/\pi)\arctan u$.

The existence of asymptotes for a diagram $\omega \in \mathcal{D}$,

(7.2.8) $$\lim_{x \to -\infty} (\omega(x)+x) = c, \qquad \lim_{x \to +\infty} (\omega(x)-x) = d,$$

is equivalent to the existence of the first moment p_1 of the corresponding Rayleigh function F. In this case one can replace the last formula in equation (7.2.6) by

$$\omega(x) = \int_{-\infty}^{x} F(u)\,du + \int_{x}^{\infty} (1-F(u))\,du$$

(see Figure 12 on the next page). Then $-d = c = p_1$ in equation (7.2.8), the asymptotes cross at zero level, and the region $\{(x,y) : |x-c| \leqslant y \leqslant \omega(x)\}$ bounded by the asymptotes and the graph of $y = \omega(x)$ generalizes the shape of a Young diagram. The area of this shape (i.e., "the number of boxes") equals one half of

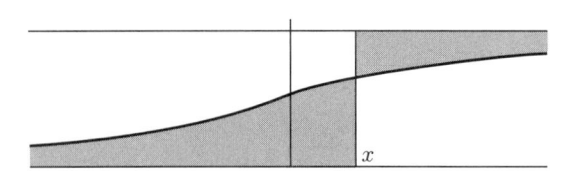

FIGURE 12. The value $\omega(x)$ via a Rayleigh function.

the variance $p_2 - p_1^2$ of the corresponding Rayleigh function F, and also equals the variance $h_2 - h_1^2$ of the measure μ associated with F by equation (2.3.2).

(7.2.9) EXAMPLE. Take for F the distribution function of the arcsine law

$$\tau(du) = \frac{1}{\pi} \frac{du}{\sqrt{4-u^2}}, \qquad |u| \leqslant 2.$$

Then the function

$$(7.2.10) \qquad \Omega(x) = \begin{cases} (2/\pi)(u \arcsin(u/2) + \sqrt{4-u^2}), & |u| \leqslant 2, \\ |u|, & |u| \geqslant 2, \end{cases}$$

represents the corresponding diagram (note that $\Omega'(u) = (2/\pi) \arcsin(u/2)$ for $|u| \leqslant 2$). The graph of the function $y = \Omega(x)$ is shown in Figure 13, along with that of the rectangular diagram of Figure 10. One can observe the convergence of diagrams ω_n corresponding to the interlacing zeros of Chebyshev polynomials to the diagram (7.2.10). See [21, 23] for more details. The limiting shape (7.2.10) was first found in the papers [31, 41, 42] in connection with the asymptotics of large Young diagrams.

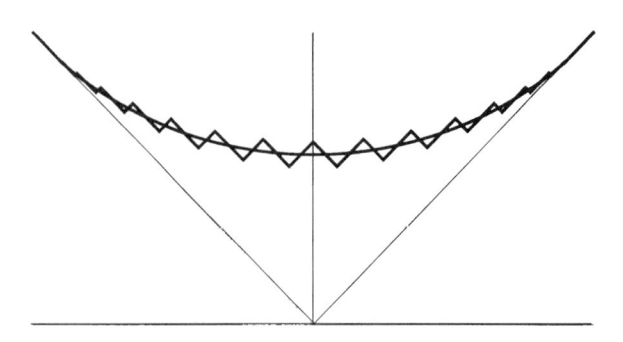

FIGURE 13. The diagram of the arcsine law.

§8. Exponential representations related to Gaussian measure

8.1. The exponential representation of the Gaussian distribution. Let τ be an absolutely continuous measure with the density

$$(8.1.1) \qquad \frac{d\tau(u)}{du} = \frac{1}{\sqrt{2\pi}} \frac{1}{|D_{-1}(iu)|^2}, \qquad -\infty < u < \infty,$$

where

$$D_{-1}(z) = e^{z^2/4} \int_z^\infty e^{-x^2/2} dx = \sqrt{\frac{\pi}{2}} e^{z^2/4} \operatorname{Erfc}\left(\frac{z}{\sqrt{2}}\right)$$

is a parabolic cylinder function. It is shown in [**5**] that τ is a probability distribution, and that its Stieltjes transform $T(z)$ has a simple continued fraction decomposition,

$$(8.1.2) \qquad T(z) = \int_{-\infty}^\infty \frac{d\tau(u)}{z-u} = \cfrac{1}{z - \cfrac{1}{z - \cfrac{2}{z - \cfrac{3}{z - \cdots}}}}.$$

(8.1.3) THEOREM. *The standard Gaussian measure with the density*

$$\frac{d\mu(u)}{du} = \frac{1}{\sqrt{2\pi}} e^{-u^2/2}, \qquad -\infty < u < \infty,$$

admits the exponential representation

$$(8.1.4) \qquad \frac{1}{\sqrt{2\pi}} \int_{-\infty}^\infty \frac{e^{-u^2/2}\, du}{z-u} = \exp\left(-\frac{1}{\sqrt{2\pi}} \int_{-\infty}^\infty \frac{\ln(z-u)\, du}{|D_{-1}(iu)|^2}\right).$$

Equivalently, μ is the Markov transform of the measure τ.

PROOF. Recall that the Stieltjes transform of the Gaussian measure may be written as

$$(8.1.5) \quad R(z) = \frac{1}{\sqrt{2\pi}} \int_{-\infty}^\infty \frac{e^{-u^2/2}}{z-u} = \cfrac{1}{z - \cfrac{1}{z - \cfrac{2}{z - \cfrac{3}{z - \cdots}}}} = \frac{1}{z} + \sum_{n=1}^\infty \frac{(2n-1)!!}{z^{2n-1}}.$$

Considering the master identity in its form (2.4.3), we must show that $R'(z) + R(z)T(z) = 0$, i.e., that

$$-\frac{d}{dz}\ln \cfrac{1}{z - \cfrac{1}{z - \cfrac{2}{z - \cdots}}} = \cfrac{1}{z - \cfrac{2}{z - \cfrac{3}{z - \cdots}}}.$$

Taking the derivative of the formal power series in (8.1.5) one readily obtains

$$(8.1.6) \qquad -R'(z) = \frac{1}{z^2} + \frac{1\cdot 3}{z^4} + \frac{1\cdot 3\cdot 5}{z^6} + \cdots = zR(z) - 1.$$

On the other hand, it follows from the continued fraction representation of $R(z)$ that

(8.1.7) $$(z - T(z))R(z) = 1.$$

The identity $R'(z) + R(z)T(z) = 0$ is an immediate consequence of equations (8.1.6) and (8.1.7). □

One can feel rather awkward about taking the derivative of a diverging series in (8.1.6), and desire a more combinatorial proof. Then the argument can be as follows. Our aim is to prove that the moments $\{p_{2n}\}$, $\{h_{2n}\}$ of the distributions τ, μ are related by formula (3.3.3), which can also be written as

(8.1.8) $$(2n+1)h_{2n} = \sum_{k=0}^{n} h_{2k} p_{2n-2k}, \qquad n = 1, 2, \ldots.$$

Identity (8.1.7) implies that $h_0 = 1$ and

(8.1.9) $$h_{2n+2} = \sum_{k=0}^{n} h_{2k} p_{2n-2k}, \qquad n = 1, 2, \ldots.$$

Since $h_{2n+2} = (2n+1)!! = (2n+1)h_{2n}$, equation (8.1.8) follows.

(8.1.10) REMARK. The first few moments of the distribution (8.1.1) are

$$p_0 = 1, \quad p_2 = 2, \quad p_4 = 10, \quad p_6 = 74, \quad p_8 = 706, \quad p_{10} = 8162.$$

Touchard [38] suggests the following combinatorial interpretation of these numbers.

Consider a system s of nonintersecting semi-circular arcs in the upper half plane, with the endpoints in the set $I_n = \{1, \ldots, 2n\}$ (see Figure 14). If the set I_n can be split in two nonempty subintervals, $I_n = I_k \cup \{2k+1, \ldots, 2n\}$, in such a way that no arc in the system s has its endpoints in different intervals, then Touchard calls the system *decomposable*. If no such decomposition exists, the system s is *indecomposable*, and we denote by $\tilde{p}_{2n}$ the number of indecomposable systems on the set I_{2n+2}.

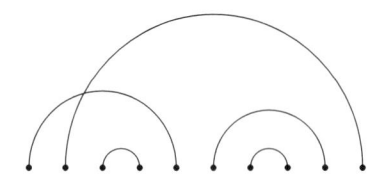

FIGURE 14. An indecomposable system of 5 arcs.

Every system of arcs splits uniquely as a union of an indecomposable subsystem on a subinterval $I_{2k+2} \subset I_{2n+2}$, and the remaining system of arcs on the complementary interval. As a result, there is a recurrence relation

(8.1.11) $$(2n+1)!! = \sum_{k=0}^{n} (2k-1)!! \, \tilde{p}_{2n-2k}.$$

Comparing (8.1.11) and (8.1.9), one derives that $\tilde{p}_{2n} = p_{2n}$, so that the number p_{2n} enumerates the indecomposable systems of $(n+1)$ arcs.

(8.1.12) REMARK. Let $Q_n(x)$, $n = 0, 1, 2, \ldots$, be the sequence of unital polynomials orthogonal with respect to the measure τ. Then $Q_0(x) = 1$, $Q_1(x) = x$, and other polynomials can be obtained from the recurrence relation

$$(8.1.13) \qquad Q_{n+1}(x) = xQ_n(x) - (n+1)Q_{n-1}(x), \qquad n = 1, 2, \ldots.$$

The coefficients of the recurrence relation coincide with the elements of the continued fraction (8.1.2). One has the following combinatorial description of these polynomials in the spirit of [43].

Let D and R be two finite sets with the total number of $|D|+|R| = n$ elements. Call an *arc* any pair of distinct points in $D \cup R$. By definition, an *n-chip* is a system s of arcs in $D \cup R$ with disjoint endpoints such that
1. every point in D belongs to some arc;
2. every arc has at least one point in R.

Denote by C_n the set of all n-chips. For instance, all three 2-chips are shown in Figure 15. Let $a(s)$ denote the number of arcs in an n-chip $s \in C_n$. Then

$$(8.1.14) \qquad Q_n(x) = \sum_{s \in C_n} (-1)^{a(s)} x^{n-2a(s)}.$$

The formula follows easily from equation (4.18) in [5].

FIGURE 15. All three chips with two vertices.

8.2. A generalization of Theorem 8.1.3.

Askey and Wimp have found in [5] the explicit orthogonality relations for associated Hermite polynomials. Using their result, we shall find the exponential representations (2.4.2) for the corresponding distributions.

Recall that the *parabolic cylinder function* $y = D_\nu(z)$ can be defined as the solution of the Weber equation

$$\frac{d^2 y}{dz^2} + \left(\nu + \frac{1}{2} - \frac{z^2}{4}\right) y = 0$$

with the initial conditions

$$D_\nu(0) = \frac{\Gamma(1/2) \, 2^{\nu/2}}{\Gamma((1-\nu)/2)}, \qquad D'_\nu(0) = \frac{\Gamma(-1/2) \, 2^{(\nu-1)/2}}{\Gamma(-\nu/2)}$$

(see [33, p. 324]). If $\nu = -c < 0$, there is an integral representation (see [33, equation (8.1.4)])

$$D_{-c}(z) = \frac{e^{-z^2/4}}{\Gamma(c)} \int_0^\infty e^{-zx} x^{c-1} e^{-x^2/2} dx,$$

so that up to a constant factor the function $e^{z^2/4} D_{-c}(z)$ is the Laplace transform of the so called χ-distribution with c degrees of freedom (see [13]).

Consider a measure μ_c with the density

$$(8.2.1) \qquad \frac{d\mu_c(u)}{du} = W(u,c) \equiv \frac{1}{\sqrt{2\pi}\,\Gamma(c+1)} \frac{1}{|D_{-c}(iu)|^2}, \qquad -\infty < u < \infty.$$

We use the following result of [5] (slightly adapted for our purposes).

(8.2.2) THEOREM (Askey, Wimp). *For every $c > -1$, equation (8.2.1) determines a probability distribution. The Cauchy–Stieltjes transform of the measure μ_c has a simple continued fraction expansion, namely*

$$(8.2.3) \qquad R_c(z) \equiv \int_{-\infty}^{\infty} \frac{d\mu_c(u)}{z-u} = \cfrac{1}{z - \cfrac{c+1}{z - \cfrac{c+2}{z - \cdots}}}.$$

(Note that the continued fraction decomposition carries the same information as the recurrence relation for orthogonal polynomials; see equation (1.7) in [5].)

We show that the exponential representation of the measure μ_c is provided by the absolutely continuous probability distribution τ_c with density

$$(8.2.4) \qquad \frac{d\tau_c(u)}{du} = (c+1)\,W(u,c+1) - c\,W(u,c), \qquad -\infty < u < \infty.$$

(8.2.5) THEOREM. *For each $c > -1$, the measures (8.2.1) and (8.2.4) are related by the identity*

$$-\frac{d}{dz} \ln \int_{-\infty}^{\infty} \frac{d\mu_c(u)}{z-u} = \int_{-\infty}^{\infty} \frac{d\tau_c(u)}{z-u}.$$

PROOF. By definition (8.2.4), the Cauchy–Stieltjes transform of the measure τ_c equals

$$T_c(z) \equiv \int_{-\infty}^{\infty} \frac{d\tau_c(u)}{z-u} = (c+1)\,R_{c+1}(z) - c\,R_c(z).$$

It also follows directly from (8.2.3) that $zR_c(z) - 1 = (c+1)\,R_c(z)\,R_{c+1}(z)$. If we show that $y = R_c(z)$ satisfies the Riccati equation $R'(z) = cR_c^2(z) - zR_c(z) + 1$, this would yield the master identity in the form $R_c'(z) + R_c(z)\,T_c(z) = 0$.

It has been known since Euler and Lagrange that the continued fraction (8.2.3) is a formal solution of the Riccati equation. In order to see this, note that the substitution

$$(8.2.6) \qquad R_c(z) = -\frac{1}{c}\frac{\varphi'(z)}{\varphi(z)}$$

transforms the Riccati equation into a second order linear equation

$$(8.2.7) \qquad \varphi''(z) + z\varphi'(z) + c\varphi(z) = 0.$$

Taking consecutive derivatives of (8.2.7), one obtains

$$(8.2.8) \qquad \varphi^{(n+2)}(z) + z\varphi^{(n+1)}(z) + (c+n)\varphi^{(n)}(z) = 0.$$

Now write equations (8.2.7), (8.2.8) in the form

$$-\frac{1}{c}\frac{\varphi'(z)}{\varphi(z)} = \frac{1}{z - (-\varphi''(z)/\varphi'(z))}, \qquad -\frac{\varphi^{(n+1)}(z)}{\varphi^{(n)}(z)} = \frac{c+n}{z - (-\varphi^{(n+2)}(z)/\varphi^{(n+1)}(z))}.$$

We get, at least formally,

$$R_c(z) = \cfrac{1}{z - \cfrac{c+1}{z - \cfrac{c+2}{z - \cdots}}}.$$

Though there is a vast literature on solving Riccati equations in terms of continued fractions (see [19] and the references therein), we could not find an appropriate reference. For the reader's convenience we prove the convergence, too. In the course of the proof we use the associated Hermite polynomials introduced in [5].

8.3. Associated Hermite polynomials. Denote by $H_n(x;c)$ the polynomials defined by the recurrence relation

(8.3.1) $$H_{n+1}(x;c) = xH_n(x;c) - (c+n)H_{n-1}(x;c)$$

and the usual initial conditions. The first examples are

$$H_0(x;c) = 1$$
$$H_1(x;c) = x$$
$$H_2(x;c) = x^2 - (c+1)$$
$$H_3(x;c) = x^3 - (2c+3)x$$
$$H_4(x;c) = x^4 - (3c+6)x^2 + (c^2 + 4c + 3).$$

It follows from [5] that the polynomials $H_n(x;c)$ (called there the *associated Hermite polynomials*) are orthogonal with respect to the measure μ_c,

$$\int_{-\infty}^{\infty} H_m(x;c)H_n(x;c)\,d\mu_c(x) = \delta_{mn}.$$

In the particular case of $c = 0$ one obtains the ordinary Hermite polynomials, and the polynomials $H_n(x;1) = Q_n(x)$ coincide with those of Remark 8.1.12. Note that the combinatorial description (8.1.14) of the latter polynomials can be extended to the case of general $c > -1$:

(8.3.2) $$H_n(x;c) = \sum_{s \in C_n} (-1)^{a(s)} c^{m(s)} x^{n-2a(s)}.$$

Here $m(s)$ denotes the number of cycles in a permutation generated by the vertical arcs of an n-chip $s \in C_n$ (see Figure 16 on the next page). An arc is called *vertical*, if it has a point in D. The sets D and R are assumed to be linearly ordered, so that the vertical arcs determine a permutation in $\mathfrak{S}_d$, $d = |D|$.

From the recurrence relation (8.3.1) one can easily derive that

(8.3.3) $$H_n(x;c) = H_n(x;c+1) + H'_{n-1}(x;c+1),$$
(8.3.4) $$H_{n+1}(x;c-1) - xH_n(x;c) + cH_{n-1}(x;c+1) = 0.$$

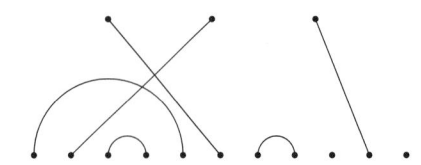

FIGURE 16. Vertical arcs of a chip determine a permutation.

Applying the elementary theory of continued fractions, one can write

$$(8.3.5) \qquad \frac{H_{n-1}(z;c+1)}{H_n(z;c)} = \cfrac{1}{z - \cfrac{c+1}{z - \cfrac{c+2}{\cdots - \cfrac{c+n-1}{z}}}} \longrightarrow R_c(z),$$

as long as $n \to \infty$, and

$$\frac{H_n(z;c)}{H_{n+1}(z;c)} = \cfrac{1}{z - \cfrac{c+n}{z - \cfrac{c+n-1}{\cdots - \cfrac{c+1}{z}}}}, \qquad \frac{H_{n-1}(z;c+1)}{H_n(z;c+1)} = \cfrac{1}{z - \cfrac{c+n}{z - \cfrac{c+n-1}{\cdots - \cfrac{c+2}{z}}}}.$$

It follows from the Flajolet Theorem 5.1.8 that the power series expansions of these functions at infinity have identical terms up to the order z^{-2n}.

Add the ratio $H'_n(z;c)/H_n(z;c)$ to the inverses of the latter fractions. Then the formulas (8.3.3), (8.3.4) yield

$$(8.3.6) \qquad \frac{H_{n+1}(z;c)}{H_n(z;c)} + \frac{H'_n(z;c)}{H_n(z;c)} = \frac{H_{n+1}(z;c-1)}{H_n(z;c)} = z - c\frac{H_{n-1}(z;c+1)}{H_n(z;c)}$$

and it follows that

$$(8.3.7)$$
$$\frac{H_n(z;c+1)}{H_{n-1}(z;c+1)} + \frac{H'_n(z;c)}{H_n(z;c)}$$
$$= \frac{H_n(z;c)}{H_{n-1}(z;c+1)} - \left(\frac{H'_{n-1}(z;c+1)}{H_{n-1}(z;c+1)} - \frac{H'_n(z;c)}{H_n(z;c)}\right)$$
$$= \frac{H_n(z;c)}{H_{n-1}(z;c+1)} - \left(\frac{H_{n-1}(z;c+1)}{H_n(z;c)}\right)' \left(\frac{H_{n-1}(z;c+1)}{H_n(z;c)}\right)^{-1}.$$

By (8.3.5), the limit of (8.3.6), as $n \to \infty$, is $z - cR_c(z)$, and (8.3.7) converges to

$$\frac{1}{R_c(z)} - \frac{R'_c(z)}{R_c(z)}.$$

Since both expressions (8.3.6), (8.3.7) have n identical first terms in their power series expansions at infinity, the limiting asymptotic series coincide, and

$$z - cR_c(z) = \frac{1}{R_c(z)} - \frac{R'_c(z)}{R_c(z)} \quad \text{for } z \in \mathbb{C} \setminus \mathbb{R}.$$

This is exactly the Riccati equation for $R_c(z)$, and the proof of Theorem 8.2.5 is completed. □

8.4. A deformation of the Gaussian distribution to the semi-circle distribution. Using the affine invariance of the master identity, one can slightly generalize Theorem 8.2.5. To this end, we denote by $\mu_{c,\beta}$ the distribution with the density

$$(8.4.1) \qquad w_{c,\beta}(u) = \frac{1}{\sqrt{\beta}} W\left(\frac{u}{\sqrt{\beta}}, \frac{c}{\beta}\right), \qquad \beta > 0.$$

In words, $\mu_{c,\beta}$ is the image of the measure $\mu_{c/\beta}$ under the dilatation $u \mapsto u\sqrt{\beta}$. In a similar way, we introduce the measure $\tau_{c,\beta}$ with the density

$$(8.4.2) \qquad \frac{d\tau_{c,\beta}(u)}{du} = \left(1 + \frac{c}{\beta}\right) w_{c+\beta,\beta}(u) - c w_{c,\beta}(u).$$

Once again, $\tau_{c,\beta}$ is the image of $\tau_{c/\beta}$ under the dilatation $v = u\sqrt{\beta}$.

(8.4.3) COROLLARY. *The measures $\mu_{c,\beta}$ and $\tau_{c,\beta}$ are related by the identity*

$$\int_{-\infty}^{\infty} \frac{d\mu_{c,\beta}(u)}{z-u} = \exp \int_{-\infty}^{\infty} \ln \frac{1}{z-u} \, d\tau_{c,\beta}(u),$$

for any choice of the parameters $c > -1$, $\beta > 0$.

PROOF. This is a direct consequence of Theorem 8.2.5. □

We complete this section with the description of the moments $h_n(c, \beta)$, $p_n(c, \beta)$ of the measures $\mu_{c,\beta}$, $\tau_{c,\beta}$. The first few examples are as follows:

$h_0(c, \beta) = 1,$ $\qquad\qquad p_0(c, \beta) = 1,$
$h_2(c, \beta) = (c + \beta),$ $\qquad\qquad p_2(c, \beta) = 2(c + \beta),$
$h_4(c, \beta) = (c + \beta)(2c + 3\beta),$ $\qquad\qquad p_4(c, \beta) = 2(c + \beta)(3c + 5\beta),$
$h_6(c, \beta) = (c + \beta)(5c^2 + 17c\beta + 15\beta^2),$ $\qquad p_6(c, \beta) = 2(c + \beta)(10c^2 + 38c\beta + 37\beta^2).$

(8.4.4) LEMMA. *The moments $h_{2n}(c, \beta)$, $p_{2n}(c, \beta)$ are homogeneous polynomials of degree n in the parameters c, β. The coefficient of c^n is the Catalan number $\binom{2n}{n}/(n+1)$ for $h_{2n}(c, \beta)$, and the binomial coefficient $\binom{2n}{n}$ for $p_{2n}(c, \beta)$.*

PROOF. Taking into account the general relations (3.3.3) between the moments h_n and p_n, and Example 5.2.7, it suffices to prove our claim for h_n. By the definition of the measure $\mu_{c,\beta}$, its Cauchy–Stieltjes transform $R_{c,\beta}(z)$ is related to that of the measure μ_c by the equation

$$R_{c,\beta}(z) = \frac{1}{\sqrt{\beta}} R_{c/\beta}\left(\frac{z}{\sqrt{\beta}}\right).$$

Using the continued fraction expansion (8.2.3) of $R_c(z)$, one derives that

$$(8.4.5) \qquad R_{c,\beta}(z) = \cfrac{1}{z - \cfrac{c+\beta}{z - \cfrac{c+2\beta}{z - \cfrac{c+3\beta}{z - \cdots}}}},$$

so that the elements $\lambda_n = c + n\beta$ of the continued fraction are linear and homogeneous in c and β. The Lemma now follows from the Flajolet Theorem 5.1.8. $\square$

(8.4.6) COROLLARY. *If $\beta = 0$, then $\mu_{c,0}$ is the semi-circle distribution*

$$\frac{d\mu_{c,0}(u)}{du} = \frac{1}{2\pi c}\sqrt{4c - u^2}, \qquad -\infty < u < \infty,$$

with the variance c. Hence, the family $\mu_{c,\beta}$ provides a deformation between the Gaussian measure $\mu_{0,1}$ and the semi-circle law $\mu_{1,0}$.

References

1. N. I. Akhiezer, *The classical moment problem*, Oliver & Boyd, Edinburgh and London, 1965.
2. V. I. Arnold, *Mathematical methods of classical mechanics*, Springer-Verlag, New York–Berlin–London–Tokyo, 1989.
3. N. Aronszajn and W. F. Donoghue, *On exponential representations of analytic functions in the upper half-plane with positive imaginary part*, J. Analyse Math. **5** (1956/57), 321–388.
4. _____, *On exponential representations of analytic functions in the upper half-plane with positive imaginary part*, J. Analyse Math. **12** (1964), 113–127.
5. R. Askey and J. Wimp, *Associated Laguerre and Hermite polynomials*, Proc. Roy. Soc. Edinburgh Sect. A **96** (1984), 15–37.
6. F. V. Atkinson, *Discrete and continuous boundary problems*, Academic Press, New York–London, 1964.
7. M. Sh. Birman and D. R. Yafaev, *Spectral properties of scattering matrix*, Algebra i Analiz **4** (1992), no. 5, 1–44; English transl. in St. Petersburg. Math. J. **4** (1992), no. 5, 1–27.
8. D. M. Cifarelli and E. Regazzini, *Some remarks on the distribution functions of means of a Dirichlet process*, Ann. Statist. **18** (1990), 429–442.
9. H. Delange, *On two theorems of S. Verblunsky*, Proc. Cambridge Phil. Soc. **46** (1950), 57–66.
10. P. Diaconis and J. Kemperman, *Some new tools for Dirichlet priors*, Preprint (1994), 1–12.
11. R. Donaghey and L. W. Shapiro, *Motzkin numbers*, J. Combin. Theory Ser. A **23** (1977), 291–301.
12. A. Erdélyi, *Asymptotic expansions*, Dover, New York, 1956.
13. W. Feller, *An introduction to probability theory and its applications*, vol. II, Wiley, New York–London–Sydney–Toronto, 1966.
14. P. Flajolet, *Combinatorial aspects of continued fractions*, Discrete Math. **32** (1980), 125–161.
15. F. Harary and Ed. M. Palmer, *Graphical enumeration*, Academic Press, New York and London, 1973.
16. F. Hausdorff, *Momentenprobleme für ein endliches Interval*, Math. Z. **16** (1923), 220–248.
17. G. James and A. Kerber, *The representation theory of the symmetric group*, Addison-Wesley, London, 1981.
18. W. B. Jones and W. J. Thorn, *Continued fractions. Analytic theory and applications*, Addison-Wesley, London, 1980.
19. _____, *Continued fractions in numerical analysis*, in: Continued Fractions and Padé Approximations, Elsevier Science Publishers, North-Holland, 1990, pp. 169–256.
20. I. S. Kac and M. G. Krein, *R-functions: analytic functions mapping upper half-plane to itself*, in: Discrete and Continuous Boundary Problems by F. V. Atkinson, addendum to the Russian translation, 1968.

21. S. V. Kerov, *Separation of roots of orthogonal polynomials and the limiting shape of generic large Young diagrams*, Preprint of the Trondheim Univ. **8** (1992), Trondheim, Norway, 1–23.
22. _____, *q-Analog of the hook walk algorithm and random Young tableaux*, Funktsional. Anal. i Prilozhen. **26** (1992), no. 3, 35–45; English transl., Functional Anal. Appl. **26** (1992), no. 3, 179–187.
23. _____, *The asymptotics of interlacing roots of orthogonal polynomials*, Algebra i Analiz **5** (1993), no. 5, 68–86; English transl., St. Petersburg Math. J. **5** (1994), 925–941.
24. _____, *Transition probabilities of continuous Young diagrams and Markov moment problem*, Funktsional. Anal. i Prilozhen. **27** (1993), no. 2, 32–49; English transl. in Functional Anal. Appl. **27** (1993), no. 2, 104–117.
25. _____, *The differential model of growth of Young diagrams*, Proc. St. Petersburg Math. Soc. **4** (1996), 167–194.
26. J. F. C. Kingman, *Random partitions in population genetics*, Proc. Roy. Soc. London Ser. A **361** (1978), 1–20.
27. _____, *Poisson processes*, Clarendon Press, Oxford, 1993.
28. M. G. Krein and A. A. Nudelman, *The Markov moment problem and extremal problems*, "Nauka", Moscow, 1973; English transl., Translations of Mathematical Monographs, vol. 50, Amer. Math. Soc., Providence, RI, 1977.
29. M. G. Krein, *On certain new studies in the perturbation theory for selfadjoint operators*, Topics in Differential and Integral Equations and Operator Theory (Operator Theory: Advances and Applications, Vol. 7), Birkhäuser, Basel–Boston–Stuttgart, 1983.
30. I. M. Lifshits, *On a problem of perturbation theory related to quantum statistics*, Uspekhi Mat. Nauk **7** (1952), no. 1, 171–180.
31. B. F. Logan and L. A. Shepp, *A variational problem for random Young tableaux*, Adv. in Math. **26** (1977), 206–222.
32. I. G. Macdonald, *Symmetric functions and Hall polynomials*, Clarendon Press, Oxford, 1979.
33. W. Magnus, F. Oberhettinger, and R. P. Soni, *Formulas and theorems for the special functions of mathematical physics*, Springer-Verlag, New York, 1966.
34. A. A. Markov, *Nouvelles applications des fractions continues*, Math. Ann. **47** (1896), 579–597.
35. L. W. Shapiro et al., *The Riordan group*, Discrete Appl. Math. **34** (1991), 229–239.
36. J. A. Shohat and J. D. Tamarkin, *The problem of moments*, Amer. Math. Soc., New York, 1943.
37. T. J. Stieltjes, *Recherches sur les fractions continues*, Mémoire presentés par divers Savants à, l'Académie des Sciences de l'Institut de France, Sciences et mathématiques **32** (1892), no. 2, 1–196.
38. J. Touchard, *Sur un problème de configurations et sur les fractions continues*, Canad. J. Math. **4** (1952), 2–25.
39. S. Verblunsky, *Two moment problems for bounded functions*, Proc. Cambridge Phil. Soc. **42** (1946), 189–196.
40. _____, *On the initial moments of a bounded function*, Proc. Cambridge Phil. Soc. **43** (1947), 275–279.
41. A. M. Vershik and S. V. Kerov, *Asymptotics of the Plancherel measure of the symmetric group and the limiting form of Young tableax*, Dokl. Akad. Nauk SSSR **18** (1977), 527–531; English transl. in Soviet Math. Dokl. **233** (1977), 1024–1027.
42. A. M. Vershik and S. V. Kerov, *Asymptotic of the largest and the typical dimensions of irreducible representations of a symmetric group*, Funktsional. Anal. i Prilozhen. **19** (1985), no. 1, 25–36; English transl., Functional Anal. Appl. **19** (1985), no. 1, 21–31.
43. G. Viennot, *Une theorie combinatoire des polynomes orthogonaux generaux*, Notes de conférences données au Département de mathématiques et d'informatique, Montréal, Québec (1983).
44. H. Yamato, *Characteristic functions of means of distributions chosen from a Dirichlet process*, Ann. Probab. **12** (1984), 262–267.

Steklov Mathematical Institute, St. Peterburg Branch (POMI), Fontanka 27, St. Petersburg, 191011, Russia
E-mail address: kerov@pdmi.ras.ru

Amer. Math. Soc. Transl.
(2) Vol. **181**, 1998

Quasicommuting Families of Quantum Plücker Coordinates

Bernard Leclerc and Andrei Zelevinsky

To A. A. Kirillov on his 60th birthday

§1. Introduction

Let $\mathbf{Q}_q[\mathcal{F}]$ be the q-deformation of the coordinate ring of the flag variety of type A_r, as defined in [**LR**, **TT**]. This is the algebra with unit over the field of rational functions $\mathbf{Q}(q)$ generated by $2^{r+1} - 1$ generators $[J]$ labeled by nonempty subsets $J \subset [1, r+1] := \{1, \ldots, r+1\}$, subject to the q-deformed Plücker relations (see §2 below). We refer to the generators $[J]$ as *quantum flag minors* (they can be identified with q-minors of a generic q-matrix whose row set consists of several initial rows).

We say that two quantum flag minors $[I]$ and $[J]$ *quasicommute* if $[J][I] = q^n[I][J]$ for some integer n. In this paper, we deal with the following problem.

PROBLEM A. Describe all families of quasicommuting quantum flag minors.

The motivation for Problem A comes from the study of canonical bases for quantum groups of type A_r (see [**BZ**, **K**]). It was proved in [**BZ**] that the *dual canonical basis* (= Kashiwara's *upper global crystal basis*) in $\mathbf{Q}_q[\mathcal{F}]$ contains all generators $[J]$. The main conjecture of [**BZ**] claims that the dual canonical basis is stable under the taking of quasicommutative products. This makes Problem A quite natural.

Our first main result is a combinatorial criterion for the quasicommutativity of two quantum flag minors $[I]$ and $[J]$. We shall write $I \prec J$ if $i < j$ for any $i \in I$, $j \in J$ (in particular, if one or both of I, J are empty). We write $I - J := \{i : i \in I, i \notin J\}$, and write $|I|$ for the size of I. We say that I and J are *weakly separated* if at least one of the following two conditions holds:
  (1) $|I| \geq |J|$, and $J - I$ can be partitioned into a disjoint union $J - I = J' \cup J''$ so that $J' \prec I - J \prec J''$;
  (2) $|J| \geq |I|$, and $I - J$ can be partitioned into a disjoint union $I - J = I' \cup I''$ so that $I' \prec J - I \prec I''$.

1991 *Mathematics Subject Classification*. Primary 14M15, 17B37.
A. Zelevinsky's research was partially supported by NSF grant DMS-9304247.

©1998 American Mathematical Society

THEOREM 1.1. *Two quantum flag minors $[I]$ and $[J]$ quasicommute if and only if the sets I and J are weakly separated. In case (1) above, we have*

(1.1) $$[J][I] = q^{|J''|-|J'|}[I][J].$$

The proof will be given in §2. The "if" part was essentially proved in [**KL**]. Our proof of the "only if" part uses a method similar to the one in [**BZ**].

In view of Theorem 1.1, Problem A becomes equivalent to the combinatorial problem of describing all collections of weakly separated subsets of $[1, r+1]$. Our first result in this direction gives the maximum possible size of such a collection.

THEOREM 1.2. *The maximum possible size of a collection of weakly separated subsets of $[1, r+1]$ is $\binom{r+2}{2} + 1$.*

This theorem will be proved in §3. Note that $\binom{r+2}{2} = \binom{r+1}{2} + r + 1$ is the dimension of the affine cone over the flag variety $\mathcal{F}$ of $G = SL_{r+1}$ in its Plücker embedding. It follows that Theorem 1.2 would be a natural consequence of the above-mentioned conjecture in which the dual canonical basis in $\mathbf{Q}_q[\mathcal{F}]$ contains all quasicommuting monomials in quantum flag minors. (The appearance of the extra summand 1 in Theorem 1.2 is due to the fact that the empty subset is allowed to be a member of a weakly separated collection.)

An example of a weakly separated collection of maximum possible size is given by the collection $\mathcal{C}_{\min}$ consisting of the empty set and all intervals $[a,b]$ ($1 \leq a \leq b \leq r+1$); another example is the collection $\mathcal{C}_{\max}$ consisting of complements of subsets from $\mathcal{C}_{\min}$. (The notation $\mathcal{C}_{\min}$ and $\mathcal{C}_{\max}$ refers to a natural partial order on the set of all weakly separated collections that will be discussed later.)

We shall generalize Theorem 1.2 in two different directions. First, we impose some restrictions on the size of subsets in a weakly separated collection.

THEOREM 1.3. *Let k and l be any two indices such that $0 \leq k \leq l \leq r+1$. The maximum possible size of a collection $\mathcal{C}$ of weakly separated subsets of $[1, r+1]$ such that $k \leq |I| \leq l$ for all $I \in \mathcal{C}$ is $\binom{r+2}{2} - \binom{r+2-l}{2} - \binom{k+1}{2} + 1$.*

To formulate the second generalization, consider any permutation $w \in S_{r+1}$. As usual, $l(w)$ will denote the length of w (i.e., the number of inversions). Following [**BFZ**, subsection 5.3], we shall say that a subset $I \subset [1, r+1]$ is a *w-chamber set* if for each $j \in I$ the set I also contains all indices i such that $i < j$ and $w(i) < w(j)$ (this terminology will be explained in §4).

THEOREM 1.4. *The maximum possible size of a weakly separated collection of w-chamber subsets of $[1, r+1]$ is $l(w) + r + 2$.*

This theorem includes Theorem 1.2 as a special case corresponding to $w = w_0$, the longest permutation (clearly, every subset of $[1, r+1]$ is a w_0-chamber set).

Both Theorems 1.3 and 1.4 will also be proved in §3. Geometrically, in Theorem 1.3 we think of the flag variety $\mathcal{F}$ as a member of the family of incomplete flag varieties, while in Theorem 1.4 we include $\mathcal{F}$ into the family of Schubert varieties.

We conjecture that Theorem 1.4 can be sharpened as follows.

CONJECTURE 1.5. *All maximal (by inclusion) weakly separated collections of w-chamber subsets of $[1, r+1]$ are of the same size $l(w) + r + 2$.*

The conjecture was computer checked for $r \leq 4$ and in some other special cases. In §4 we shall prove it for a special class of *strongly separated collections* that are the second main object of study in this paper.

We say that subsets I and J of $[1, r+1]$ are *strongly separated* if either $I - J \prec J - I$ or $J - I \prec I - J$ (in particular, this is so if one of the sets I and J contains the other). Clearly, any two strongly separated subsets are weakly separated. A collection of subsets of $[1, r+1]$ is *strongly separated* if the same is also true of any two members of this collection. An algebraic interpretation of strongly separated subsets is given by Corollary 2.10 below: two subsets I and J are strongly separated if and only if one of the products $[I][J]$ and $[J][I]$ is invariant under the natural involution $x \mapsto \overline{x}$ of $\mathbf{Q}_q[\mathcal{F}]$ that interchanges q and q^{-1} (see [**L, K**]).

Strongly separated collections arise naturally in several interesting combinatorial and geometric contexts. The first such context is the study of *commutation classes* of reduced words for permutations. For a permutation $w \in S_{r+1}$ let $R(w)$ denote the set of *reduced words* for w, that is, the set of sequences $\mathbf{h} = (h_1, \ldots, h_l)$ of minimal length $l = l(w)$ such that $w = s_{h_1} \cdots s_{h_l}$ (as usual, s_i stands for the simple transposition $(i, i+1)$). Recall that a *2-move* for a reduced word $\mathbf{h} \in R(w)$ consists in interchanging two consecutive entries i and j in $\mathbf{h}$ provided $|i - j| \geqslant 2$. Two reduced words obtained from each other by a sequence of 2-moves are said to belong to the same *commutation class*. Let $R^c(w)$ denote the set of commutation classes in $R(w)$. These classes were studied recently in [**E, F, S**], but the following problem remains open.

PROBLEM B. Describe the set $R^c(w)$ for all permutations $w \in S_{r+1}$.

The relationship between commutation classes and strongly separated collections is given by the following construction. Following [**BFZ**], we associate to any reduced word $\mathbf{h} = (h_1, \ldots, h_l) \in R(w)$ a collection $\mathcal{C}(\mathbf{h})$ of $l+r+2$ nonempty subsets of $[1, r+1]$ called *chamber sets* for $\mathbf{h}$. The collection $\mathcal{C}(\mathbf{h})$ consists of the empty set, the $r+1$ intervals $[1, a]$, $a = 1, \ldots, r+1$ and the l subsets $s_{h_l} s_{h_{l-1}} \cdots s_{h_k}([1, h_k])$, $k = 1, \ldots, l$. For example, if $r = 3$ and $\mathbf{h} = (3, 2, 1, 3)$, then

$$\mathcal{C}(\mathbf{h}) = \{\varnothing, 1, 2, 12, 24, 123, 124, 234, 1234\}.$$

(The members of $\mathcal{C}(\mathbf{h})$ are called chamber sets because in [**BFZ**] they were defined in terms of the geometry of the *pseudo-line arrangement* naturally associated with $\mathbf{h}$ (see §4 below). The connection between pseudo-line arrangements, commutation classes, and other combinatorial structures like sorting networks or CC systems is discussed in the monograph [**Kn**].) By [**BFZ**, Proposition 5.3.1], any chamber set for some $\mathbf{h} \in R(w)$ is a w-chamber set, as defined above. It is also easy to see that $\mathcal{C}(\mathbf{h}) = \mathcal{C}(\mathbf{h}')$ if and only if reduced words $\mathbf{h}$ and $\mathbf{h}'$ belong to the same commutation class. In §4 we shall prove the following

THEOREM 1.6. *For any permutation $w \in S_{r+1}$, the correspondence $\mathbf{h} \mapsto \mathcal{C}(\mathbf{h})$ is a bijection between the set $R^c(w)$ of commutation classes and the set of all maximal (by inclusion) strongly separated collections of w-chamber sets.*

In particular, Theorem 1.6 implies Conjecture 1.5 for strongly separated collections.

The relationship between (commutation classes of) reduced words and strongly separated collections has an interesting geometric application that we learned from Peter Magyar. Namely, every $\mathbf{h} \in R(w)$ gives rise to a Bott–Samelson desingularization of the Schubert variety corresponding to w. As shown by Magyar, this Bott–Samelson variety has a nice explicit description in terms of the collection $\mathcal{C}(\mathbf{h})$.

The quasicommuting family of quantum Plücker coordinates $\{[I] : I \in \mathcal{C}(\mathbf{h})\}$ is closely related with the "quantization" of the corresponding Bott–Samelson variety, recently constructed by B. Feigin (this construction is discussed in [**IM**] (for type A_r) and [**J**] (for an arbitrary semisimple Lie algebra); recently, A. Berenstein has extended it to any Kac–Moody Lie algebra). The relationship between quantum Plücker coordinates and quantum "Bott–Samelson coordinates" will be studied in a separate publication; in the classical case $q = 1$, this relationship is given by Theorem 1.4 in [**BFZ**] (for type A_r). Note that the geometric meaning of weakly but not strongly separated collections has yet to be understood.

Let $\mathbf{W} = \mathbf{W}_r$ be the set of all weakly separated collections of subsets in $[1, r+1]$ that have maximum possible size $\binom{r+2}{2} + 1$. Let $\mathbf{S} = \mathbf{S}_r$ be the subset of $\mathbf{W}$ formed by strongly separated collections. Using Theorem 1.6, we can identify $\mathbf{S}$ with $R^c(w_0)$. The set $R^c(w_0)$ has an important poset structure via the *second Bruhat order* introduced in [**MS**]. This partial order is the transitive closure of the following relation on reduced words: $\mathbf{h}' \geqslant \mathbf{h}$ if $\mathbf{h}'$ is obtained from $\mathbf{h}$ by a *raising 3-move* replacing three consecutive indices i, $i+1$, i by $i+1$, i, $i+1$. Using the above identification of $R^c(w_0)$ with $\mathbf{S}$, we can transfer the second Bruhat order to $\mathbf{S}$. The geometric interpretation of 3-moves in terms of pseudo-line arrangements given in [**BFZ**, §2], implies the following description of the transferred partial order on $\mathbf{S}$. If three indices $i < j < k$ and a subset $L \subset [1, r+1]$ disjoint from $\{i, j, k\}$ are such that a strongly separated collection $\mathcal{C}$ contains seven subsets L, Li, Lj, Lk, Lij, Ljk, and $Lijk$, then a raising 3-move on $\mathcal{C}$ replaces Lj by Lik (here and in the sequel, we write $Li, Lj, \ldots$ for subsets $L \cup \{i\}, L \cup \{j\}, \ldots$). We have $\mathcal{C}' \geqslant \mathcal{C}$ if $\mathcal{C}'$ is obtained from $\mathcal{C}$ by a sequence of raising 3-moves. Under this partial order, the collection $\mathcal{C}_{\min}$ described above is the unique minimal element of $\mathbf{S}$, while $\mathcal{C}_{\max}$ is the unique maximal element.

In §5, we introduce a natural extension of raising 3-moves and the partial order from $\mathbf{S}$ to $\mathbf{W}$. It is based on the following result.

THEOREM 1.7. *Let $i < j < k$, and let $L \subset [1, r+1]$ be a subset disjoint from $\{i, j, k\}$. Suppose that $\mathcal{C}$ is a weakly separated collection containing five subsets Li, Lj, Lk, Lij, and Ljk. Then the collection obtained from $\mathcal{C}$ by replacing Lj with Lik is also weakly separated.*

The operation in Theorem 1.7, i.e., replacing Lj with Lik in the presence of four "witnesses" Li, Lk, Lij, and Ljk, will be called a *weak raising flip*. We define the poset structure on $\mathbf{W}$ as follows: $\mathcal{C}' \geqslant \mathcal{C}$ if $\mathcal{C}'$ is obtained from $\mathcal{C}$ by a sequence of weak raising flips. It is not hard to show that the restriction of this partial order to $\mathbf{S}$ coincides with the transferred second Bruhat order.

For instance, the picture in Figure 1 shows the poset $\mathbf{W}_3$. (For convenience, the subsets of the form $[1, a]$ or $[a, 4]$, which belong to all collections, have been omitted.) Removing in it the two vertices printed in bold type and the four dashed edges, one gets the usual poset $\mathbf{S}_3$.

CONJECTURE 1.8. *The collection $\mathcal{C}_{\min}$ is the unique minimal element of the poset $\mathbf{W}$, and $\mathcal{C}_{\max}$ is the unique maximal element.*

In other words, we conjecture that every collection $\mathcal{C} \in \mathbf{W}$ different from $\mathcal{C}_{\max}$ admits a weak raising flip, and every $\mathcal{C} \in \mathbf{W}$ different from $\mathcal{C}_{\min}$ admits a weak lowering flip. Some partial results in this direction are obtained in §5.

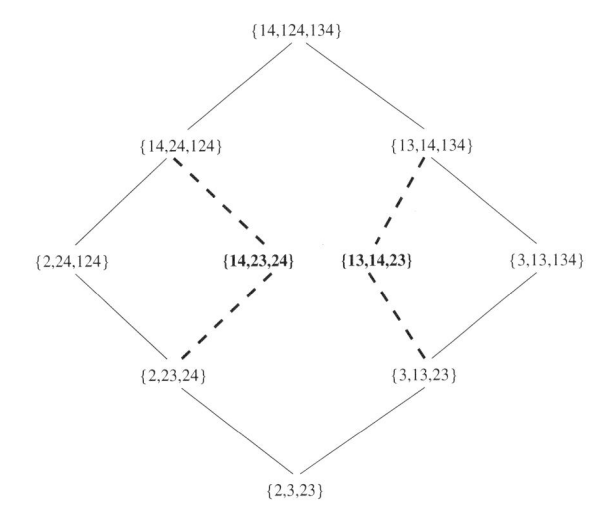

FIGURE 1

Note that the definition of a weak raising flip was motivated by the Chamber Ansatz in [**BFZ**], more specifically, by the observation that the 3-term relation $M_{Lik}M_{Lj} = M_{Lij}M_{Lk} + M_{Ljk}M_{Li}$ in [**BFZ** (2.5.4)], does not involve subsets L and $Lijk$. Using this observation, we shall show in §5 that every collection obtained from $\mathcal{C}_{\min}$ by a sequence of weak raising flips gives rise to a total positivity criterion for upper-triangular matrices (in the sense of [**BFZ,** subsection 3.2]). Thus, Conjecture 1.8 would imply that every collection $\mathcal{C} \in \mathbf{W}$ gives rise to a total positivity criterion.

The material of this paper is organized as follows. Section 2 is devoted to the proof of Theorem 1.1. For the reader's convenience, we recall there the necessary definitions and results from [**TT, FRT, BZ**].

Theorems 1.2, 1.3, and 1.4 are proved in §3, and Theorem 1.6 is proved in §4. As a by-product of these proofs, we obtain an inductive procedure for generating the set $\mathbf{W}_r$ and $\mathbf{S}_r$. This procedure is based on the following result. For a collection $\mathcal{C} \in \mathbf{W}_r$, let $\mathcal{C}^-$ denote the collection of all subsets $J \subset [1, r]$ such that $J = I \cap [1, r]$ for some $I \in \mathcal{C}$.

THEOREM 1.9. *The correspondence $\mathcal{C} \mapsto \mathcal{C}^-$ is a surjection $\mathbf{W}_r \to \mathbf{W}_{r-1}$, and its restriction to $\mathbf{S}_r$ is a surjection $\mathbf{S}_r \to \mathbf{S}_{r-1}$.*

Note that Knuth [**Kn**] found and implemented an algorithm for enumerating commutation classes. It is essentially equivalent to our inductive procedure for $\mathbf{S}_r$.

Finally, in §5 we prove Theorem 1.7 and some other results about the extended second Bruhat order.

This work was partly done during the visit of one of the authors (A. Z.) to the Fourier Institute in Grenoble, France, in June 1995. He is grateful to M. Brion and L. Manivel for their hospitality and stimulating discussions.

§2. A criterion for the quasicommutativity of quantum flag minors

In this section we prove Theorem 1.1. First, let us recall the defining relations of the algebra $\mathbf{Q}_q[\mathcal{F}]$ given in [**TT**]. Given two subsets I and J of $[1, r+1]$, we shall denote by inv(I, J) the number of pairs $(i,j) \in I \times J$ such that $i > j$. The relations from [**TT**] can be written in the following form:

$R_1(I, J)$: for all subsets I and J such that $|I| \leq |J|$, we have

$$[I][J] = \sum_M (-q)^{\text{inv}(J-M,M)-\text{inv}(I,M)} [I \cup M][J - M],$$

where the sum is over all $M \subset J - I$ with $|M| = |J| - |I|$;

$R_2(I, J)$: for all subsets I and J such that $|I| - 1 \geq |J| + 1$, we have

$$\sum_{i \in I-J} (-q)^{\text{inv}(i,I-i)-\text{inv}(i,J)}[I - \{i\}][J \cup \{i\}] = 0.$$

We shall use the following notation. If the quantum minors $[I]$ and $[J]$ quasi-commute, then we define the integer $c(I, J)$ from

(2.1) $$[J][I] = q^{c(I,J)}[I][J].$$

Clearly, we have

(2.2) $$c(J, I) = -c(I, J).$$

We can now prove the "if" part of Theorem 1.1, which takes the following form.

LEMMA 2.1. *Let I and J be weakly separated sets satisfying condition* (1) *in the Introduction. Then $[I]$ and $[J]$ quasicommute, and $c(I, J) = |J''| - |J'|$. In particular, if $J \subset I$, then $c(I, J) = 0$.*

PROOF. In the case when $I \cap J = \emptyset$, Lemma 2.1 is an immediate consequence of [**KL**, Lemmas 3.6, 3.7]. The general case can be reduced to this particular one with the help of the following observation.

LEMMA 2.2. *The quantum flag minors $[I]$ and $[J]$ quasicommute if and only if $[I - J]$ and $[J - I]$ quasicommute. Moreover, we have $c(I, J) = c(I - J, J - I)$.*

Lemma 2.2 is an immediate consequence of the following

LEMMA 2.3. *For a subset L of $[1, r+1]$, let $\mathcal{A}_L$ (resp., $\mathcal{B}_L$) denote the subalgebra of $\mathbf{Q}_q[\mathcal{F}]$ generated by the quantum flag minors $[I]$ such that $I \cap L = \emptyset$ (resp., $L \subset I$). Then there is an algebra isomorphism $\mathcal{A}_L \to \mathcal{B}_L$ that sends each $[I]$ to $[L \cup I]$.*

PROOF OF LEMMA 2.3. The defining relations of $\mathcal{A}_L$ (resp., $\mathcal{B}_L$) are just relations $R_1(I, J)$ and $R_2(I, J)$ with $L \cap (I \cup J) = \emptyset$ (resp., $L \subset I \cap J$). Suppose that $L \cap (I \cup J) = \emptyset$. Then, formally replacing every minor $[P]$ occurring in $R_1(I, J)$ (resp., $R_2(I, J)$) by $[L \cup P]$, we exactly get the relation $R_1(L \cup I, L \cup J)$ (resp., $R_2(L \cup I, L \cup J)$). Conversely, if $L \subset I \cap J$, the substitution of $[P - L]$ in place of $[P]$ in $R_1(I, J)$ (resp., $R_2(I, J)$) gives the relation $R_1(I - L, J - L)$ (resp., $R_2(I - L, J - L)$). This completes the proof of Lemma 2.3 and of the "if" part of Theorem 1.1.

To prove the "only if" part of Theorem 1.1, we shall use another interpretation of $\mathbf{Q}_q[\mathcal{F}]$. Recall that the entries of a *generic q-matrix* $X = (x_{ij})$, $i, j \in [1, r+1]$, are indeterminates over $\mathbf{Q}(q)$ subject to the following relations: every 2×2 submatrix $\begin{pmatrix} a & b \\ c & d \end{pmatrix}$ of X satisfies

(2.3)
$$ba = qab, \quad ca = qac, \quad bc = cb,$$
$$db = qbd, \quad dc = qcd, \quad da - ad = (q - q^{-1})bc.$$

The $\mathbf{Q}(q)$-algebra $\mathbf{Q}_q[\text{Mat}]$ generated by the x_{ij} is the q-deformation of the ring of polynomials on the matrix algebra studied in [**FRT**]. For $J = \{j_1 < \cdots < j_l\} \subset [1, r+1]$, the generator $[J]$ of $\mathbf{Q}_q[\mathcal{F}]$ can be identified with the *quantum flag minor* of X on rows $1, \ldots, l$ and columns $j_1, \ldots, j_l$ given by

(2.4)
$$[J] = \sum_{\sigma \in S_l} (-q)^{-l(\sigma)} x_{1, j_{\sigma(1)}} \cdots x_{l, j_{\sigma(l)}}$$

(see [**LR, TT**]). Thus, $\mathbf{Q}_q[\mathcal{F}]$ can be identified with the subalgebra of $\mathbf{Q}_q[\text{Mat}]$ generated by quantum flag minors.

Let U_+ denote the algebra with unit over $\mathbf{Q}(q)$ generated by the elements $E_1, \ldots, E_r$ subject to the *quantum Serre relations*

(2.5)
$$E_i E_j = E_j E_i \quad \text{for } |i - j| > 1,$$
$$E_i^2 E_j - (q + q^{-1}) E_i E_j E_i + E_j E_i^2 = 0 \quad \text{for } |i - j| = 1.$$

This is the quantized universal enveloping algebra of the Lie algebra $\text{Lie}(N)$, where N is the maximal unipotent subgroup of upper unitriangular matrices in SL_{r+1}. Both algebras $\mathbf{Q}_q[\text{Mat}]$ and U_+ are graded via

(2.6)
$$\deg x_{ij} = \varepsilon_j, \quad \deg E_j = \alpha_j = \varepsilon_j - \varepsilon_{j+1},$$

where $\varepsilon_1, \ldots, \varepsilon_{r+1}$ is the standard basis in the lattice $\mathbb{Z}^{r+1}$. We shall use a q-deformation of the action of $\text{Lie}(N)$ on the matrix space by infinitesimal right translations. This quantized action is given by the following proposition, which can be proved by a straightforward calculation.

PROPOSITION 2.4. *There is a unique action of U_+ on $\mathbf{Q}_q[\text{Mat}]$ satisfying*

(2.7) $$E_j(x_{kl}) = \delta_{j, l-1} x_{k, l-1},$$
(2.8) $$E_j(fg) = E_j(f) g + q^{-(\gamma, \alpha_j)} f E_j(g),$$

where in (2.8), *γ denotes the degree of a homogeneous element f, and (γ, α) is the standard scalar product defined by $(\varepsilon_i, \varepsilon_j) = \delta_{ij}$.*

The formulas (2.7), (2.8), and (2.4) readily imply the following

LEMMA 2.5. *For any subset $I \subset [1, r+1]$, we have*

(2.9) $$E_j([I]) = [I \cup \{j\} - \{j+1\}]$$

if $j \notin I$ and $j + 1 \in I$; otherwise, $E_j([I]) = 0$.

We shall denote by δ_I the indicator function of a set I given by $\delta_I(i) = 1$ if $i \in I$, and $\delta_I(i) = 0$, otherwise.

LEMMA 2.6. *The property that $[I]$ and $[J]$ quasicommute is preserved by the following transformations of a pair (I, J):*
  (1) *replacing (I, J) by $(I \cup \{i\} - \{i+1\}, J)$ for some index i such that $i+1 \in I - J$ and $i \notin I$; in this case, $c(I \cup \{i\} - \{i+1\}, J) = c(I, J) + \delta_J(i)$;*
  (2) *replacing (I, J) by $(I, J \cup \{j\} - \{j+1\})$ for some index j such that $j+1 \in J - I$ and $j \notin J$; in this case, $c(I, J \cup \{j\} - \{j+1\}) = c(I, J) - \delta_I(j)$.*

PROOF. Part (2) follows from (1) and (2.2), so it is enough to prove part (1). Note that $[J]$ is a homogeneous element of degree $\sum_{j \in J} \varepsilon_j$. Writing $I' = I \cup \{i\} - \{i+1\}$ for short, and applying (2.8) and (2.9), we obtain

$$E_i([J][I]) = q^{-\delta_J(i)} [J][I'], \qquad E_i([I][J]) = [I'][J],$$

which implies the desired statement.

LEMMA 2.7. *Suppose I and J are two disjoint sets that are not weakly separated, but become weakly separated after any transformation as in Lemma 2.6. Interchanging I and J if necessary, the pair (I, J) assumes one of the following forms*:

(2.10) $\qquad I = [1, j-1] \cup [j+1, s], \quad J = \{j\} \qquad (1 < j < s),$

(2.11) $\qquad I = [1, j-2] \cup \{j\}, \quad J = \{j-1\} \cup [j+1, s] \qquad (2 < j < s).$

PROOF. The conditions readily imply that $I \cup J = [1, s]$ for some s. Interchanging I and J if necessary, we can assume that $1 \in I$. The sets I and J subdivide $[1, s]$ into the union of disjoint intervals $I_1 \cup J_1 \cup I_2 \cup J_2 \cdots$ (read in increasing order) so that we have $I = I_1 \cup \cdots \cup I_k$, $J = J_1 \cup \cdots \cup J_{k-1}$ if $s \in I$, and $I = I_1 \cup \cdots \cup I_k$, $J = J_1 \cup \cdots \cup J_k$ if $s \in J$. Let A and B be two consecutive intervals in this subdivision, i.e., $(A, B) = (I_p, J_p)$ or $(A, B) = (J_p, I_{p+1})$. By an (A, B)-*reduction* we mean the following operation on a pair (I, J): removing the last point of A from one of these sets and the first point of B from the other one. The conditions of the lemma mean that any (A, B)-reduction transforms (I, J) into a weakly separated pair, in particular, leaves at most three disjoint intervals. Hence, the total number of intervals in I and J is at most five. Let us consider three cases.

Case 1. $I = I_1 \cup I_2 \cup I_3$, $J = J_1 \cup J_2$. Applying the (I_1, J_1)-reduction and the (J_2, I_3)-reduction, we conclude that $|J_1| = |J_2| = 1$. But then the (J_1, I_2)-reduction will not lead to a weakly separated pair, so this case is impossible.

Case 2. $I = I_1 \cup I_2$, $J = J_1 \cup J_2$. Suppose that $|J_1| > 1$. Applying the (I_1, J_1)-reduction, we see that $|I_1| = 1$, and $|I_2| \geq |J_1| + |J_2| - 1 \geq 2$. On the other hand, applying the (J_1, I_2)-reduction yields $|I_2| = 1$. This contradiction shows that $|J_1| = 1$. The same argument proves that $|J_2| = 1$, i.e., we are in the situation (2.11).

Case 3. $I = I_1 \cup I_2$, $J = J_1$. Suppose $|J_1| > 1$. Since I and J are not weakly separated, at least one of I_1 and I_2 must have more than one point. But then the reduction involving this interval together with J_1 will also give a pair that is not weakly separated. This contradiction shows that $|J_1| = 1$, i.e., we are in the situation (2.10). This completes the proof of Lemma 2.7.

Now everything is ready for the proof of the "only if" part of Theorem 1.1. Assume, on the contrary, that there exist two nonweakly separated sets I and J such that $[I]$ and $[J]$ quasicommute. Consider such a pair with minimal possible value of $\sum_{i\in I} i + \sum_{j\in J} j$. Using Lemmas 2.2, 2.6, and 2.7, we can assume that (I, J) is one of the pairs given in (2.10) and (2.11).

First, suppose I and J are given by (2.10). Using Lemmas 2.1 and 2.6 (1), we obtain

$$(2.12) \qquad c(I,J) = c([1,j] \cup [j+2,s], \{j\}) + 1 = 1.$$

On the other hand, using Lemmas 2.1 and 2.6 (2), we obtain

$$(2.13) \qquad c(I,J) = c([1,j-1] \cup [j+1,s], \{j-1\}) - 1 = -1.$$

Comparing (2.12) and (2.13) yields a contradiction.

The case when I and J are given by (2.11) is treated similarly. On one hand, Lemmas 2.1 and 2.6 (1) imply that

$$(2.14) \qquad \begin{aligned} c(I,J) &= c([1,j-1], \{j-1\} \cup [j+1,s]) + 1 \\ &= c([1,j-2], [j+1,s]) + 1 = \min(j-2, s-j) + 1. \end{aligned}$$

On the other hand, Lemmas 2.1 and 2.6 (2) imply that

$$(2.15) \qquad \begin{aligned} c(I,J) &= c([1,j-2] \cup \{j\}, \{j-2\} \cup [j+1,s]) - 1 \\ &= c([1,j-3] \cup \{j\}, [j+1,s]) - 1 = \min(j-2, s-j) - 1. \end{aligned}$$

Comparing (2.14) and (2.15) again yields a contradiction, thus completing the proof of Theorem 1.1.

We conclude this section with some comments and complementary results. First, let us discuss the relationship between the present setting and that of [**BZ**]. The main object of study in [**BZ**] is the algebra $\mathcal{A}_r$ of noncommutative polynomials in the matrix entries t_{ij} ($i < j$) of the matrix T described in [**BZ**, §1]. Note that T is not a q-matrix, i.e., its entries do not satisfy the commutation relations (2.3). To make our notation more unified, we shall rename $\mathcal{A}_r$ to $\mathbf{Q}_q[N]$, where N is the group of unipotent upper triangular matrices in $G = SL_{r+1}$; the algebra $\mathbf{Q}_q[N]$ is indeed a natural q-deformation of the ring of polynomial functions on N.

Both algebras $\mathbf{Q}_q[\mathcal{F}]$ and $\mathbf{Q}_q[N]$ can be naturally embedded as subalgebras into the algebra $\mathbf{Q}_q[B]$, the q-deformation of the ring of regular functions on the upper triangular group B. To construct $\mathbf{Q}_q[B]$, we first consider the quotient algebra of $\mathbf{Q}_q[\text{Mat}]$ modulo the two-sided ideal generated by the x_{ij}, $i > j$. We denote this algebra by $\mathbf{Q}_q[\text{Upp}]$. For $i \leqslant j$ let b_{ij} be the image of x_{ij} in $\mathbf{Q}_q[\text{Upp}]$. It is natural to call the matrix (b_{ij}) a *generic upper triangular q-matrix*. Note that the diagonal entries b_{ii} commute with each other in $\mathbf{Q}_q[\text{Upp}]$. Now $\mathbf{Q}_q[B]$ can be defined as the algebra obtained from $\mathbf{Q}_q[\text{Upp}]$ by adjoining the inverses b_{ii}^{-1} of diagonal entries.

Finally, let θ be the $\mathbf{Q}$-algebra automorphism of $\mathbf{Q}_q[N]$ defined by $\theta(t_{i,i+1}) = t_{i,i+1}$ and $\theta(q) = q^{-1}$.

The relationship between $\mathbf{Q}_q[\mathcal{F}]$ and $\mathbf{Q}_q[N]$ can be now summarized as follows.

PROPOSITION 2.8. (a) *The restriction of the projection $\mathbf{Q}_q[\text{Mat}] \to \mathbf{Q}_q[B]$ to the subalgebra $\mathbf{Q}_q[\mathcal{F}]$ is injective.*

(b) *There is a natural algebra embedding $\mathbf{Q}_q[N] \to \mathbf{Q}_q[B]$ which sends $\theta(t_{ij})$ for $i < j$ to $b_{ij} b_{ii}^{-1}$.*

(c) *For every subset $J \subset [1, r+1]$ of size l, the embedding in* (b) *sends $\theta(\Delta([1,l]; J))$ to $[J][[1,l]]^{-1}$, where $\Delta([1,l]; J)$ is the minor defined in* [**BZ**, (1.6)].

This proposition is an easy consequence of the definitions. It implies that two generators of $\mathbf{Q}_q[\mathcal{F}]$ quasicommute if and only if the corresponding flag minors of T do also. Thus, Theorem 1.1 provides the following criterion.

COROLLARY 2.9. *Two flag minors $\Delta([1, |I|]; I)$ and $\Delta([1, |J|]; J)$ quasicommute in $\mathbf{Q}_q[N]$ if and only if the sets I and J are weakly separated.*

Finally, let us briefly discuss the algebraic meaning of strong separation (see the introduction for the definition of strongly separated sets). There is a natural involution $x \mapsto \overline{x}$ defined on $\mathbf{Q}_q[\mathrm{Mat}]$, which sends q to q^{-1} and preserves the canonical basis (see, e.g., [**K**]). It is known that all quantum minors are fixed under this involution, and that it satisfies

$$\overline{uv} = q^{(\lambda_r, \mu_r) - (\lambda_l, \mu_l)} \overline{v}\,\overline{u}$$

for all weight vectors $u, v \in \mathbf{Q}_q[\mathrm{Mat}]$ [**K**]. Here $\mathbf{Q}_q[\mathrm{Mat}]$ is regarded as a bimodule over $U_q(sl_{r+1})$ and λ_r, λ_l (resp., μ_r, μ_l) denote the respective weights of u (resp., v) under the right and left actions (the right and left weights of x_{ij} are ε_j and ε_i, respectively). It then follows from an easy calculation that

(2.16) $$\overline{[I][J]} = q^{d(I,J)}[J][I]$$

for $|I| \geqslant |J|$, where

(2.17) $$d(I, J) = \min(|I - J|, |J - I|) = \min(|I|, |J|) - |I \cap J|.$$

The following characterization of strongly separated subsets is easily obtained by comparing (2.16) with (2.1).

COROLLARY 2.10. *Two subsets I and J are strongly separated if and only if one of the products $[I][J]$ and $[J][I]$ is invariant under the involution $x \mapsto \overline{x}$.*

§3. Weakly separated collections of maximum size

Our first goal in this section is to prove Theorem 1.2. The following lemma is an immediate consequence of the definition of weak separation.

LEMMA 3.1. *Let I and J be two subsets of $[1, r]$. Then the following conditions are equivalent*:
(1) *I is weakly separated from $J \cup \{r+1\}$*;
(2) *either $|I| > |J|$ and I is weakly separated from J, or $|I| \leqslant |J|$ and $I - J \prec J - I$.*

For a subset $I \subset [1, r+1]$, we shall use the notation $I^- = I \cap [1, r]$. Lemma 3.1 implies the following.

LEMMA 3.2. *If $I, J \subset [1, r+1]$ are weakly separated, then I^- and J^- are also weakly separated.*

Now let $\mathcal{C}$ be a weakly separated collection of subsets in $[1, r+1]$. Let $\mathcal{C}^-$ denote the collection of all subsets $J \subset [1, r]$ such that $J = I^-$ for some $I \in \mathcal{C}$. Thus, the correspondence $I \mapsto I^-$ is a surjective map $\mathcal{C} \to \mathcal{C}^-$. Obviously, the fiber of this

map over each subset $J \in \mathcal{C}^-$ has at most two elements, namely J and $J \cup \{r+1\}$. We shall use the notation
$$\mathcal{C}_p^- = \{J \in \mathcal{C}^- : |J| = p\}.$$

LEMMA 3.3. *For any $p = 0, 1, \ldots, r$, there is at most one $J \in \mathcal{C}_p^-$ such that both J and $J \cup \{r+1\}$ belong to $\mathcal{C}$.*

PROOF. Suppose that $I, J \in \mathcal{C}_p^-$ are such that I, $I \cup \{r+1\}$, J, and $J \cup \{r+1\}$ all belong to $\mathcal{C}$. Since $\mathcal{C}$ is weakly separated, Lemma 3.1 implies that $I - J \prec J - I$ and $J - I \prec I - J$. Hence $I = J$, as desired.

Now everything is ready for the proof of Theorem 1.2. The collection $\mathcal{C}_{\min}$ of all intervals provides an example of a weakly separated collection of size $\binom{r+2}{2} + 1$. It remains to show that every weakly separated collection $\mathcal{C}$ of subsets in $[1, r+1]$ has size $\leqslant \binom{r+2}{2} + 1$. The proof is by induction on r. For $r = 0$ or $r = 1$, we have $\binom{r+2}{2} + 1 = 2^{r+1}$, so our statement is obvious. So we shall assume that $r \geqslant 2$, and that every weakly separated collection of subsets in $[1, r]$ is known to have size $\leqslant \binom{r+1}{2} + 1$. In view of Lemma 3.3 and the inductive assumption, we have

$$(3.1) \qquad |\mathcal{C}| \leqslant r + 1 + |\mathcal{C}^-| \leqslant r + 1 + \binom{r+1}{2} + 1 = \binom{r+2}{2} + 1.$$

Theorem 1.2 is proved.

Let $\mathbf{W}_r$ denote the set of all weakly separated collections $\mathcal{C}$ of subsets of $[1, r+1]$, that have maximum possible size $|\mathcal{C}| = \binom{r+2}{2} + 1$. The estimate in (3.1) implies the following

COROLLARY 3.4. *For every $\mathcal{C} \in \mathbf{W}_r$, the collection $\mathcal{C}^- = \{I^- : I \in \mathcal{C}\}$ belongs to $\mathbf{W}_{r-1}$. Furthermore, for each $p = 0, 1, \ldots, r$, there is exactly one subset $J \in \mathcal{C}_p^-$ such that both J and $J \cup \{r+1\}$ belong to $\mathcal{C}$.*

Thus, the correspondence $\mathcal{C} \mapsto \mathcal{C}^-$ is a map $\mathbf{W}_r \to \mathbf{W}_{r-1}$. Let us describe the fibers of this map.

For $\mathcal{C} \in \mathbf{W}_r$, let $\Sigma(\mathcal{C})$ denote the family of subsets $J \in \mathcal{C}^-$ such that both J and $J \cup \{r+1\}$ belong to $\mathcal{C}$. By Corollary 3.4, for each $p = 0, 1, \ldots, r$, the family $\Sigma(\mathcal{C})$ contains a unique subset of size p.

Now let $\mathcal{C}' \in \mathbf{W}_{r-1}$. By a *section* of $\mathcal{C}'$ we shall mean a family of subsets $\Sigma = (J_0, J_1, \ldots, J_r)$ satisfying the following two conditions:
(1) each subset J_p belongs to $\mathcal{C}'_p$ and is strongly separated from all subsets $J \in \mathcal{C}'_p$;
(2) $J_p - J_{p+1} \prec J_{p+1} - J_p$ for $p < r$.

The following result is a refinement of the first statement in Theorem 1.9.

THEOREM 3.5. *For any $\mathcal{C} \in \mathbf{W}_r$, the family $\Sigma(\mathcal{C})$ is a section of $\mathcal{C}^-$. Furthermore, for any $\mathcal{C}' \in \mathbf{W}_{r-1}$, the correspondence $\mathcal{C} \mapsto \Sigma(\mathcal{C})$ is a bijection between the set of collections $\mathcal{C} \in \mathbf{W}_r$ such that $\mathcal{C}^- = \mathcal{C}'$ and the set of sections Σ of $\mathcal{C}'$. The inverse bijection sends a section $\Sigma = (J_0, J_1, \ldots, J_r)$ of $\mathcal{C}'$ to the collection $\mathcal{C}$ given by*

$$(3.2) \qquad \begin{aligned} \mathcal{C}_p = &\{J \in \mathcal{C}'_p : J - J_p \prec J_p - J\} \\ &\cup \{J \cup \{r+1\} : J \in \mathcal{C}'_{p-1}, J_{p-1} - J \prec J - J_{p-1}\}. \end{aligned}$$

PROOF. Let $\mathcal{C} \in \mathbf{W}_r$, and $\Sigma(\mathcal{C}) = (J_0, J_1, \ldots, J_r)$. Let $J \in \mathcal{C}_p^-$ be a subset different from J_p. Then exactly one of the subsets J and $J \cup \{r+1\}$ belongs to $\mathcal{C}$. Using Lemma 3.1, we see that if $J \in \mathcal{C}$ then $J - J_p \prec J_p - J$; similarly, if $J \cup \{r+1\} \in \mathcal{C}$ then $J_p - J \prec J - J_p$. In both cases, J_p is strongly separated from J, so $\Sigma(\mathcal{C})$ satisfies the first condition in the definition of a section. The second condition follows from Lemma 3.1 applied to $I = J_p$ and $J = J_{p+1}$. Thus, $\Sigma(\mathcal{C})$ is a section of $\mathcal{C}^-$. Furthermore, the above argument shows that the collection $\mathcal{C}$ is obtained from $\mathcal{C}' = \mathcal{C}^-$ and $\Sigma = \Sigma(\mathcal{C}^-)$ by means of (3.2).

It remains to show that, conversely, for any $\mathcal{C}' \in \mathbf{W}_{r-1}$ and any section Σ of $\mathcal{C}'$, the collection $\mathcal{C}$ defined by (3.2) is weakly separated. In view of Lemma 3.1, we only need to show the following: if $I \in \mathcal{C}'_p$ and $J \in \mathcal{C}'_q$ are such that $p \leqslant q$, $I - J_p \prec J_p - I$ and $J_q - J \prec J - J_q$, then $I - J \prec J - I$. This is a consequence of the following lemma.

LEMMA 3.6. *Suppose that $I, J, K \subset [1, r]$ are such that $|I| \leqslant |J| \leqslant |K|$, $I - J \prec J - I$, $J - K \prec K - J$, and I is weakly separated from K. Then $I - K \prec K - I$.*

Before proving Lemma 3.6, we state the following obvious reformulations of strong and weak separation.

LEMMA 3.7. *Two sets I and J are strongly separated if and only if there do not exist three indices $a < b < c$ such that either $I \cap \{a, b, c\} = \{b\}$, $J \cap \{a, b, c\} = \{a, c\}$ or $I \cap \{a, b, c\} = \{a, c\}$, $J \cap \{a, b, c\} = \{b\}$.*

LEMMA 3.8. (a) *Suppose $|I| < |J|$. Then I and J are weakly separated if and only if there do not exist three indices $a < b < c$ such that $I \cap \{a, b, c\} = \{b\}$, $J \cap \{a, b, c\} = \{a, c\}$.*

(b) *Suppose $|I| = |J|$. Then I and J are weakly separated if and only if there do not exist four indices $a < b < c < d$ such that either $I \cap \{a, b, c, d\} = \{b, d\}$, $J \cap \{a, b, c, d\} = \{a, c\}$ or $I \cap \{a, b, c, d\} = \{a, c\}$, $J \cap \{a, b, c, d\} = \{b, d\}$.*

PROOF OF LEMMA 3.6. Suppose there exist $a \in K - I$ and $b \in I - K$ such that $a < b$. We shall show that this leads to a contradiction. First, the following two chains of implications show that either $\{a, b\} \subset J$ or $J \cap \{a, b\} = \varnothing$:

$$a \in J \implies a \in J - I \implies b \notin I - J \implies b \in J,$$
$$b \in J \implies b \in J - K \implies a \notin K - J \implies a \in J.$$

Suppose $\{a, b\} \subset J$. Since $b \in J - K$ and $|J| \leqslant |K|$, it follows that $K - J \neq \varnothing$. Choose $c \in K - J$; since $J - K \prec K - J$, it follows that $b < c$. Since $a \in J - I$ and $I - J \prec J - I$, it follows that $c \notin I$. Thus, $I \cap \{a, b, c\} = \{b\}$ and $K \cap \{a, b, c\} = \{a, c\}$. By Lemma 3.8 (a), we must have $|I| \geqslant |K|$, hence $|I| = |J| = |K|$. It follows that $I - J \neq \varnothing$. Choose $d \in I - J$; since $I - J \prec J - I$, it follows that $d < a$. But then both possibilities $d \in K$ and $d \notin K$ lead to a contradiction: if $d \in K$, then $d < b$ contradicts $J - K \prec K - J$; if $d \notin K$, then $I \cap \{d, a, b, c\} = \{d, b\}$ and $K \cap \{d, a, b, c\} = \{a, c\}$, which contradicts Lemma 3.8 (b).

The case $J \cap \{a, b\} = \varnothing$ is treated in the same way; we leave the details to the reader.

Lemma 3.6 and Theorem 3.5 are proved.

Theorem 3.5 provides an inductive procedure for generating all the collections $\mathcal{C} \in \mathbf{W}_r$. Let us illustrate this procedure by some examples. In listing the subsets in a collection $\mathcal{C} \in \mathbf{W}_r$, we shall omit the subsets $[1,a]$ and $[a+1,r+1]$, that always belong to $\mathcal{C}$. The remaining subsets will be called *bounded*. (This terminology comes from the geometric interpretation of strongly separated subsets as chamber sets of an arrangement of pseudo-lines, which will be discussed later.) Every $\mathcal{C} \in \mathbf{W}_r$ contains $\binom{r}{2}$ bounded subsets. We shall also write a section Σ of $\mathcal{C}' \in \mathbf{W}_{r-1}$ as $(J_1, \ldots, J_{r-1})$, omitting the subsets $J_0 = \varnothing$ and $J_r = [1,r]$.

There is only one collection in $\mathbf{W}_1$ consisting of all four subsets in $[1,2]$. It has two sections: $\Sigma = (1)$ and $\Sigma' = (2)$. The corresponding two collections in $\mathbf{W}_2$ are (listing only bounded sets) $\mathcal{C} = \{13\}$ and $\mathcal{C}' = \{2\}$. The collection $\mathcal{C} = \{13\}$ has five sections:
$$(1,12),\ (1,13),\ (1,23),\ (3,13),\ (3,23);$$
the corresponding five collections in $\mathbf{W}_3$ are
$$\{14,124,134\},\ \{14,13,134\},\ \{14,13,23\},\ \{3,13,134\},\ \{3,13,23\}.$$

The collection $\mathcal{C}' = \{2\}$ also has five sections:
$$(1,12),\ (1,23),\ (2,12),\ (2,23),\ (3,23);$$
the corresponding five collections in $\mathbf{W}_3$ are
$$\{14,24,124\},\ \{14,24,23\},\ \{2,24,124\},\ \{2,24,23\},\ \{2,3,23\}.$$

Thus, $|\mathbf{W}_3| = 10$.

Let us turn to the proof of Theorem 1.3. We shall need the following lemma, which is an immediate consequence of Lemma 3.8.

LEMMA 3.9. *Any interval $I = [a,b] \subset [1,r+1]$ is weakly separated from all subsets $J \subset [1,r+1]$ with $|J| \leq |I|$. Any subset $I = [1,r+1] - [a,b]$, which is the complement of an interval, is weakly separated from all subsets $J \subset [1,r+1]$ with $|J| \geq |I|$.*

PROOF OF THEOREM 1.3. Let $\mathcal{C}$ be a weakly separated collection of subsets in $[1,r+1]$ such that $k \leq |I| \leq l$ for all $I \in \mathcal{C}$. Let $\widetilde{\mathcal{C}}$ be the collection obtained from $\mathcal{C}$ by adjoining all intervals $[a,b]$ in $[1,r+1]$ with $|[a,b]| > l$, and all complements of intervals $[1,r+1] - [a,b]$ with $|[1,r+1] - [a,b]| < k$. By Lemma 3.9, $\widetilde{\mathcal{C}}$ is a weakly separated collection. By Theorem 1.2, $|\widetilde{\mathcal{C}}| \leq \binom{r+2}{2} + 1$. Now the desired upper bound
$$|\mathcal{C}| \leq \binom{r+2}{2} - \binom{r+2-l}{2} - \binom{k+1}{2} + 1$$
becomes a consequence of the following obvious statement.

LEMMA 3.10. *The number of intervals $[a,b]$ in $[1,r+1]$ with $|[a,b]| > l$ is equal to $\binom{r+2-l}{2}$.*

To complete the proof of Theorem 1.3, it remains to construct a weakly separated collection $\mathcal{C}$ as above, of size $\binom{r+2}{2} - \binom{r+2-l}{2} - \binom{k+1}{2} + 1$. We define

(3.3) $\mathcal{C}_k = \{[1,k]\} \cup \{[1,a] \cup [b+1, b+k-a] : 0 \leq a \leq k-1,\ a+1 \leq b \leq a+r+1-k\},$

and, for $k < j \leq l$, we define $\mathcal{C}_j$ to be the collection of all intervals of size j in $[1,r+1]$. Using Lemmas 3.8, 3.9, and 3.10, it is easy to check that $\mathcal{C} = \bigcup_{j=k}^{l} \mathcal{C}_j$ is a weakly separated collection of the required size. Theorem 1.3 is proved.

We now turn to the proof of Theorem 1.4. The above proof of Theorem 1.2 extends to this more general case with the following modifications. Fix $w \in S_{r+1}$ and let $s = w(r+1)$. We associate to w a permutation $w^- \in S_r$ as follows: for $i \in [1, r]$ set $w^-(i) = w(i)$ if $w(i) < s$, and $w^-(i) = w(i) - 1$ if $w(i) > s$. Counting inversions of w and w^-, we see that

$$(3.4) \qquad l(w) = l(w^-) + r + 1 - s.$$

The next lemma follows at once from definitions.

LEMMA 3.11. *Let $J \subset [1, r]$. Then J is a w-chamber set if and only if J is a w^--chamber set. The subset $J \cup \{r+1\}$ is a w-chamber set if and only if J is a w^--chamber set and $w^{-1}([1, s-1]) \subset J$.*

Now let $\mathcal{C}$ be a weakly separated collection of w-chamber sets. By Lemma 3.11, the collection $\mathcal{C}^-$ consists of w^--chamber sets. Furthermore, if $J \in \mathcal{C}_p^-$ and $p < s-1$, then $(J \cup \{r+1\}) \notin \mathcal{C}$. Combining this with Lemma 3.3 and using induction on r, we obtain the following generalization of (3.1):

$$(3.5) \qquad |\mathcal{C}| \leqslant |\mathcal{C}^-| + r + 2 - s \leqslant l(w^-) + r + 1 + r + 2 - s = l(w) + r + 2$$

(the last equality in (3.5) follows from (3.4)).

To complete the proof of Theorem 1.4, it remains to give an example of a collection that turns (3.5) into an equality. This can be easily done "by hand" but we prefer to give a procedure generating all such collections; this procedure generalizes Theorem 3.5.

Let $\mathbf{W}(w)$ denote the set of all weakly separated collections $\mathcal{C}$ of w-chamber sets that have maximum possible size $|\mathcal{C}| = l(w) + r + 2$. The following statement generalizes Corollary 3.4 and is proved in the same way.

COROLLARY 3.12. *For every $\mathcal{C} \in \mathbf{W}(w)$, the collection $\mathcal{C}^-$ belongs to $\mathbf{W}(w^-)$. Furthermore, for each $p = s-1, s, \ldots, r$, there is exactly one subset $J \in \mathcal{C}_p^-$ such that both J and $J \cup \{r+1\}$ belong to $\mathcal{C}$.*

Thus, the correspondence $\mathcal{C} \mapsto \mathcal{C}^-$ is a map $\mathbf{W}(w) \to \mathbf{W}(w^-)$. Generalizing Theorem 3.5, we shall describe fibers of this map. By a *section* of a collection $\mathcal{C}' \in \mathbf{W}(w^-)$ we shall mean a family of subsets $\Sigma = (J_{s-1}, J_s, \ldots, J_r)$ satisfying conditions (1) and (2) before Theorem 3.5 and such that $J_{s-1} = w^{-1}([1, s-1])$.

THEOREM 3.13. *For any $\mathcal{C}' \in \mathbf{W}(w^-)$, there is a bijection between sections Σ of $\mathcal{C}'$ and collections $\mathcal{C} \in \mathbf{W}(w)$ such that $\mathcal{C}^- = \mathcal{C}'$. Under this bijection, the collection $\mathcal{C}$ corresponding to a section $\Sigma = (J_{s-1}, J_s, \ldots, J_r)$ of $\mathcal{C}'$, has $\mathcal{C}_p$ given by (3.2) for $p \geqslant s$, and $\mathcal{C}_p = \mathcal{C}_p'$ for $0 \leqslant p \leqslant s-1$.*

This is proved by the same arguments as Theorem 3.5. The only additional statement to check is the following: if $J \in \mathcal{C}_p'$ for some $p \geqslant s$ is such that $J_p - J \prec J - J_p$ then $J_{s-1} \subset J$ (by Lemma 3.11, this is needed to guarantee that $J \cup \{r+1\}$ is a w-chamber set). To prove this, we set $I = J_{s-1} = w^{-1}([1, s-1])$ and $K = (w^-)^{-1}([1, p]) = w^{-1}([1, p] - \{s\})$. Then I, J, and K satisfy the assumptions of the next lemma, which implies the desired statement.

LEMMA 3.14. *Suppose that $I, J, K \subset [1, r]$ are such that $|J| = |K|$, $I - J \prec J - I$, $J - K \prec K - J$, and $I \subset K$. Then $I \subset J$.*

PROOF. Suppose our statement is not true, i.e., $I - J \neq \varnothing$. Let $b \in I - J$. Since $I \subset K$, we also have $b \in K$, so $b \in K - J$. Since $|J| = |K|$, it follows that $J - K \neq \varnothing$. Let $a \in J - K$. Since $J - K \prec K - J$, it follows that $a < b$. However, since $I \subset K$, we also have $a \in J - I$, which contradicts the condition that $I - J \prec J - I$.

Lemma 3.14 and Theorem 3.13 are proved. In view of Theorem 3.13, an inductive argument shows that $\mathbf{W}(w) \neq \varnothing$ for $w \in S_{r+1}$. This completes the proof of Theorem 1.4.

§4. Maximal strongly separated collections

In this section we prove Theorem 1.6. Our proof will be based on the geometric description of chamber sets given in [**BFZ**]. Let us recall this description. We fix a permutation $w \in S_{r+1}$ of length $l(w) = l$. Let $\mathbf{h} = (h_1, \ldots, h_l) \in R(w)$ be a reduced word for w. We associate with $\mathbf{h}$ a *pseudo-line arrangement* Arr$(\mathbf{h})$. This is a configuration of $r + 1$ pseudo-lines in the plane (modulo natural isotopy) which may be defined recursively as follows. If $\mathbf{h}$ is the empty word, i.e., the unique element of $R(\text{id})$, then Arr$(\mathbf{h})$ consists of $r + 1$ parallel horizontal lines labelled $1, \ldots, r+1$ from bottom to top. Otherwise, Arr$(\mathbf{h})$ is obtained from Arr$(\mathbf{h}^-)$, where $\mathbf{h}^- = (h_2, \cdots, h_l)$, by adding on the left a crossing between the h_1th and $(h_1 + 1)$th pseudo-lines from the bottom. For example, if $r = 3$ and $\mathbf{h} = (3, 2, 1, 3)$, then Arr$(\mathbf{h})$ is the configuration shown in Figure 2.

The *chambers* of Arr$(\mathbf{h})$ are the connected components of the complement of the union of all pseudo-lines. It is easy to see that there are $l + r + 2$ of them. We label every chamber A by the set $L(A)$ of indices of all pseudo-lines that pass below A. The sets $L(A)$ are the *chamber sets* of Arr$(\mathbf{h})$ (or simply of $\mathbf{h}$). We denote by $\mathcal{C}(\mathbf{h})$ the collection of all chamber sets of $\mathbf{h}$. For instance, we see from Figure 2 that

$$\mathcal{C}(3, 2, 1, 3) = \{\varnothing, 1, 2, 12, 24, 123, 124, 234, 1234\}.$$

It is easy to see that this description is equivalent to the one in the Introduction.

As a first step toward Theorem 1.6, we prove the following lemma.

LEMMA 4.1. *For any reduced word $\mathbf{h} \in R(w)$, the collection $\mathcal{C}(\mathbf{h})$ is a strongly separated collection of w-chamber sets.*

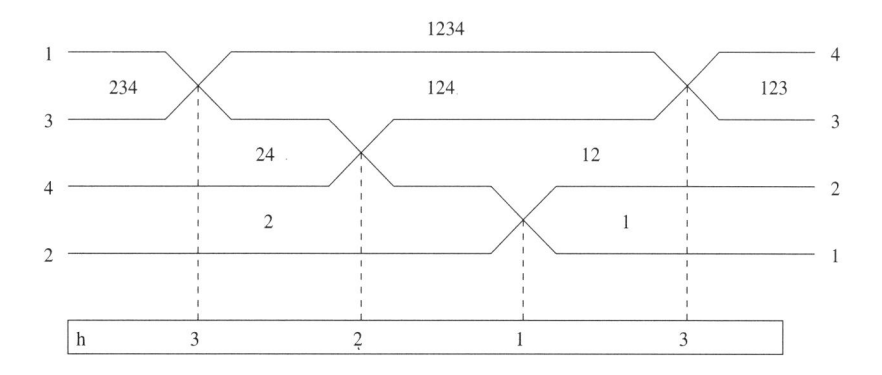

FIGURE 2

PROOF. Recall that a subset $I \subset [1, r+1]$ is a *w-chamber set* if for each $j \in I$ the set I also contains all indices i such that $i < j$ and $w(i) < w(j)$. The definition of Arr(**h**) implies easily that the pseudo-lines Line$_i$ and Line$_j$ with $i < j$ cross each other if and only if $w(i) > w(j)$. It follows that every chamber set of **h** is a w-chamber set.

To prove that $\mathcal{C}(\mathbf{h})$ is strongly separated, consider two chamber sets $I = L(A)$ and $J = L(B)$. Suppose $I - J \not\prec J - I$. This means that there exist indices $i < j$ such that the pseudo-line Line$_i$ passes above the chamber A but below B, while Line$_j$ passes below A but above B. In other words, the chambers A and B lie inside two opposite angles formed by Line$_i$ and Line$_j$ at their crossing point: the chamber A is to the left and B is to the right of this crossing point. In particular, this implies that every point of A is to the left of every point of B. Similarly, the assumption $J - I \not\prec I - J$ implies that every point of B is to the left of every point of A. It follows that at least one of the conditions $I - J \prec J - I$ and $J - I \prec I - J$ must hold, i.e., I and J are strongly separated. Lemma 4.1 is proved.

Recall from §3 that $\mathbf{W}(w)$ denotes the set of all weakly separated collections of w-chamber sets in $[1, r+1]$ that have maximum possible size $l + r + 2$. Let $\mathbf{S}(w)$ denote the subset of $\mathbf{W}(w)$ formed by strongly separated collections. By Lemma 4.1, for every $\mathbf{h} \in R(w)$, the chamber set collection $\mathcal{C}(\mathbf{h})$ belongs to $\mathbf{S}(w)$. As a next step in the proof of Theorem 1.6, we now establish the reverse inclusion.

LEMMA 4.2. *Every collection $\mathcal{C} \in \mathbf{S}(w)$ is of the form $\mathcal{C}(\mathbf{h})$ for some $\mathbf{h} \in R(w)$.*

PROOF. We shall use the surjection $\mathcal{C} \mapsto \mathcal{C}^-$ from $\mathbf{W}(w)$ to $\mathbf{W}(w^-)$ and the description of its fibers given by Theorem 3.13.

LEMMA 4.3. *A collection $\mathcal{C} \in \mathbf{W}(w)$ is strongly separated if and only if $\mathcal{C}^-$ is strongly separated and the section $\Sigma = \Sigma(\mathcal{C}) = (J_{s-1}, J_s, \ldots, J_r)$ forms a flag, i.e., $J_{p-1} \subset J_p$ for $s \leq p \leq r$.*

PROOF OF LEMMA 4.3. Clearly, if $\mathcal{C}$ is strongly separated, then so is $\mathcal{C}^-$. It is also clear that if $J_{p-1} \not\subset J_p$ then J_p is not strongly separated from $J_{p-1} \cup \{r+1\}$. This proves the "only if" part of our lemma. To prove the "if" part assume that $\mathcal{C}^-$ is strongly separated and the section $\Sigma = \Sigma(\mathcal{C}) = (J_{s-1}, J_s, \ldots, J_r)$ forms a flag. It is enough to check the following: if $s - 1 \leq p < p' \leq r$, and $J \in \mathcal{C}_p^-$, $J' \in \mathcal{C}_{p'}^-$ are such that $J_p - J \prec J - J_p$, $J' - J_{p'} \prec J_{p'} - J'$, then $J' - J \prec J - J'$ (this guarantees that J' is strongly separated from $J \cup \{r+1\}$). Since $\mathcal{C}^-$ is strongly separated, we can assume that $J - J' \prec J' - J$; our goal then is to prove that $J \subset J'$. By Lemma 3.6, the conditions $J_p - J \prec J - J_p$ and $J - J' \prec J' - J$ imply that $J_p - J' \prec J' - J_p$. Applying Lemma 3.14 to the triple $(I, J, K) = (J_p, J', J_{p'})$, we obtain that $J_p \subset J'$. But then the conditions $J_p - J \prec J - J_p$ and $J - J' \prec J' - J$ imply that $J \subset J'$, as desired (this is proved by exactly the same argument as Lemma 3.14). Lemma 4.3 is proved.

Note that Lemma 4.3 establishes the following generalization of the second statement in Theorem 1.9.

COROLLARY 4.4. *The correspondence $\mathcal{C} \mapsto \mathcal{C}^-$ is a surjection $\mathbf{S}(w) \to \mathbf{S}(w^-)$.*

Continuing the proof of Lemma 4.2, let us assume that $\mathcal{C} \in \mathbf{S}(w)$. By Lemma 4.3, $\mathcal{C}^- \in \mathbf{S}(w^-)$. Using induction on r, we can assume that $\mathcal{C}^- = \mathcal{C}(\mathbf{h}')$

for some reduced word $\mathbf{h}'$ of w^-. The condition $J_{p-1} \subset J_p$ means that the corresponding chambers in the arrangement $\mathrm{Arr}(\mathbf{h}')$ are adjacent to each other (they are separated by the pseudo-line Line_j where $\{j\} = J_p - J_{p-1}$). Therefore, we can extend the arrangement $\mathrm{Arr}(\mathbf{h}')$ to an arrangement of $r+1$ pseudo-lines, by adding one more pseudo-line Line_{r+1} that passes through all chambers in $\mathrm{Arr}(\mathbf{h}')$ with chamber sets $J_{s-1}, J_s, \ldots, J_r$. Clearly, this new arrangement is isotopic to $\mathrm{Arr}(\mathbf{h})$ for some reduced word $\mathbf{h}$ of w, and $\mathcal{C}$ is exactly the collection of its chamber sets. Lemma 4.2 is proved.

To complete the proof of Theorem 1.6, it remains to show the following

LEMMA 4.5. *Any strongly separated collection $\mathcal{C}$ of w-chamber sets in $[1, r+1]$ is contained in $\mathcal{C}(\mathbf{h})$ for some $\mathbf{h} \in R(w)$.*

PROOF. Let $s = w(r+1)$ and let $u = w^- \in S_r$. We shall write the collection $\mathcal{C}^-$ of subsets in $[1, r]$ as $\mathcal{C}^- = \mathcal{C}' \cup \mathcal{C}''$, where

$$\mathcal{C}' = \{I \subset [1, r] : I \in \mathcal{C}\}, \qquad \mathcal{C}'' = \{I \subset [1, r] : I \cup \{r+1\} \in \mathcal{C}\}.$$

Clearly, $\mathcal{C}^-$ is a strongly separated collection of u-chamber sets. Using induction on r, we can assume that $\mathcal{C}^- \subset \mathcal{C}(\mathbf{h}')$ for some $\mathbf{h}' \in R(u)$. Furthermore, the subcollections $\mathcal{C}'$ and $\mathcal{C}''$ are easily seen to have the following property:

(1) for any $I \in \mathcal{C}'$ and $J \in \mathcal{C}''$, we have $I - J \prec J - I$, and $u^{-1}[1, s-1] \subset J$.

We shall say that a subset $K \subset [1, r]$ *separates* $\mathcal{C}'$ and $\mathcal{C}''$ if $I - K \prec K - I$ and $K - J \prec J - K$ for all $I \in \mathcal{C}'$, $J \in \mathcal{C}''$. Taking Lemmas 4.2 and 4.3 into account, we see that Lemma 4.5 becomes a consequence of the following

LEMMA 4.6. *Let $\mathbf{h}' \in R(u)$ and suppose that $\mathcal{C}'$ and $\mathcal{C}''$ are two subcollections of the chamber set collection $\mathcal{C}(\mathbf{h}')$ that satisfy (1) above. There exists a flag $(u^{-1}[1, s-1] = J_{s-1} \subset J_s \subset \cdots \subset J_r = [1, r])$ in $\mathcal{C}(\mathbf{h}')$ such that each J_p separates $\mathcal{C}'$ and $\mathcal{C}''$.*

PROOF OF LEMMA 4.6. First of all, it is clear that $u^{-1}[1, s-1] = J_{s-1}$ separates $\mathcal{C}'$ and $\mathcal{C}''$ (we have $I - J_{s-1} \prec J_{s-1} - I$ for all u-chamber sets I).

Now suppose we have already constructed the chamber sets $J_{s-1} \subset J_s \subset \cdots \subset J_p$ separating $\mathcal{C}'$ and $\mathcal{C}''$, for some p with $s - 1 \leqslant p < r$. To conclude the proof, it remains to construct the chamber set J_{p+1} of size $p+1$ such that $J_p \subset J_{p+1}$, and J_{p+1} separates $\mathcal{C}'$ and $\mathcal{C}''$. In view of Lemma 3.6, we can and will assume that all subsets from $\mathcal{C}'$ and $\mathcal{C}''$ have size $> p$.

Let B be a chamber in the arrangement $\mathrm{Arr}(\mathbf{h}')$. By the *shadow* of B we shall mean the union of all chambers below B (including B itself) that can be joined with B by a polygonal line that crosses only horizontal segments of $\mathrm{Arr}(\mathbf{h}')$. It is easy to see that a chamber A belongs to the shadow of B if and only if $L(A) \subset L(B)$. We shall use the following geometrically obvious lemma.

LEMMA 4.7. *Suppose $I = L(A)$ and $J = L(B)$ are two chamber sets of $\mathrm{Arr}(\mathbf{h}')$ such that $|I| \leqslant |J|$ (i.e., the chamber A lies below B or on the same horizontal level with B). Then $I - J \prec J - I$ if and only if A either belongs to the shadow of B or lies to the right of the shadow.*

Now let A be the chamber in the arrangement $\mathrm{Arr}(\mathbf{h}')$ with the chamber set $L(A) = J_p$. Lemma 4.7 implies easily that for every two chambers B' and B'' such that $L(B') \in \mathcal{C}'$, $L(B'') \in \mathcal{C}''$, there exists a chamber B adjacent from above to A

and lying inside or to the left of the shadow of B' and inside or to the right of the shadow of B''. Let us choose B' so that the right end of its shadow on the level immediately above A is as far to the left as possible; similarly, choose B'' so that the left end of its shadow on the level immediately above A is as far to the right as possible. Taking the corresponding chamber B as above, we conclude that the chamber set $J_{p+1} = L(B)$ contains J_p and separates $\mathcal{C}'$ and $\mathcal{C}''$. This completes the proof of Lemmas 4.5 and 4.6 and Theorem 1.6.

§5. Extension of the second Bruhat order

In this section we prove Theorem 1.7. Recall from [MS] the definition of the second Bruhat order on the set $R^c(w_0)$ of commutation classes of reduced words for w_0. This partial order is the transitive closure of the following relation on reduced words: $\mathbf{h}' \geqslant \mathbf{h}$ if $\mathbf{h}'$ is obtained from $\mathbf{h}$ by a *raising 3-move* replacing three consecutive indices $i, i+1, i$ by $i+1, i, i+1$. By Theorem 1.6, the correspondence $\mathbf{h} \to \mathcal{C}(\mathbf{h})$ that associates to each reduced word $\mathbf{h}$ the chamber set collection of the pseudo-line arrangement $\mathrm{Arr}(\mathbf{h})$, allows us to identify $R^c(w_0)$ with the set $\mathbf{S}_r$ of all maximal strongly separated collections of subsets in $[1, r+1]$. Let us describe how this identification transforms the second Bruhat order.

In terms of the pseudo-line arrangements $\mathrm{Arr}(\mathbf{h})$ and $\mathrm{Arr}(\mathbf{h}')$, a raising 3-move translates into the following *Yang–Baxter raising flip*. If $\mathrm{Arr}(\mathbf{h})$ has a "triangular" chamber bounded by three pseudo-lines, Line_i, Line_j, and Line_k, such that Line_j passes below the crossing of Line_i and Line_k, then $\mathrm{Arr}(\mathbf{h}')$ is obtained from $\mathrm{Arr}(\mathbf{h})$ by deforming Line_j so that it will now pass above the crossing of Line_i and Line_k (and keep the same position with respect to all other crossings).

It is geometrically obvious that $\mathrm{Arr}(\mathbf{h})$ has a "triangular" chamber A as above if and only if its chamber set collection $\mathcal{C}(\mathbf{h})$ contains seven subsets, L, Li, Lj, Lk, Lij, Ljk, and $Lijk$, where the chamber set $L(A)$ is equal to $Lj := L \cup \{j\}$. Furthermore, $\mathcal{C}(\mathbf{h}')$ is obtained from $\mathcal{C}(\mathbf{h})$ by replacing Lj by Lik. We see that a raising 3-move translates into the following operation on a maximal strongly separated collection $\mathcal{C}$: replacing $Lj \in \mathcal{C}$ by Lik for some indices $i < j < k$ and a subset $L \subset [1, r+1]$ disjoint from $\{i, j, k\}$, provided that $\mathcal{C}$ contains six "witnesses", L, Li, Lk, Lij, Ljk, and $Lijk$. We call this operation a *strong raising flip*. Thus, $\mathcal{C}' \geqslant \mathcal{C}$ in $\mathbf{S}_r$ if $\mathcal{C}'$ is obtained from $\mathcal{C}$ by a sequence of strong raising flips. These definitions are illustrated by Figure 3.

We now define a *weak raising flip* on a collection of subsets $\mathcal{C}$ as the operation of replacing $Lj \in \mathcal{C}$ by Lik for some indices $i < j < k$ and a subset $L \subset [1, r+1]$ disjoint from $\{i, j, k\}$, provided that $\mathcal{C}$ contains four "witnesses", Li, Lk, Lij, and Ljk. The inverse operation will be called a *weak lowering flip*.

Theorem 1.7 is a consequence of the following statement.

THEOREM 5.1. *Weak raising and lowering flips preserve the set of weakly separated collections.*

PROOF OF THEOREM 5.1. It suffices to prove the following

LEMMA 5.2. *Suppose $i < j < k$ and $L \cap \{i, j, k\} = \varnothing$. If a subset $I \subset [1, r+1]$ is weakly separated from each of the four subsets Li, Lk, Lij, Ljk and is different from Lj and Lik, then I is weakly separated from Lj and Lik.*

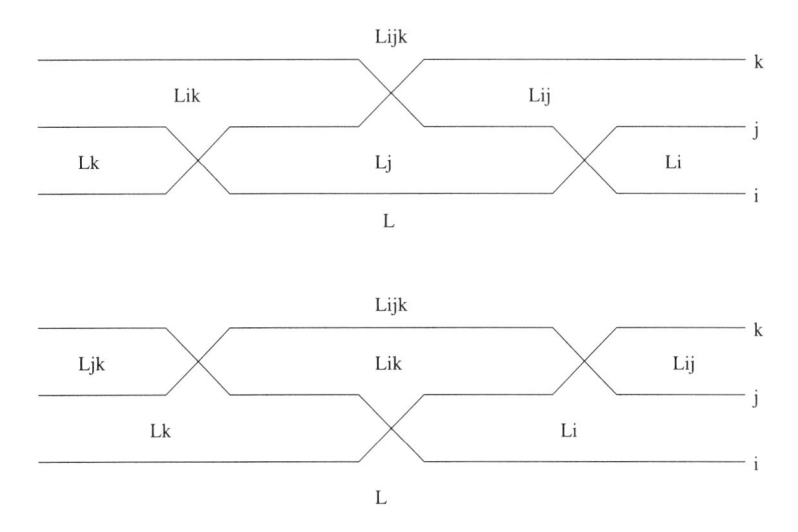

FIGURE 3

PROOF OF LEMMA 5.2. Suppose that I is not weakly separated from Lj. Consider two cases.

Case 1. Suppose $|I| \leqslant |Lj|$. By Lemma 3.8, there exist three indices $a < b < c$ such that $I \cap \{a,b,c\} = \{b\}$, $Lj \cap \{a,b,c\} = \{a,c\}$. Since at least one of the indices i and k is different from b, at least one of the intersections $Lij \cap \{a,b,c\}$ and $Ljk \cap \{a,b,c\}$ is equal to $\{a,c\}$. Hence at least one of Lij and Ljk is not weakly separated from I, which contradicts our assumptions.

Case 2. Suppose $|I| > |Lj|$. By Lemma 3.8, there exist three indices $a < b < c$ such that $I \cap \{a,b,c\} = \{a,c\}$, $Lj \cap \{a,b,c\} = \{b\}$.

Subcase 2.1. Suppose $b = j$. Then $j \notin I$. We claim that in this case I contains i and k. Suppose that $i \notin I$. Then $i \neq a$, so we have either $i < a$, or $a < i < j = b$. If $i < a$, then $I \cap \{i,a,j,c\} = \{a,c\}$, $Lij \cap \{i,a,j,c\} = \{i,j\}$, which contradicts the assumption that I and Lij are weakly separated. If $a < i < j$, then $I \cap \{a,i,c\} = \{a,c\}$, $Li \cap \{a,i,c\} = \{i\}$, which contradicts the assumption that I and Li are weakly separated. We see that $i \in I$. The fact that $k \in I$ is proved in a similar way. Thus, $I = I'ik$ with $I' \cap \{i,j,k\} = \varnothing$. We have $|I'| \geqslant |L|$ and $I' \neq L$, because I is assumed to be different from Lik. Thus, $I' - L \neq \varnothing$.

Subsubcase 2.1.1. Suppose $L \subset I'$. Then $Lij - I = Ljk - I = \{j\}$. Since $I - Lij = (I' - L)k$, $I - Ljk = (I' - L)i$, the conditions that I is weakly separated from Lij and Ljk imply that $j \prec I' - L$ and $I' - L \prec j$, a contradiction.

Subsubcase 2.1.2. Suppose $L \not\subset I'$, i.e., $L - I' \neq \varnothing$. Since $Li - I = Lk - I = L - I'$ and $I - Li = (I' - L)k$, $I - Lk = (I' - L)i$, the conditions that I is weakly separated from Li and Lk imply that the set $(I' - L)ik$ lies between two consecutive elements of $L - I'$. Therefore, $L - I'$ either contains some l such that $(I' - L)k \prec l$, or contains some h such that $h \prec (I' - L)i$. Suppose, $L - I'$ contains some l such that $(I' - L)k \prec l$. Then $Lij \cap \{j,k,l\} = \{j,l\}$ and $I \cap \{j,k,l\} = \{k\}$, so the fact that I and Lij are weakly separated, implies that $j \prec I - Lij = (I' - L)k \prec l$. Hence $I' - L$ contains some m such that $j < m < l$. But then we have $Ljk \cap \{i,j,m,l\} = \{j,l\}$

and $I \cap \{i,j,m,l\} = \{i,m\}$; in view of Lemma 3.8, this contradicts the assumption that I and Ljk are weakly separated. The case when $L - I'$ contains some h such that $h \prec (I' - L)i$ is treated in the same way.

Subcase 2.2. Suppose $b \neq j$, i.e., $b \in L - I$. Then the condition that I is weakly separated from Li and Lk, can be satisfied only if $a = i$ and $c = k$, i.e., we again have $I = I'ik$ for some subset I' such that $|I'| \geq |L|$ and $I' \neq L$. Now we have $b \in L - I' = Li - I$ and $k \in (I' - L)k = I - Li$, so the condition that I and Li are weakly separated implies that $b \prec I' - L$. However, by a similar argument, the condition that I and Lk are weakly separated, implies that $I' - L \prec b$. This contradiction completes the proof of the fact that I is weakly separated from Lj. The remaining part of Lemma 5.2 (that I is weakly separated from Lik) can be proved in a similar way, or deduced from the first part by passing to complementary subsets. Lemma 5.2 and Theorem 5.1 are proved.

Recall that $\mathbf{W} = \mathbf{W}_r$ is the set of all weakly separated collections of subsets in $[1, r+1]$ that have maximum possible size $\binom{r+2}{2} + 1$. We introduce a partial order on $\mathbf{W}_r$ as follows: $\mathcal{C}' \geq \mathcal{C}$ if $\mathcal{C}'$ is obtained from $\mathcal{C}$ by a sequence of weak raising flips. This definition makes $\mathbf{W}_r$ a ranked poset with respect to the rank function

(5.1) $$r(\mathcal{C}) = \sum_{I \in \mathcal{C}} |I|$$

(this is obvious, since any weak raising flip increases rank by 1).

Clearly, the collection $\mathcal{C}_{\min}$ of all intervals in $[1, r+1]$ is a minimal element of the poset $\mathbf{W}_r$ because every weak lowering flip involves a subset of the form Lik which is not an interval. Similarly, the collection $\mathcal{C}_{\max}$ of all interval complements is a maximal element of $\mathbf{W}_r$. This is also a special case of the following statement, which is an immediate consequence of the definitions.

PROPOSITION 5.3. *For a collection $\mathcal{C}$ of subsets in $[1, r+1]$, let $\overline{\mathcal{C}}$ denote the collection of complements of subsets from $\mathcal{C}$. Then the correspondence $\mathcal{C} \to \overline{\mathcal{C}}$ is an order-reversing involution of $\mathbf{W}_r$.*

Conjecture 1.8 claims that $\mathcal{C}_{\min}$ is the only minimal element of $\mathbf{W}_r$, while $\mathcal{C}_{\max}$ is the only maximal element. Let us present some partial results in this direction.

Consider an increasing filtration

$$\{\mathcal{C}_{\min}\} = \mathbf{W}^1 \subset \mathbf{W}^2 \subset \cdots \subset \mathbf{W}^r = \mathbf{W}$$

on $\mathbf{W} = \mathbf{W}_r$, where $\mathbf{W}^p$ consists of collections $\mathcal{C} \in \mathbf{W}$ such that all subsets $I \in \mathcal{C}$ with $|I| > p$ are intervals. Clearly, each $\mathbf{W}^p$ is invariant under weak lowering flips, i.e., is an order ideal in $\mathbf{W}$. The following easy lemma gives several equivalent descriptions of these order ideals.

LEMMA 5.4. *Let $\mathcal{C} \in \mathbf{W}_r$, and $p = 1, \ldots, r$. The following conditions are equivalent:*
  (1) $\mathcal{C} \in \mathbf{W}^p$;
  (2) $\mathcal{C}$ *contains all intervals of size* $\geq p$;
  (3) $\mathcal{C}$ *contains all intervals of size* p.

PROOF. Since $\mathcal{C}$ is a maximal (by inclusion) weakly separated collection, the implication (1) $\Rightarrow$ (2) follows from Lemma 3.9. The implication (2) $\Rightarrow$ (3) is trivial, and (3) $\Rightarrow$ (1) follows from the easy observation that every noninterval set I is not weakly separated from some interval $[a, b]$ of smaller size.

The following conjecture sharpens Conjecture 1.8.

CONJECTURE 5.5. *For $p \geqslant 2$, every collection $\mathcal{C} \in \mathbf{W}^p - \mathbf{W}^{p-1}$ admits a weak lowering flip $Lik \to Lj$ with $|Lik| = p$.*

This conjecture was computer checked for $r \leqslant 4$ and all $p \geqslant 2$, and for $r = 5$, $p \leqslant 3$. We shall prove it for $p = 2$ and r arbitrary, by giving an explicit combinatorial description of the poset $\mathbf{W}^2$.

Our answer will be given in terms of parenthesizings of strings of letters. They are defined recursively as follows. For one letter a there is only one parenthesizing (a). Now a parenthesizing of a string $a_1 \cdots a_n$ for $n > 1$ is obtained by splitting it into two nonempty substrings $a_1 \cdots a_k$ and $a_{k+1} \cdots a_n$ and taking parenthesizings of each of them plus the pair of parentheses surrounding the whole string. Thus, each parenthesizing of $a_1 \cdots a_n$ has $2n - 1$ pairs of matching opening and closing parentheses. It is well known that the number of such parenthesizings is the Catalan number $\frac{1}{n}\binom{2n-2}{n-1}$. For example, the string $a_1 a_2 a_3 a_4$ has five different parenthesizings:

$$((a_1)((a_2)((a_3)(a_4)))), \quad ((a_1)(((a_2)(a_3))(a_4))), \quad (((a_1)(a_2))((a_3)(a_4))),$$
$$(((a_1)((a_2)(a_3)))(a_4)), \quad ((((a_1)(a_2))(a_3))(a_4)).$$

We shall also need *partial parenthesizings*: a partial parenthesizing of a string $a_1 \cdots a_n$ is obtained by splitting it into any number of substrings and taking a parenthesizing of each of these substrings; in particular, any parenthesizing is considered as a partial parenthesizing. In depicting partial parenthesizings, we put substrings into which our string is split in brackets. For example, a string $a_1 a_2 a_3$ has five partial parenthesizings:

(5.2)
$$[((a_1)((a_2)(a_3)))], \quad [(((a_1)(a_2))(a_3))], \quad [((a_1)(a_2))][(a_3)],$$
$$[(a_1)][((a_2)(a_3))], \quad [(a_1)][(a_2)][(a_3)].$$

THEOREM 5.6. *The subset $\mathbf{W}^2 \subset \mathbf{W}_r$ is in a bijection with all partial parenthesizings of the string $a_{12} a_{23} \cdots a_{r,r+1}$. The collection $\mathcal{C} \in \mathbf{W}^2$ corresponding to a partial parenthesizing π is described as follows. One-element members of $\mathcal{C}$ are $\{1\}, \{r+1\}$ and all $\{i\}$ such that $a_{i-1,i}$ and $a_{i,i+1}$ belong to two different brackets in π. Two-element members of $\mathcal{C}$ are $\{i < j\}$ such that $a_{i,i+1}$ and $a_{j-1,j}$ belong to the same bracket of π, and the parenthesizing of this bracket in π has a matching pair of parentheses with the opening one right before $a_{i,i+1}$ and the closing one right after $a_{j-1,j}$.*

For example, if $r = 3$, there are five collections in $\mathbf{W}^2$ corresponding to five partial parenthesizings in (5.2) (where each a_i is replaced by $a_{i,i+1}$). Skipping the subsets $\{1\}, \{4\}, \{1,2\}, \{3,4\}$ that are always present, the lists of remaining one- and two-element subsets in these collections are the following:

$$\{14, 24, 23\}, \quad \{14, 13, 23\}, \quad \{3, 13, 23\}, \quad \{2, 24, 23\}, \quad \{2, 3, 23\}.$$

PROOF. Let $\mathcal{C} \in \mathbf{W}^2$. By Lemma 5.4, $\mathcal{C}$ contains all subsets $\{i, i+1\}$ for $i = 1, \ldots, r$. Let $\{i_0\}, \{i_1\}, \ldots, \{i_{s+1}\}$ be all one-element subsets in $\mathcal{C}$, where $s \geqslant 1$ and $1 = i_0 < i_1 < \cdots < i_s < i_{s+1} = r+1$. Clearly, for every two-element subset $\{i < j\} \in \mathcal{C}$, the interval $[i, j]$ is contained in one of $[i_k, i_{k+1}]$, $k = 0, \ldots, s$. Since $\mathcal{C}$ is a maximal weakly separated collection, it follows that $\mathcal{C}$ contains all subsets

$\{i_k, i_{k+1}\}$. Furthermore, we claim that if $i_{k+1} - i_k > 1$ for some k, then there exists a j such that $i_k < j < i_{k+1}$ and both $\{i_k, j\}$ and $\{j, i_{k+1}\}$ belong to $\mathcal{C}$. Indeed, take
$$j = \max\{i : i_k < i < i_{k+1}, \{i_k, i\} \in \mathcal{C}\};$$
the fact that $\{j, i_{k+1}\} \in \mathcal{C}$ again follows from the maximality of $\mathcal{C}$. Repeating this argument, we conclude that the subsets $\{i, j\} \in \mathcal{C}$ contained in $[i_k, i_{k+1}]$ are exactly the ones corresponding to some parenthesizing of the string $a_{i_k, i_k+1} \cdots a_{i_{k+1}-1, i_{k+1}}$, as described in the theorem.

COROLLARY 5.7. *Conjecture 5.5 holds for* $p = 2$. *Lowering flips on a collection* $\mathcal{C} \in \mathbf{W}^2 - \mathbf{W}^1$ *correspond to the following operations on a partial parenthesizing:*
$$[((A)(B))] \to [(A)][(B)]$$
(*splitting one bracket into two*).

Let C_r be the cardinality of the subset $\mathbf{W}^2 \subset \mathbf{W}_r$ (we set $C_0 = 1$). Theorem 5.6 easily implies the following expression for the generating function of the sequence (C_r):
$$\sum_{r \geq 0} C_r x^r = \frac{2}{1 + (1 - 4x)^{1/2}}. \tag{5.3}$$

We leave the proof of (5.3) as an exercise for the reader.

Note also that a collection $\mathcal{C} \in \mathbf{W}^2$ is strongly separated if and only if the corresponding partial parenthesizing has at most two symbols $a_{i,i+1}$ inside each bracket. The number F_r of such collections is easily seen to be the Fibonacci number (i.e., we have $F_0 = F_1 = 1$ and $F_r = F_{r-1} + F_{r-2}$ for $r \geq 2$).

Now let us consider another interesting subposet of $\mathbf{W}_r$. Namely, let $\mathcal{C}' \in \mathbf{W}_{r-1}$ be the collection of all intervals in $[1, r]$, and let
$$\mathbf{M} = \mathbf{M}_r = \{\mathcal{C} \in \mathbf{W}_r : \mathcal{C}^- = \mathcal{C}'\} \tag{5.4}$$
be the fiber over $\mathcal{C}'$ of the natural projection $\mathbf{W}_r \to \mathbf{W}_{r-1}$. According to Theorem 3.5, the collections from $\mathbf{M}$ are in a bijection with sections $(J_0, J_1, \ldots, J_r)$ of $\mathcal{C}'$; this bijection is given by (3.2).

The definitions in §3 readily imply that every section of $\mathcal{C}'$ has the form ($J_p = [b_p - p + 1, b_p]$), where $(b_0, b_1, \ldots, b_r)$ is a sequence of integers satisfying
$$0 = b_0 \leq b_1 \leq \cdots \leq b_r = r, \qquad b_p \geq p. \tag{5.5}$$

We shall denote the corresponding collection $\mathcal{C} \in \mathbf{M}$ by $\mathcal{C}(b_1, \ldots, b_{r-1})$. Thus, the collection $\mathcal{C}(b_1, \ldots, b_{r-1})$ consists of intervals $[b - p + 1, b]$ with $0 \leq p \leq b \leq b_p$ and of subsets $[b - p + 1, b] \cup \{r + 1\}$ with $b_p \leq b \leq r$. In particular, this implies that the intersections of $\mathbf{M}$ with the terms of the filtration $(\mathbf{W}^p)$ in $\mathbf{W}_r$ are given by
$$\mathbf{M} \cap \mathbf{W}^p = \{\mathcal{C}(b_1, \ldots, b_{r-1}) : b_p = r\}. \tag{5.6}$$

The cardinality of $\mathbf{M}$ is easily seen to be the Catalan number $\frac{1}{r+1}\binom{2r}{r}$. One way to see this is to establish a bijection between sequences satisfying (5.5) and lattice paths from $(0, 0)$ to (r, r) lying below the diagonal $x = y$; to construct such a bijection, we associate to $(b_0, b_1, \ldots, b_r)$ the lattice path whose horizontal segments are $[(b_p, p), (b_{p+1}, p)]$ for $p = 0, 1, \ldots, r - 1$.

The proof of the following proposition is straightforward.

PROPOSITION 5.8. *The subset* **M** *is an order ideal in* $\mathbf{W}_r$. *Furthermore, a collection* $\mathcal{C}(b'_1, \ldots, b'_{r-1})$ *is obtained from* $\mathcal{C}(b_1, \ldots, b_{r-1})$ *by a weak lowering flip if and only if*

$$(b'_1, \ldots, b'_{r-1}) = (b_1, \ldots, b_{p-1}, b_p + 1, b_{p+1}, \ldots, b_{r-1})$$

for some p such that $b_p < b_{p+1}$. *This flip replaces* Lik *with* Lj, *where* $i = b_p - p + 1, j = b_p + 1, k = r + 1$ *and* $L = [b_p - p + 2, b_p]$.

Comparing this description of weak lowering flips with (5.6), we obtain the following corollary.

COROLLARY 5.9. *Conjecture 5.5 holds for any collection* $\mathcal{C}(b_1, \ldots, b_{r-1}) \in \mathbf{M}$ *different from* $\mathcal{C}_{\min} = \mathcal{C}(r, r, \ldots, r)$.

Note that, in view of Lemma 4.3, a collection $\mathcal{C}(b_1, \ldots, b_{r-1})$ is strongly separated if and only if $b_{p+1} \leqslant b_p + 1$ for $1 \leqslant p \leqslant r - 1$. Such sequences $(b_1, \ldots, b_{r-1})$ are in bijection with binary strings of length $r - 1$ via the correspondence

$$(b_1, \ldots, b_{r-1}) \mapsto (b_2 - b_1, b_3 - b_2, \ldots, b_r - b_{r-1}).$$

In particular, the number of strongly separated collections in **M** is 2^{r-1}.

We conclude this paper by a "classical" application of weak flips to total positivity criteria. Consider the subgroup $B \subset GL_{r+1}$ of upper-triangular matrices. For a subset $J \subset [1, r+1]$ of size l let $\Delta^J(x)$ denote the minor of x with the row set $[1, l]$ and column set J; we shall think of Δ^J as an element of the algebra $\mathbf{Q}[B]$ of polynomial functions on B. A collection $\mathcal{C}$ of nonempty subsets in $[1, r+1]$ will be called a *total positivity base* for B if $|\mathcal{C}| = \dim(B) = \binom{r+2}{2}$ and every flag minor $\Delta^I \in \mathbf{Q}[B]$ is a subtraction-free rational expression in the minors Δ^J ($J \in \mathcal{C}$). In view of [**BFZ**, Theorem 3.2.1], every such collection $\mathcal{C}$ gives rise to a total positivity criterion: a matrix $x \in B$ is totally positive (i.e., all minors not identically equal to 0 on B take positive real values at x) if and only if $\Delta^J(x) > 0$ for all $J \in \mathcal{C}$. The classical Fekete criterion (see [**BFZ**, Theorem 3.2.2]) says that the collection $\mathcal{C}_{\min} - \{\varnothing\}$ of all nonempty intervals in $[1, r+1]$ is a total positivity base for B.

PROPOSITION 5.10. *Weak raising and lowering flips preserve the set of all totally positivity bases for* B.

PROOF. Suppose $i < j < k$ and $L \cap \{i, j, k\} = \varnothing$. Our statement follows at once from the fact that the flag minors satisfy the 3-term Plücker relation

(5.7) $$\Delta^{Lj} \Delta^{Lik} = \Delta^{Li} \Delta^{Ljk} + \Delta^{Lk} \Delta^{Lij}$$

(this relation is obtained from the relation $R_2(Lijk, L)$ in §2 by the specialization $q = 1$).

In particular, combining this proposition with the Fekete criterion, we obtain that all maximal strongly separated collections are total positivity bases (cf. [**BFZ**, Proposition 3.2.3]). Conjecture 1.8 would imply that the same is true for all maximal weakly separated collections. In any case, in view of Corollaries 5.7 and 5.9, all the collections in Theorem 5.6 and Proposition 5.8 are total positivity bases, hence give rise to new criteria for total positivity.

References

[BFZ] A. Berenstein, S. Fomin, and A. Zelevinsky, *Parametrizations of canonical bases and totally positive matrices*, Adv. Math. **122** (1996), 49–149.

[BZ] A. Berenstein and A. Zelevinsky, *String bases for quantum groups of type A_r*, I. M. Gelfand Seminar, Advances in Soviet Math., vol. 16, Part 1, Amer. Math. Soc., Providence, RI, 1993, pp. 51–89.

[E] S. Elnitsky, *Rhombic tilings of polygons and classes of reduced words in Coxeter groups*, Ph.D. Dissertation (1993), University of Michigan.

[FRT] L. D. Faddeev, N. Y. Reshetikhin, and L. A. Takhtadzhyan, *Quantization of Lie groups and Lie algebras*, Algebra i Analiz **1** (1989), 178–206; English transl., Leningrad Math. J. **1** (1990), 193–226.

[F] C. K. Fan, *A Hecke algebra quotient and some combinatorial applications*, J. Alg. Comb. **5** (1996), 175–189.

[IM] K. Iohara and F. Malikov, *Rings of skew polynomials and Gelfand-Kirillov conjecture for quantum groups*, Comm. Math. Phys. **164** (1994), 217–237.

[J] A. Joseph, *Sur une conjecture de Feigin*, C. R. Acad. Sci. Paris Sér. I **320** (1995), 1441–1444.

[K] M. Kashiwara, *Global crystal bases of quantum groups*, Duke Math. J. **69** (1993), 455–485.

[Kn] D. E. Knuth, *Axioms and hulls*, Springer Lecture Notes in Computer Science, vol. 606, 1992.

[KL] D. Krob and B. Leclerc, *Minor identities for quasi-determinants and quantum determinants*, Comm. Math. Phys. **169** (1995), 1–23.

[LR] V. Lakshmibai and N. Yu. Reshetikhin, *Quantum deformations of SL_n/B and its Schubert varieties*, Special Functions (Okayama, 1990), ICM-90 Satell. Conf. Proc., Springer-Verlag, Tokyo, 1991, pp. 149–168.

[L] G. Lusztig, *Introduction to quantum groups*, Birkhäuser, Boston, 1993.

[MS] Yu. I. Manin and V. V. Schechtman, *Arrangements of hyperplanes, higher braid groups and higher Bruhat orders*, Adv. Stud. Pure Math. **17** (1989), 289–308.

[S] J. Stembridge, *On the fully commutative elements of Coxeter groups*, J. Alg. Comb. **5** (1996), 353–385.

[TT] E. Taft and J. Towber, *Quantum deformation of flag schemes and Grassmann schemes I — A q-deformation of the shape-algebra for $GL(n)$*, J. Algebra **142** (1991), 1–36.

DÉPARTEMENT DE MATHÉMATIQUES, UNIVERSITÉ DE CAEN, 14032 CAEN CEDEX, FRANCE
E-mail address: Bernard.Leclerc@litp.ibp.fr

DEPARTMENT OF MATHEMATICS, NORTHEASTERN UNIVERSITY, BOSTON MA 02115, USA
E-mail address: andrei@neu.edu

Factorial Supersymmetric Schur Functions and Super Capelli Identities

Alexander Molev

Dedicated to Professor A. A. Kirillov on his 60th birthday

ABSTRACT. A factorial analog of the supersymmetric Schur functions is introduced. It is shown that factorial versions of the Jacobi–Trudi and Sergeev–Pragacz formulas hold. The results are applied to construct a linear basis in the center of the universal enveloping algebra for the Lie superalgebra $\mathfrak{gl}(m|n)$ and to obtain super-analogs of the higher Capelli identities.

§0. Introduction

For a partition λ of length $\leq m$ the factorial Schur function $s_\lambda(x|a)$ in the variables $x = (x_1, \ldots, x_m)$ depending on an arbitrary numerical sequence $a = (a_i)$, $i \in \mathbb{Z}$, may be defined as follows [20, p. 54]. Let

$$(x|a)^k = (x - a_1) \cdots (x - a_k)$$

for each $k \geq 0$. Then

(0.1) $$s_\lambda(x|a) = \frac{\det[(x_j|a)^{\lambda_i + m - i}]_{1 \leq i,j \leq m}}{\Delta(x)},$$

where $\Delta(x)$ stands for the Vandermonde determinant,

$$\Delta(x) = \prod_{i<j}(x_i - x_j) = \det[(x_j|a)^{m-i}]_{1 \leq i,j \leq m}.$$

Note that the usual Schur function $s_\lambda(x)$ coincides with $s_\lambda(x|a)$ for the zero sequence a. The factorial Schur functions admit many of the classical properties of $s_\lambda(x)$ (see [4–6, 9, 10, 21, 29, 31, 32]), as well as some new ones, e.g., the characterization theorem [29] which plays an important role in the proofs of analogs of the Capelli identity [24, 29]. (The term "factorial" was primarily used only for the case of the sequence a with $a_i = i - 1$; we use it in the present paper in a broader sense, for an arbitrary a). In particular, the functions $s_\lambda(x|a)$ may be equivalently

---

1991 *Mathematics Subject Classification.* 05A15, 05E05, 17A70.

©1998 American Mathematical Society

defined in terms of tableaux, which also enables one to introduce the skew factorial Schur functions $s_{\lambda/\mu}(x|a)$ for each pair of partitions $\mu \subset \lambda$.

Super-analogs of the Schur functions can be defined by using some specializations of the usual symmetric functions in infinitely many variables (see [20, p. 58]). This approach goes back to Littlewood [19] (see also [23]). On the other hand, they naturally emerge in the representation theory of Lie superalgebras [15, 16] and have been studied by several authors; see, e.g., [2, 3, 7, 9, 13, 33–35, 38–40].

For a pair of partitions λ and μ with $\mu \subset \lambda$, the supersymmetric skew Schur function $s_{\lambda/\mu}(x/y)$ in the variables $x = (x_1, \ldots, x_m)$ and $y = (y_1, \ldots, y_n)$ can be defined by the formula

$$(0.2) \qquad s_{\lambda/\mu}(x/y) = \sum_{\mu \subset \nu \subset \lambda} s_{\lambda/\nu}(x) s_{\nu'/\mu'}(y),$$

where λ' denotes the partition conjugate to λ.

The linear span of the functions (0.2) consists of all supersymmetric polynomials P in x and y, that is, of those polynomials that are symmetric in x and y separately and possess the following *cancellation property*: the result of setting $x_m = -y_n = z$ in P is independent of z.

In this paper we introduce factorial analogs $s_{\lambda/\mu}(x/y|a)$ of the supersymmetric Schur functions parametrized by arbitrary numerical sequences $a = (a_i)$, $i \in \mathbb{Z}$ (see §1). For the zero sequence a, this function coincides with $s_{\lambda/\mu}(x/y)$, and the functions $s_\lambda(x/y|a)$ with $\lambda_{m+1} \leqslant n$ form a (nonhomogeneous) basis in the space of supersymmetric polynomials. We prove that many properties of the supersymmetric Schur functions have their factorial analogs.

First, we find the generating series for the corresponding elementary and complete functions and check that they are supersymmetric.

Then, using a modified Gessel–Viennot method [8], we establish an analog of the Jacobi–Trudi formula and thus prove that the polynomials $s_{\lambda/\mu}(x/y|a)$ are also supersymmetric.

Further, we prove a characterization theorem for the polynomials $s_\lambda(x/y|a)$ analogous to the corresponding theorem for the factorial Schur polynomials [29] (see also [37]).

Using this theorem, we prove a factorial analog of the Sergeev–Pragacz formula. For the usual supersymmetric Schur functions, this formula can be proved in several different ways; see, e.g., [3, 13, 22, 33, 34]. However, these proofs cannot be easily carried over to the case of the functions $s_\lambda(x/y|a)$, because for a general sequence a they lose both the symmetry property of the functions (0.2)

$$(0.3) \qquad s_{\lambda/\mu}(x/y) = s_{\lambda'/\mu'}(y/x)$$

and the specialization property with respect to x

$$(0.4) \qquad s_\lambda(x/y)|_{x_m=0} = s_\lambda(x'/y),$$

where $x' = (x_1, \ldots, x_{m-1})$.

In particular, in the case of $a = (0)$, we obtain one more proof of the Sergeev–Pragacz formula.

As a corollary of the factorial Sergeev–Pragacz formula, we get an analog of the Berele–Regev factorization theorem for the polynomials $s_\lambda(x/y|a)$ [2] (see also [9, 35]).

A special case of the factorization theorem yields an analog of the dual Cauchy formula, which gives a decomposition of the double product $\prod\prod(x_i + y_j)$ into a sum of products of the factorial Schur functions. Two other proofs of this formula are given in [17, 18, 21].

Goulden and Greene [9] and Macdonald [21] found a new tableau representation for the functions $s_{\lambda/\mu}(x/y)$ in the case of infinite sets of variables $x = (x_i)$ and $y = (y_i)$, $i \in \mathbb{Z}$. Using this representation, we prove that the corresponding function $s_{\lambda/\mu}(x/y|a)$ does not depend on a (here $a = (a_i)$ is regarded as a sequence of independent variables) and coincides with $s_{\lambda/\mu}(x/y)$.

Finally, we show that many results of [28–31] concerning higher Capelli identities and shifted Schur functions have their natural super-analogs. In particular, a basis in the center of the universal enveloping algebra $U(\mathfrak{gl}(m|n))$ is explicitly constructed. The eigenvalues of the basis elements in highest weight representations are super-analogs of the shifted Schur functions. We outline two proofs of the super-versions of the higher Capelli identities. The first proof uses the characterization theorem for the factorial supersymmetric Schur polynomials, while the second one is based on the properties of the Jucys–Murphy elements in the group algebra for the symmetric group. These identities include the super Capelli identity found by Nazarov [27].

I would like to thank A. Lascoux, M. Nazarov, A. Okounkov, G. Olshanski, and P. Pragacz for useful remarks and discussions.

§1. Definition and combinatorial interpretation of the functions $s_{\lambda/\mu}(x/y|a)$

We shall suppose that $a = (a_i)$, $i \in \mathbb{Z}$, is a fixed sequence of complex numbers.

For a pair of partitions $\mu \subset \lambda$, the skew factorial Schur function can be defined by the formula (see, e.g., [9, 21])

$$(1.1) \quad s_{\lambda/\mu}(x|a) = \sum_T \prod_{\alpha \in \lambda/\mu} (x_{T(\alpha)} - a_{T(\alpha)+c(\alpha)}),$$

with summation over all semistandard skew tableaux T of shape λ/μ with entries in the set $\{1, \ldots, m\}$, where $T(\alpha)$ is the entry of T in the cell α and $c(\alpha) = j - i$ is the content of $\alpha = (i, j)$. The entries of a semistandard tableau are assumed weakly increasing along rows and strictly increasing down columns. It can be verified directly (see [9]) that the polynomials (1.1) are symmetric in x. It was proved in [21] that for a standard (nonskew) shape λ formulas (0.1) and (1.1) define the same function, which makes the symmetry property obvious in this case.

Due to (1.1), the factorial elementary and complete symmetric polynomials are given by

$$(1.2) \quad e_k(x|a) = s_{(1^k)}(x|a) = \sum_{i_1 < \cdots < i_k} (x_{i_1} - a_{i_1})(x_{i_2} - a_{i_2-1}) \cdots (x_{i_k} - a_{i_k-k+1}),$$

$$(1.3) \quad h_k(x|a) = s_{(k)}(x|a) = \sum_{i_1 \leq \cdots \leq i_k} (x_{i_1} - a_{i_1})(x_{i_2} - a_{i_2+1}) \cdots (x_{i_k} - a_{i_k+k-1}).$$

It is immediate from (1.1) that the highest component of $s_{\lambda/\mu}(x\,|\,a)$ is the usual skew Schur polynomial $s_{\lambda/\mu}(x)$. This implies that the polynomials $s_\lambda(x\,|\,a)$ with $l(\lambda) \leqslant m$ form a basis in the symmetric polynomials in x.

DEFINITION 1.1. Let $x = (x_1, \ldots, x_m)$ and $y = (y_1, \ldots, y_n)$ be two families of variables. Given a sequence a, denote by $a^* = (a_i^*)$ another sequence, defined by $a_i^* = -a_{n-i+1}$, and introduce the *factorial supersymmetric Schur polynomials* (*functions*) by the formula

$$(1.4) \qquad s_{\lambda/\mu}(x/y\,|\,a) = \sum_{\mu \subset \nu \subset \lambda} s_{\lambda/\nu}(x\,|\,a) s_{\nu'/\mu'}(y\,|\,a^*).$$

We shall prove in §3 that these polynomials are indeed supersymmetric, so their name will be justified.

Comparing (0.2) and (1.4), we see that the highest homogeneous component of the polynomial $s_{\lambda/\mu}(x/y\,|\,a)$ is $s_{\lambda/\mu}(x/y)$ if the latter is nonzero, and that $s_{\lambda/\mu}(x/y\,|\,a)$ coincides with $s_{\lambda/\mu}(x/y)$ for the zero sequence a.

Using formula (1.1), we can reformulate definition (1.4) in terms of tableaux. To distinguish the indices of x and y, let us identify the indices of y with the symbols $1', \ldots, n'$. Consider the diagram of shape λ/μ and fill it with the indices $1', \ldots, n', 1, \ldots, m$ so that:
  (a) in each row (resp., column) each primed index is to the left of (resp., above) each unprimed index;
  (b) primed indices strictly decrease along the rows and weakly decrease down the columns;
  (c) unprimed indices weakly increase along the rows and strictly increase down the columns.

Denote the resulting tableau by T.

PROPOSITION 1.2. *One has the formula*
$$(1.5) \qquad s_{\lambda/\mu}(x/y\,|\,a) = \sum_T \prod_{\substack{\alpha \in \lambda/\mu \\ T(\alpha) \text{ unprimed}}} (x_{T(\alpha)} - a_{T(\alpha)+c(\alpha)}) \prod_{\substack{\alpha \in \lambda/\mu \\ T(\alpha) \text{ primed}}} (y_{T(\alpha)} + a_{T(\alpha)+c(\alpha)}).$$

PROOF. For each tableau T the cells of the diagram λ/μ occupied by primed indices form a subdiagram ν/μ, where $\mu \subset \nu \subset \lambda$. Let us fix such a partition ν and sum in (1.5) first over the tableaux T whose primed part forms a subtableau of shape ν/μ. The part of such a tableau T formed by unprimed indices is a semistandard subtableau of shape λ/ν, and taking the sum over these subtableaux we get by (1.1) that

$$\sum_T \prod_{\alpha \in \lambda/\nu} (x_{T(\alpha)} - a_{T(\alpha)+c(\alpha)}) = s_{\lambda/\nu}(x\,|\,a).$$

It remains to verify that

$$(1.6) \qquad \sum_T \prod_{\alpha \in \nu/\mu} (y_{T(\alpha)} + a_{T(\alpha)+c(\alpha)}) = s_{\nu'/\mu'}(y\,|\,a^*),$$

with summation over ν/μ-tableaux T with entries from $\{1,\ldots,n\}$ whose rows strictly decrease and columns weakly decrease. Indeed, since $s_{\nu'/\mu'}(y|a^*)$ is symmetric in y, setting $\tilde{y} = (y_n,\ldots,y_1)$ and using (1.1), we may write

$$
\begin{aligned}
s_{\nu'/\mu'}(y|a^*) = s_{\nu'/\mu'}(\tilde{y}|a^*) &= \sum_{T'} \prod_{\alpha' \in \nu'/\mu'} (\tilde{y}_{T'(\alpha')} - a^*_{T'(\alpha')+c(\alpha')}) \\
&= \sum_{T'} \prod_{\alpha' \in \nu'/\mu'} (y_{n-T'(\alpha')+1} + a_{n-T'(\alpha')+1-c(\alpha')}),
\end{aligned}
\tag{1.7}
$$

with summation over semistandard ν'/μ'-tableaux T' with entries from $\{1,\ldots,n\}$. Note that the map

$$T'(\alpha') \to T(\alpha) = n - T'(\alpha') + 1,$$

where $\alpha = (i,j) \in \nu/\mu$ and $\alpha' = (j,i) \in \nu'/\mu'$, is a bijection between the set of semistandard ν'/μ'-tableaux and the set of ν/μ-tableaux whose rows strictly decrease and columns weakly decrease. Obviously, $c(\alpha) = -c(\alpha')$; hence, (1.7) coincides with the left-hand side of (1.6), which completes the proof.

§2. Generating series for the elementary and complete factorial supersymmetric polynomials

Now we introduce the *elementary* and *complete factorial supersymmetric polynomials* as special cases of $s_\lambda(x/y|a)$ in which λ is a column or row partition, respectively:

$$e_k(x/y|a) = s_{(1^k)}(x/y|a) \quad \text{and} \quad h_k(x/y|a) = s_{(k)}(x/y|a).$$

Using definition (1.4), we can express them in terms of the polynomials (1.2) and (1.3) as follows:

$$e_k(x/y|a) = \sum_{p+q=k} e_p(x|\tau^{-q}a) h_q(y|a^*), \tag{2.1}$$

$$h_k(x/y|a) = \sum_{p+q=k} h_p(x|\tau^q a) e_q(y|a^*), \tag{2.2}$$

where τ is the shift operator acting on sequences a by replacing each a_i by a_{i+1}. Proposition 1.2 yields the following explicit formulas for $e_k(x/y|a)$ and $h_k(x/y|a)$:

$$
e_k(x/y|a) = \sum_{\substack{p+q=k}} \sum_{\substack{i_1<\cdots<i_p \\ j_1\geqslant\cdots\geqslant j_q}} (y_{j_1}+a_{j_1})\cdots(y_{j_q}+a_{j_q-q+1}) \\
\times (x_{i_1}-a_{i_1-q})\cdots(x_{i_p}-a_{i_p-k+1}), \tag{2.3}
$$

$$
h_k(x/y|a) = \sum_{\substack{p+q=k}} \sum_{\substack{i_1\leqslant\cdots\leqslant i_p \\ j_1>\cdots> j_q}} (y_{j_1}+a_{j_1})\cdots(y_{j_q}+a_{j_q+q-1}) \\
\times (x_{i_1}-a_{i_1+q})\cdots(x_{i_p}-a_{i_p+k-1}). \tag{2.4}
$$

We shall suppose that $e_k(x/y|a) = h_k(x/y|a) = 0$ if $k < 0$. Note that formulas (2.3) and (2.4) imply the following symmetry property:

$$h_k(x/y|a) = e_k(y/x|-\tau^{k-1}a).$$

THEOREM 2.1. *One has the following generating series for the polynomials* $e_k(x/y|a)$ *and* $h_k(x/y|a)$:

(2.5)
$$1 + \sum_{k=1}^{\infty} \frac{(-1)^k e_k(x/y|a)}{(t-a_{m-k+1})\cdots(t-a_m)} = \frac{(t-x_1)\cdots(t-x_m)(t-a_1)\cdots(t-a_n)}{(t-a_1)\cdots(t-a_m)(t+y_1)\cdots(t+y_n)},$$

(2.6)
$$1 + \sum_{k=1}^{\infty} \frac{h_k(x/y|a)}{(t-a_{m+1})\cdots(t-a_{m+k})} = \frac{(t-a_1)\cdots(t-a_m)(t+y_1)\cdots(t+y_n)}{(t-x_1)\cdots(t-x_m)(t-a_1)\cdots(t-a_n)}.$$

PROOF. We shall use induction on m. The generating series for the complete factorial symmetric polynomials $h_k(y|a)$ is given by

(2.7)
$$1 + \sum_{k=1}^{\infty} \frac{h_k(y|a)}{(t-a_{n+1})\cdots(t-a_{n+k})} = \frac{(t-a_1)\cdots(t-a_n)}{(t-y_1)\cdots(t-y_n)}.$$

This formula was proved in [**31, 32**] in the special case of the sequence a with $a_i = i - 1$, and this proof works in the general case as well. Note that for $m = 0$ we have $e_k(x/y|a) = h_k(y|a^*)$ by (2.1), and (2.5) follows from (2.7). Suppose now that $m \geqslant 1$. Denote $x' = (x_1, \ldots, x_{m-1})$. We see from (2.3) that

(2.8)
$$e_k(x/y|a) = e_k(x'/y|a) + e_{k-1}(x'/y|a)(x_m - a_{m-k+1}).$$

So, using the induction hypotheses, we may write the right-hand side of (2.5) as

$$\sum_{k=0}^{\infty} \frac{(-1)^k e_k(x'/y|a)}{(t-a_{m-k})\cdots(t-a_{m-1})} \frac{t-x_m}{t-a_m}$$

$$= \sum_{k=0}^{\infty} \frac{(-1)^k e_k(x'/y|a)}{(t-a_{m-k+1})\cdots(t-a_m)} \frac{t-x_m}{t-a_{m-k}}$$

$$= \sum_{k=0}^{\infty} \frac{(-1)^k e_k(x'/y|a) + (-1)^k e_{k-1}(x'/y|a)(x_m - a_{m-k+1})}{(t-a_{m-k+1})\cdots(t-a_m)},$$

where we have used
$$\frac{t-x_m}{t-a_{m-k}} = 1 - \frac{x_m - a_{m-k}}{t-a_{m-k}}.$$

Due to (2.8), this proves (2.5).

The following formula for the generating series for the elementary factorial symmetric polynomials $e_k(y|a)$ is contained in [**20**, p. 55] (see also [**31, 32**]):

(2.9)
$$1 + \sum_{k=1}^{\infty} \frac{(-1)^k e_k(y|a)}{(t-a_{n-k+1})\cdots(t-a_n)} = \frac{(t-y_1)\cdots(t-y_n)}{(t-a_1)\cdots(t-a_n)}.$$

For $m = 0$ we have $h_k(x/y|a) = e_k(y|a^*)$ by (2.2), and so, for this case, (2.6) follows from (2.9). Now let $m \geqslant 1$. We see from (2.4) that

(2.10)
$$h_k(x/y|a) = \sum_{r+s=k} h_r(x'/y|a)(x_m - a_{m+k-s})\cdots(x_m - a_{m+k-1}).$$

By the induction hypotheses we can write the right-hand side of (2.6) in the form

$$\sum_{r=0}^{\infty} \frac{h_r(x'/y|a)}{(t-a_m)\cdots(t-a_{m+r-1})} \frac{t-a_m}{t-x_m}$$

$$= \sum_{r=0}^{\infty} \frac{h_r(x'/y|a)}{(t-a_{m+1})\cdots(t-a_{m+r})} \frac{t-a_{m+r}}{t-x_m}$$

$$= \sum_{r=0}^{\infty} \frac{h_r(x'/y|a)}{(t-a_{m+1})\cdots(t-a_{m+r})} \sum_{s=0}^{\infty} \frac{(x_m-a_{m+r})\cdots(x_m-a_{m+r+s-1})}{(t-a_{m+r+1})\cdots(t-a_{m+r+s})},$$

where we have applied (2.7) with $n = 1$. The latter expression can be rewritten as

$$\sum_{k=0}^{\infty} \frac{1}{(t-a_{m+1})\cdots(t-a_{m+k})} \sum_{r+s=k} h_r(x'/y|a)(x_m-a_{m+k-s})\cdots(x_m-a_{m+k-1}),$$

which coincides with the left-hand side of (2.6) by (2.10).

COROLLARY 2.2. *For any k the polynomials $e_k(x/y|a)$ and $h_k(x/y|a)$ are supersymmetric.*

PROOF. Indeed, the cancellation property is obviously satisfied by $e_k(x/y|a)$ and $h_k(x/y|a)$, because after setting $x_m = -y_n = z$ the factors on the right-hand sides of (2.5) and (2.6) containing z cancel.

REMARK. In a letter to the author, A. Lascoux pointed out that the factorial Schur functions are recovered as a special case of the double Schubert polynomials [18, 22]. In particular, using the technique of divided differences, one can prove that the complete factorial supersymmetric polynomials $h_k(x/y|a)$ coincide with the complete factorial symmetric polynomials $h_k(x \cup a^{(n)} | -y \cup a)$, $a^{(n)} = (a_1, \ldots, a_n)$, and one can also obtain the above generating series for $h_k(x/y|a)$ (see also [1]).

§3. Jacobi–Trudi formula

For the functions $s_{\lambda/\mu}(x/y|a)$, the following analog of the Jacobi–Trudi formula holds.

THEOREM 3.1. *We have*

(3.1) $$s_{\lambda/\mu}(x/y|a) = \det[h_{\lambda_i-\mu_j-i+j}(x/y|\tau^{\mu_j-j+1}a)]_{1 \leqslant i,j \leqslant l},$$

where $l = l(\lambda)$.

PROOF. We use a modified Gessel–Viennot method [8, 36] (cf. [9, 31, 32, 35]). Consider a grid consisting of two parts (see Figure 1 on the next page). The upper half of the grid is formed by m horizontal lines labelled by $1, \ldots, m$ northwards and vertical lines consequently labelled by the elements of $\mathbb{Z}$ eastwards. For each $i \in \mathbb{Z}$ the vertical line labelled by i breaks into two lines at the intersection point with the horizontal line 1. One of the two lines goes south-east and keeps the label i, and the other goes south-west and is labelled by i'.

Each vertex of the grid will be denoted by a pair of coordinates (c, i) or (c, i'), which are the labels of the lines intersecting at the vertex, so that c labels a vertical line or a line going south-east. We shall consider paths in this grid of the following kind. Each step of a path is north or east in the upper half of the grid and is

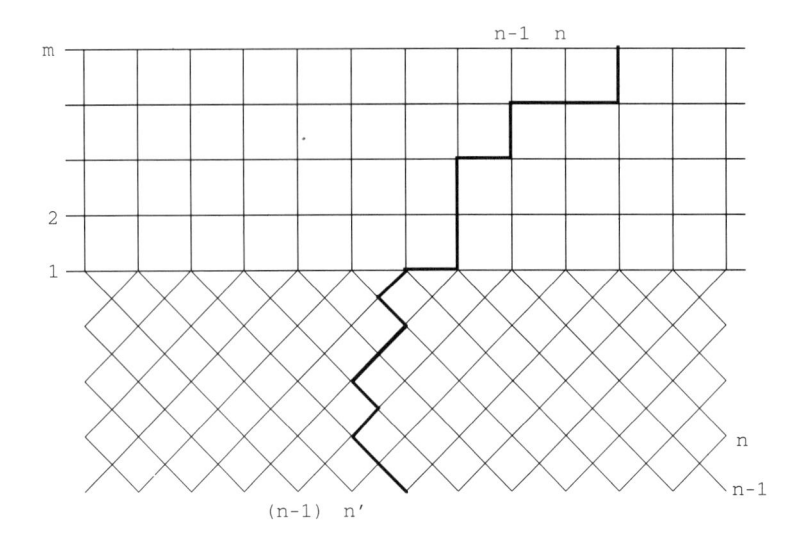

FIGURE 1

north-east or north-west in the lower part of the grid. We label each eastern step $(c, i) \to (c + 1, i)$ of a path with $x_i - a_{i+c}$ and each north-eastern step $(c, i') \to (c + 1, i')$ with $y_{i-c} + a_i$. For a path π, denote by $L(\pi)$ the product of these labels with respect to all eastern and north-eastern steps. Let us check now that

$$\sum_\pi L(\pi) = h_{s-r}(x/y \,|\, \tau^r a), \tag{3.2}$$

with summation over all paths π with the initial vertex $(r, (n+r)')$ and the final vertex (s, m). Indeed, the left-hand side of (3.2) can be calculated as follows. First fix a number q such that $0 \leqslant q \leqslant \min\{s - r, n\}$ and find the sum

$$\sum_{\pi_{\text{lower}}} L(\pi_{\text{lower}}), \tag{3.3}$$

where π_{lower} runs over the paths in the lower part of the grid with the initial vertex $(r, (n + r)')$ and the final vertex $(r + q, (r + q)')$ (which belongs to the horizontal line 1 and also has the coordinates $(r + q, 1)$ if regarded as a vertex of the upper half of the grid). Suppose that such a path π_{lower} has north-eastern steps of the form

$$(r, (j_1 + r)') \to (r + 1, (j_1 + r)'), \quad \ldots, \quad (r + q - 1, (j_q + r)') \to (r + q, (j_q + r)'),$$

where $n \geqslant j_1 \geqslant \cdots \geqslant j_q \geqslant q$. Then the product of the labels of π_{lower} with respect to the north-eastern steps equals

$$(y_{j_1} + a_{j_1+r}) \cdots (y_{j_q - q + 1} + a_{j_q + r}).$$

Hence, the sum (3.3) equals

$$\sum_{n \geqslant j_1 \geqslant \cdots \geqslant j_q \geqslant q} (y_{j_1} + a_{j_1+r}) \cdots (y_{j_q - q + 1} + a_{j_q + r})$$
$$= \sum_{n \geqslant i_1 > \cdots > i_q \geqslant 1} (y_{i_1} + a_{i_1 + r}) \cdots (y_{i_q} + a_{i_q + q - 1 + r}).$$

Applying (1.6) for $\nu/\mu = (q)$, we conclude that this expression coincides with $e_q(y|(\tau^r a)^*)$.

Similarly, one can easily check that the sum

$$\sum_{\pi_{\text{upper}}} L(\pi_{\text{upper}}),$$

where π_{upper} runs over all paths in the upper half of the grid with the initial vertex $(r+q, 1)$ and the final vertex (s, m), coincides with $h_{s-r-q}(x|\tau^{r+q} a)$.

Thus, the left-hand side of (3.2) equals

$$\sum_q h_{s-r-q}(x|\tau^{r+q} a) e_q(y|(\tau^r a)^*),$$

which coincides with $h_{s-r}(x/y|\tau^r a)$ by (2.2).

A straightforward application of Gessel–Viennot arguments with the use of (3.2) shows that the determinant on the right-hand side of (3.1) can be represented as

(3.4) $$\sum_{\mathcal{P}} L(\pi_1) \cdots L(\pi_l),$$

with summation over sets of nonintersecting paths $\mathcal{P} = (\pi_1, \ldots, \pi_l)$, where the initial vertex of π_i is $(\mu_i - i + 1, (n + \mu_i - i + 1)')$ and the final vertex is $(\lambda_i - i + 1, m)$.

It remains to prove that (3.4) coincides with $s_{\lambda/\mu}(x/y|a)$. Given a set $\mathcal{P} = (\pi_1, \ldots, \pi_l)$ of nonintersecting paths, we construct a λ/μ-tableau T as follows. Note that the total number of north-eastern and eastern steps of the path π_i equals $\lambda_i - \mu_i$. Let us number these steps starting with the initial vertex of π_i. If the kth step is north-eastern of the form $(c, j') \to (c+1, j')$, then the entry of the kth cell in the ith row of λ/μ is $(j-c)'$, and if the kth step is eastern of the form $(c, j) \to (c+1, j)$, then the entry of this cell is j. It can be easily seen that this correspondence is a bijection between the sets of nonintersecting paths and the tableaux used in the combinatorial interpretation (1.5) of the polynomials $s_{\lambda/\mu}(x/y|a)$. Moreover, the corresponding summands in (3.4) and (1.5) are clearly the same, which proves the theorem.

Note that for $n = 0$ formula (3.1) turns into the Jacobi–Trudi formula for the factorial Schur functions $s_{\lambda/\mu}(x|a)$, whereas for $m = 0$ we get a factorial analog of the Nägelsbach–Kostka formula (see [20, p. 56]), and by setting $a = (0)$ we obtain the Jacobi–Trudi formula for the functions $s_{\lambda/\mu}(x/y)$ [35].

Corollary 2.2 and Theorem 3.1 imply

COROLLARY 3.2. *The polynomials $s_{\lambda/\mu}(x/y|a)$ are supersymmetric.*

An analog of (3.1) for the factorial elementary supersymmetric polynomials can be proved in the same way by using a grid obtained from the one we have used in the above proof by interchanging its upper and lower parts. We state this formula here without proof.

THEOREM 3.3.

(3.5) $$s_{\lambda'/\mu'}(x/y|a) = \det[e_{\lambda_i - \mu_j - i + j}(x/y|\tau^{-\mu_j + j - 1} a)]_{1 \leqslant i, j \leqslant l}.$$

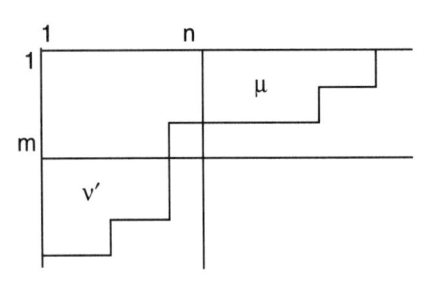

FIGURE 2

§4. Characterization theorem

Our goal in this section is to obtain characterization properties for the supersymmetric polynomials (Theorems 4.5 and 4.5′) which will play an important role in our proof of a factorial analog of the Sergeev–Pragacz formula (Theorem 5.1) and super-analogs of the Capelli identity (Theorem 8.1). We start by investigating vanishing and specialization properties of the polynomials $s_\lambda(x/y\,|\,a)$.

We say that a partition λ is *contained in the* (m,n)-*hook* if $\lambda_{m+1} \leqslant n$. Note that

(4.1) $\qquad s_\lambda(x/y\,|\,a) = 0 \quad \text{unless } \lambda \subset (m,n)\text{-hook}.$

Indeed, by definition (1.4)

(4.2) $\qquad s_\lambda(x/y\,|\,a) = \sum_{\rho \subset \lambda} s_{\rho'}(y\,|\,a^*)\, s_{\lambda/\rho}(x\,|\,a).$

Clearly, $s_{\rho'}(y\,|\,a^*) = 0$ unless $\rho_1 \leqslant n$. If λ is not contained in the (m,n)-hook, then $\lambda'_{n+1} > m$. So, if $\rho_1 \leqslant n$ then $s_{\lambda/\rho}(x\,|\,a) = 0$.

For each $\lambda \subset (m,n)$-hook we introduce two partitions $\mu = \mu(\lambda)$ and $\nu = \nu(\lambda)$ as follows. Nonzero parts of μ are defined by $\mu_i = \lambda_i - n$ for $\lambda_i > n$, and nonzero parts of ν are defined by $\nu_j = \lambda'_j - m$ for $\lambda'_j > m$, as shown in Figure 2.

For any partition α such that $l(\alpha) \leqslant l$, introduce the l-tuple

$$a_\alpha = a_\alpha^{(l)} = (a_{\alpha_1+l}, \ldots, a_{\alpha_l+1})$$

of elements of the sequence a. We shall only consider m- or n-tuples a_α as values for the variables $x = (x_1, \ldots, x_m)$ or $y = (y_1, \ldots, y_n)$, so we shall not specify the number l if it is clear from the context.

We shall often use the following vanishing properties of the factorial Schur polynomials $s_\lambda(x\,|\,a)$ (see [**29, 31**]). For any partition σ such that $l(\sigma) \leqslant m$ and $\lambda \not\subset \sigma$ one has

(4.3) $\qquad s_\lambda(a_\sigma\,|\,a) = 0.$

Moreover, if $l(\lambda) \leqslant m$, then

(4.4) $\qquad s_\lambda(a_\lambda\,|\,a) = \prod_{(i,j)\in\lambda} (a_{\lambda_i+m-i+1} - a_{m-\lambda'_j+j}).$

In particular,

$$s_\lambda(a_\lambda\,|\,a) \neq 0,$$

provided that the sequence a is multiplicity free, that is, $a_i \neq a_j$ if $i \neq j$. Relations (4.3) and (4.4) can be easily deduced from either (0.1), or (1.1). In the special case of the sequence a with $a_i = i + \text{const}$, the right-hand side of (4.4) turns into the product $H(\lambda)$ of the hook lengths of all cells of λ (see [**29**]).

PROPOSITION 4.1. *Let η be a partition of length $\leqslant n$ such that $\nu \not\subset \eta$. Then*

(4.5) $$s_\lambda(x/a_\eta^*|a) = 0.$$

PROOF. We use (4.2). Note that $s_{\lambda/\rho}(x|a) = 0$ unless $\lambda'_i - \rho'_i \leqslant m$ for any i. This means that the sum in (4.2) can be only taken over $\rho \subset \lambda$ such that $\nu' \subset \rho$. By the assumption, $\nu \not\subset \eta$; hence, $\rho' \not\subset \eta$. But in this case $s_{\rho'}(a_\eta^*|a^*) = 0$ by (4.3). Thus, all summands in (4.2) vanish for $y = a_\eta^*$, which completes the proof.

PROPOSITION 4.2. *Let γ be a partition of length $\leqslant m$ such that $\mu \not\subset \gamma$. Then*

(4.6) $$s_\lambda((\tau^n a)_\gamma/y|a) = 0.$$

PROOF. We use again the fact that $s_{\rho'}(y|a^*) = 0$ in (4.2) unless $\rho_1 \leqslant n$. Relation (4.6) will follow from the relation

(4.7) $$s_{\lambda/\rho}((\tau^n a)_\gamma|a) = 0,$$

which holds for any such ρ. This relation is a simple generalization of the vanishing property (4.3) and can be proved by similar arguments (see [**29, 31**]). Indeed, since the left-hand side of (4.7) is a polynomial in a, we may assume without loss of generality that the sequence a is multiplicity free. The polynomials $s_{\lambda/\rho}(x|a)$ are symmetric in x, so replacing x with $\tilde{x} = (x_m, \ldots, x_1)$, we can rewrite definition (1.1) in the following form:

(4.8) $$s_{\lambda/\rho}(x|a) = \sum_T \prod_{\alpha \in \lambda/\rho} (x_{m-T(\alpha)+1} - a_{T(\alpha)+c(\alpha)}),$$

where the sum is over semistandard λ/ρ-tableaux T with entries from $\{1, \ldots, m\}$. We shall verify that all summands in (4.8) vanish for $x = (\tau^n a)_\gamma$. Indeed, let us suppose that for some tableau T

$$\prod_{\alpha \in \lambda/\rho} (\{(\tau^n a)_\gamma\}_{m-T(\alpha)+1} - a_{T(\alpha)+c(\alpha)}) \neq 0$$

or, equivalently,

$$\prod_{\alpha \in \lambda/\rho} (a_{\gamma_{m-T(\alpha)+1}+T(\alpha)+n} - a_{T(\alpha)+c(\alpha)}) \neq 0.$$

Since a is multiplicity free, this implies that

(4.9) $$\gamma_{m-T(\alpha)+1} + n \neq c(\alpha)$$

for all α. For the entries of the first row of the tableau T, we have

(4.10) $$T(1, n+1) \leqslant \cdots \leqslant T(1, n+\mu_1).$$

Applying (4.9) for $\alpha = (1, n+1)$, we obtain $\gamma_{m-T(1,n+1)+1} \geqslant 1$. Further, by (4.10),

$$\gamma_{m-T(1,n+2)+1} \geqslant \gamma_{m-T(1,n+1)+1} \geqslant 1.$$

120   ALEXANDER MOLEV

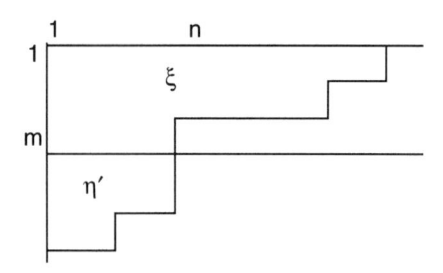

FIGURE 3

By (4.9), applied for $\alpha = (1, n+2)$, we then have $\gamma_{m-T(1,n+2)+1} \geqslant 2$. Similarly, using easy induction, we see that for any $i = 1, \ldots, \mu_1$

$$\gamma_{m-T(1,n+i)+1} \geqslant i.$$

On the other hand, for the entries of the $(n+i)$th column of T, we have

$$T(1, n+i) < \cdots < T(\mu'_i, n+i).$$

Hence,

$$\gamma_{m-T(\mu'_i,n+i)+1} \geqslant \cdots \geqslant \gamma_{m-T(1,n+i)+1} \geqslant i.$$

This means that $\gamma'_i \geqslant \mu'_i$, and so $\mu \subset \gamma$, which contradicts the assumption of the proposition. The proof is complete.

As we pointed out in the Introduction, for a general sequence a, the polynomials $s_\lambda(x/y|a)$ lose the specialization property of $s_\lambda(x/y)$ with respect to x. However, an analog of this property with respect to y still holds.

PROPOSITION 4.3. *One has*

(4.11) $$s_\lambda(x/y|a)|_{y_n=-a_n} = s_\lambda(x/y'|a),$$

where $y' = (y_1, \ldots, y_{n-1})$.

PROOF. This follows from the specialization property of the factorial Schur polynomials (see [**31, 32**]). Indeed, by (1.6) we have

(4.12) $$s_{\rho'}(y|a^*) = \sum_T \prod_{\alpha \in \rho} (y_{T(\alpha)} + a_{T(\alpha)+c(\alpha)}),$$

with summation over the ρ-tableaux T with entries from $\{1, \ldots, n\}$ whose rows strictly decrease and columns weakly decrease. If for a tableau T one has $T(1,1) = n$, then, since $c(1,1) = 0$, the corresponding summand in (4.12) vanishes for $y_n = -a_n$. So we have the property

$$s_{\rho'}(y|a^*)|_{y_n=-a_n} = s_{\rho'}(y'|a^{*'}),$$

where $a^{*'}$ is the sequence $(a_i^{*'})$ with $a_i^{*'} = -a_{n-i}$. Now (4.11) follows from (4.2).

We can now prove a vanishing theorem for the polynomials $s_\lambda(x/y|a)$ (cf. [**29, 31**]). Let ζ be a partition contained in the (m,n)-hook. Introduce two other partitions $\xi = \xi(\zeta)$ and $\eta = \eta(\zeta)$ as follows: $\xi = (\zeta_1, \ldots, \zeta_m)$, and nonzero parts of $\eta = (\eta_1, \ldots, \eta_n)$ are defined by $\eta_i = \zeta'_i - m$, if $\zeta'_i > m$, as shown in Figure 3.

In particular, $l(\xi) \leqslant m$ and $l(\eta) \leqslant n$, and we may consider the m-tuple a_ξ of elements of the sequence a and the n-tuple a^*_η of elements of the sequence a^*.

THEOREM 4.4. *Let λ, ζ be partitions contained in the (m,n)-hook.*
(i) *If $\lambda \not\subset \zeta$, then*

(4.13) $$s_\lambda(a_\xi/a_\eta^*|a) = 0.$$

(ii) *If $\lambda = \zeta$, then*

(4.14) $$s_\lambda(a_\xi/a_\eta^*|a) = \prod_{(i,j) \in \lambda} (a_{\lambda_i+m-i+1} - a_{m-\lambda_j'+j}).$$

In particular, if the sequence a is multiplicity free, then $s_\lambda(a_\xi/a_\eta^*|a) \neq 0$.

PROOF. Denote by r the length of the partition η. Then
$$a_\eta^* = (-a_{1-\eta_1}, \ldots, -a_{r-\eta_r}, -a_{r+1}, \ldots, -a_n).$$

By Proposition 4.3, the result of setting $y_i = -a_i$ for $i = r+1, \ldots, n$ in the polynomial $s_\lambda(x/y|a)$ is the polynomial $s_\lambda(x/y^{(r)}|a)$, where $y^{(r)} = (y_1, \ldots, y_r)$. So, (4.13) and (4.14) can be regarded as relations for the families of variables x and $y^{(r)}$. Moreover, due to (4.1), we need to consider only the case when λ is contained in the (r,m)-hook. In other words, we may assume without loss of generality that the length of η is n. In particular, the partition (n^m) is contained in ζ.

Now let us prove (i). By Proposition 4.1, we may assume that $\nu \subset \eta$. Since $(n^m) \subset \xi$, we may write $a_\xi = (\tau^n a)_\gamma$, where the partition γ is defined by $\gamma_i = \xi_i - n$. Hence, if $s_\lambda(a_\xi/a_\eta^*|a)$ is nonzero, then $\mu \subset \gamma$ by Proposition 4.2. Since $\lambda \subset (m,n)$-hook, this implies that $\lambda \subset \zeta$, and (i) is proved.

To prove (ii), we note that since $(n^m) \subset \zeta = \lambda$ we have

(4.15) $$a_\xi = (\tau^n a)_\mu \quad \text{and} \quad a_\eta^* = a_\nu^*.$$

As we noticed in the proof of Proposition 4.1, the sum in (4.2) can be only taken over those partitions $\rho \subset \lambda$ for which $\nu' \subset \rho$. On the other hand, using (4.3) we find that $s_{\rho'}(a_\nu^*|a^*) = 0$ unless $\rho' \subset \nu$. Hence, for $y = a_\nu^*$ relation (4.2) turns into

(4.16) $$s_\lambda(x/a_\nu^*|a) = s_\nu(a_\nu^*|a^*) s_{\lambda/\nu'}(x|a).$$

Now, using (1.1), let us write the polynomial $s_{\lambda/\nu'}(x|a)$ in terms of tableaux. By the definition of ν, we have $\lambda_i' - \nu_i = m$ for all $i = 1, \ldots n$. Since the tableaux T in (1.1) are column strict, all of them have the same entries in the first n columns, namely, the numbers $1, \ldots, m$ written in each of these columns downwards. On the other hand, the entries of the subdiagram μ can form an arbitrary semistandard μ-tableau. For a cell $(i, n+j) \in \lambda$ we obviously have $c(i, n+j) = n + c(i, j)$, so definition (1.1) for $s_{\lambda/\nu'}(x|a)$ now takes the form

$$s_{\lambda/\nu'}(x|a) = s_\mu(x|\tau^n a) \prod_{i=1}^m \prod_{j=1}^n (x_i - a_{j-\nu_j}).$$

We then obtain from (4.16) that

(4.17) $$s_\lambda((\tau^n a)_\mu/a_\nu^*|a) = s_\nu(a_\nu^*|a^*) s_\mu((\tau^n a)_\mu|\tau^n a) \prod_{i=1}^m \prod_{j=1}^n (a_{\mu_i+m+n-i+1} - a_{j-\nu_j}).$$

Now (4.14) follows from (4.4). The theorem is proved.

Theorem 4.4 implies that supersymmetric polynomials in x and y of degree $\leqslant k$ are characterized by their values at $x = a_\xi$ and $y = a_\eta^*$, where $\xi = \xi(\zeta)$ and $\eta = \eta(\zeta)$ with $|\zeta| \leqslant k$. More exactly, we have the following result (cf. [29, 31, 37]).

THEOREM 4.5. *Suppose that the sequence a is multiplicity free. Let $f(x/y)$ and $g(x/y)$ be supersymmetric polynomials of degree $\leqslant k$ such that*

(4.18) $$f(a_\xi/a_\eta^*) = g(a_\xi/a_\eta^*)$$

for any partition $\zeta \subset (m,n)$-hook with $|\zeta| \leqslant k$. Then $f(x/y) = g(x/y)$.

PROOF. It is well known that the functions $s_\lambda(x/y)$ with $\lambda \subset (m,n)$-hook form a basis in supersymmetric polynomials in x and y (see, e.g., [20, p. 61]). So do the functions $s_\lambda(x/y|a)$, because the highest term of $s_\lambda(x/y|a)$ is $s_\lambda(x/y)$. Hence, we can write the polynomial $f(x/y) - g(x/y)$ as a linear combination of the $s_\lambda(x/y|a)$:

$$f(x/y) - g(x/y) = \sum_\lambda c_\lambda s_\lambda(x/y|a).$$

Moreover, since the degree of the polynomial on the left-hand side $\leqslant k$, we may assume that $|\lambda| \leqslant k$. Introduce any total order on the set of partitions such that $|\lambda| < |\mu|$ implies $\lambda < \mu$. The condition (4.18) yields the following homogeneous system of linear equations on the coefficients c_λ:

$$\sum_\lambda c_\lambda s_\lambda(a_\xi/a_\eta^*|a) = 0, \qquad |\lambda|, |\zeta| \leqslant k.$$

Theorem 4.4 implies that the matrix $(s_\lambda(a_\xi/a_\eta^*|a))_{\lambda,\zeta}$ of this system, whose rows and columns are arranged in accordance with this order, is triangular with nonzero diagonal elements. Hence, $c_\lambda \equiv 0$, which proves the theorem.

Theorem 4.5 can be obviously reformulated in the following equivalent form.

THEOREM 4.5'. *Suppose that the sequence a is multiplicity free. Let $f(x/y)$ be a supersymmetric polynomial such that*

$$f(x/y) = s_\lambda(x/y) + \text{lower terms}$$

for some partition $\lambda \subset (m,n)$-hook and $f(a_\xi/a_\eta^*) = 0$ for any partition $\zeta \subset (m,n)$-hook with $|\zeta| < |\lambda|$. Then $f(x/y) = s_\lambda(x/y|a)$.

§5. Factorial Sergeev–Pragacz formula

In this section we apply Theorem 4.5 for the proof of an analog of the Sergeev–Pragacz formula for the polynomials $s_\lambda(x/y|a)$. In particular, for $a = (0)$, we obtain one more proof of the original formula (cf. [3, 13, 22, 33, 34]).

Suppose a partition λ is contained in the (m,n)-hook. Define the partitions μ and ν as in §4 and denote by $\rho = (\rho_1, \ldots, \rho_m)$ the part of λ which is contained in the rectangle (n^m) (see Figure 4), that is, $\rho_i = \min\{\lambda_i, n\}$.

The following analog of the Sergeev–Pragacz formula holds.

THEOREM 5.1.

(5.1) $$s_\lambda(x/y|a) = \frac{\sum_{\sigma \in S_m \times S_n} \text{sgn}(\sigma) \cdot \sigma\{f_\lambda(x/y|a)\}}{\Delta(x)\Delta(y)},$$

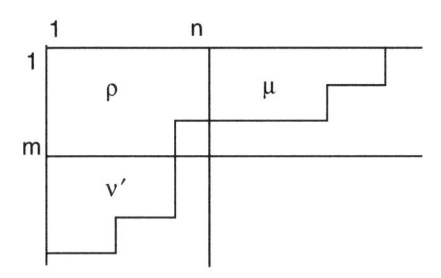

FIGURE 4

where
$$f_\lambda(x/y|a) = (x_1|\tau^{\rho_1}a)^{\mu_1+m-1}\cdots(x_m|\tau^{\rho_m}a)^{\mu_m}$$
$$\times (y_1|a^*)^{\nu_1+n-1}\cdots(y_n|a^*)^{\nu_n} \prod_{(i,j)\in\rho}(x_i+y_j).$$

PROOF. First of all, we note that both sides of (5.1) depend polynomially on a, so we may assume without loss of generality that the sequence a is multiplicity free.

Denote the right-hand side of (5.1) by $\varphi_\lambda(x/y|a)$. To apply Theorem 4.5, we must verify that this polynomial is supersymmetric and that $s_\lambda(x/y|a)$ and $\varphi_\lambda(x/y|a)$ have the same values at $x = a_\xi$ and $y = a_\eta^*$ for any $\zeta \subset (m,n)$-hook such that $|\zeta| \leqslant |\lambda|$.

The polynomials $\varphi_\lambda(x/y|a)$ are obviously symmetric in x and y, and so, to prove that they are supersymmetric, we only need to check that they satisfy the cancellation property. This can be done in exactly the same way as in the case $a = (0)$ (see, e.g., [**20** p. 61, **33**, **34**]).

Let us check now that Propositions 4.1–4.3 hold for the polynomials $\varphi_\lambda(x/y|a)$ too. To check (4.5), we represent the numerator of the right-hand side of (5.1) in the following form:
$$\sum_{\sigma\in S_m} \operatorname{sgn}(\sigma)\cdot\sigma\{(x_1|\tau^{\rho_1}a)^{\mu_1+m-1}\cdots(x_m|\tau^{\rho_m}a)^{\mu_m}g_\lambda(x/y|a)\},$$

where
$$g_\lambda(x/y|a) = \det\left[(y_j|a^*)^{\nu_i+n-i}(y_j+x_1)\cdots(y_j+x_{\rho_i'})\right]_{1\leqslant i,j\leqslant n}.$$

The condition $\nu \not\subset \eta$ means that there exists a k such that $\eta_k < \nu_k$. For $y = a_\eta^*$, the factor $(y_j|a^*)^{\nu_i+n-i}$ takes the value

(5.2) $\quad ((a_\eta^*)_j|a^*)^{\nu_i+n-i} = (a_{\eta_j+n-j+1}^* - a_1^*)\cdots(a_{\eta_j+n-j+1}^* - a_{\nu_i+n-i}^*).$

On the other hand, if $i \leqslant k \leqslant j$, then
$$1 \leqslant \eta_j+n-j+1 \leqslant \eta_k+n-k+1 \leqslant \nu_k+n-k \leqslant \nu_i+n-i.$$

This implies that (5.2) is zero and all the ijth entries of the determinant $g_\lambda(x/a_\eta^*|a)$ with $i \leqslant k \leqslant j$ are zero, and so $g_\lambda(x/a_\eta^*|a) = 0$. Since the Vandermonde determinant $\Delta(y)$ does not vanish for $y = a_\eta^*$, this proves the assertion. (These arguments are very similar to that used in [**29**, **31**] for the proof of (4.3)).

To check that the polynomials $\varphi_\lambda(x/y|a)$ satisfy (4.6), we rewrite the numerator of the right-hand side of (5.1) in the form

$$\sum_{\sigma \in S_n} \operatorname{sgn}(\sigma) \cdot \sigma\{(y_1|a^*)^{\nu_1+n-1} \cdots (y_n|a^*)^{\nu_n} h_\lambda(x/y|a)\},$$

where

$$h_\lambda(x/y|a) = \det[(x_j|\tau^{\rho_i}a)^{\mu_i+m-i}(x_j+y_1)\cdots(x_j+y_{\rho_i})]_{1 \leq i,j \leq m}.$$

The condition $\mu \not\subset \gamma$ implies that $\gamma_k < \mu_k$ for some k. In particular, this means that $\rho_k = n$ and hence $\rho_1 = \cdots = \rho_k = n$. Therefore, the ijth entry of the determinant $h_\lambda(x/y|a)$ for $i \leq k \leq j$ has the form

$$(x_j|\tau^n a)^{\mu_i+m-i}(x_j+y_1)\cdots(x_j+y_{\rho_i}).$$

Repeating the previous arguments, we conclude that $h_\lambda((\tau^n a)_\gamma/y|a) = 0$, which completes the proof.

Let us prove now that

(5.3) $$\varphi_\lambda(x/y|a)|_{y_n=-a_n} = \varphi_\lambda(x/y'|a),$$

where $y' = (y_1, \ldots, y_{n-1})$ and we define $\varphi_\lambda(x/y|a) = 0$ if $\lambda \not\subset (m,n)$-hook. Indeed, since $a_1^* = -a_n$, for $\nu_n > 0$ we obviously have

$$\varphi_\lambda(x/y|a)|_{y_n=-a_n} = 0,$$

and hence (5.3) is true because λ is not contained in the $(m, n-1)$-hook. So, we can suppose that $\nu_n = 0$. Since $\nu_i + n - i > 0$ for $i = 1, \ldots, n-1$, after setting $y_n = -a_n$, we may restrict the sum in the numerator of the right-hand side of (5.1) to the set of permutations $\sigma \in S_m \times S_{n-1}$. Further, it can be easily checked that on setting $y_n = -a_n$ in $f_\lambda(x/y|a)$, we get

$$f_\lambda(x/y|a)|_{y_n=-a_n} = f_\lambda(x/y'|a)(y_1+a_n)\cdots(y_{n-1}+a_n).$$

On the other hand,

$$\Delta(y)|_{y_n=-a_n} = \Delta(y')(y_1+a_n)\cdots(y_{n-1}+a_n).$$

The factor $(y_1+a_n)\cdots(y_{n-1}+a_n)$ is symmetric in y', so, removing it in the numerator and denominator, we see that the result is $\varphi_\lambda(x/y'|a)$, which proves (5.3).

Thus, the properties (4.5), (4.6), and (4.11) are satisfied by the polynomials $\varphi_\lambda(x/y|a)$. So, repeating the arguments used in the proof of statement (i) of Theorem 4.4, we see that these polynomials also satisfy (4.13). Hence, for any partition $\zeta \subset (m,n)$-hook such that $|\zeta| \leq |\lambda|$ and $\zeta \neq \lambda$, we have

$$s_\lambda(a_\xi/a_\eta^*|a) = \varphi_\lambda(a_\xi/a_\eta^*|a) = 0.$$

Therefore, to apply Theorem 4.5 to the polynomials $s_\lambda(x/y|a)$ and $\varphi_\lambda(x/y|a)$ it remains to check that

$$s_\lambda(a_\xi/a_\eta^*|a) = \varphi_\lambda(a_\xi/a_\eta^*|a)$$

for $\zeta = \lambda$. In this case $\eta = \nu$ and $\xi = \rho + \mu$. Due to the specialization properties (4.11) and (5.3), we may assume that $l(\nu) = n$, which implies that the partition ρ coincides with (n^m). In this case we clearly have

$$\varphi_\lambda(x/y|a) = s_\nu(y|a^*) s_\mu(x|\tau^n a) \prod_{i=1}^{m} \prod_{j=1}^{n} (x_i + y_j)$$

(see also Corollary 5.2 below). Setting $x = a_\xi = (\tau^n a)_\mu$ and $y = a_\nu^*$, we see that the result coincides with (4.17), which completes the proof of the theorem.

REMARK. As noticed in [17], (5.1) implies that the polynomials $s_\lambda(x/y|a)$ coincide with the *multi-Schur functions* (see, e.g., [22]) in the appropriate variables.

As a corollary of Theorem 5.1, we obtain the following factorization theorem for the polynomials $s_\lambda(x/y|a)$, which turns into the Berele–Regev formula for $a = (0)$ [2] (cf. [9, 35]).

COROLLARY 5.2. *If a partition λ is contained in the (m, n)-hook and contains the partition (n^m), then*

(5.4) $$s_\lambda(x/y|a) = s_\mu(x|\tau^n a) s_\nu(y|a^*) \prod_{i=1}^{m} \prod_{j=1}^{n} (x_i + y_j).$$

PROOF. Note that $\rho = (n^m)$ and that the product

$$\prod_{(i,j) \in \rho} (x_i + y_j) = \prod_{i=1}^{m} \prod_{j=1}^{n} (x_i + y_j)$$

is symmetric in x and y. So, the assertion follows from (0.1).

In the special case $\lambda = (n^m)$, formula (5.4) turns into

(5.5) $$s_{(n^m)}(x/y|a) = \prod_{i=1}^{m} \prod_{j=1}^{n} (x_i + y_j).$$

Using definition (1.4) we derive from (5.5) the following analog of the dual Cauchy formula, which is proved in [21, (6.17)] and can be also deduced from the Cauchy formula for the double Schubert polynomials [17, 18].

COROLLARY 5.3.

(5.6) $$\prod_{i=1}^{m} \prod_{j=1}^{n} (x_i + y_j) = \sum_\lambda s_{\tilde\lambda}(x|a) s_{\lambda'}(y|-a),$$

with summation over all partitions $\lambda \subset (n^m)$, where $\tilde\lambda = (n - \lambda_m, \ldots, n - \lambda_1)$ and a is an arbitrary sequence.

PROOF. By (1.4),

$$s_{(n^m)}(x/y|-a^*) = \sum_{\lambda \subset (n^m)} s_{(n^m)/\lambda}(x|-a^*) s_{\lambda'}(y|-a). \tag{5.7}$$

Since $s_{(n^m)/\lambda}(x|-a^*)$ is symmetric in x, replacing x by $\widetilde{x} = (x_m, \ldots, x_1)$ and using (1.1), we may write

$$\begin{aligned} s_{(n^m)/\lambda}(x|-a^*) &= s_{(n^m)/\lambda}(\widetilde{x}|-a^*) \\ &= \sum_T \prod_{\alpha \in (n^m)/\lambda} (x_{m-T(\alpha)+1} - a_{n-T(\alpha)-c(\alpha)+1}), \end{aligned} \tag{5.8}$$

with summation over semistandard $(n^m)/\lambda$-tableaux T. Consider the bijection between the cells of the diagram $(n^m)/\lambda$ and the cells of the diagram $\widetilde{\lambda}$ such that $\alpha = (i,j) \in (n^m)/\lambda$ corresponds to $\beta = (m-i+1, n-j+1) \in \widetilde{\lambda}$. Obviously, the content $c(\beta)$ of the cell $\beta \in \widetilde{\lambda}$ is related with $c(\alpha)$ by $c(\beta) = n - m - c(\alpha)$. Moreover, the map

$$T(\alpha) \to \widetilde{T}(\beta) = m - T(\alpha) + 1$$

is a bijection between the semistandard $(n^m)/\lambda$-tableaux and the semistandard $\widetilde{\lambda}$-tableaux. So, (5.8) gives

$$s_{(n^m)/\lambda}(x|-a^*) = \sum_{\widetilde{T}} \prod_{\beta \in \widetilde{\lambda}} (x_{\widetilde{T}(\beta)} - a_{\widetilde{T}(\beta)+c(\beta)}) = s_{\widetilde{\lambda}}(x|a).$$

Using (5.7) and (5.5), we complete the proof.

§6. Macdonald–Goulden–Greene formula

Formula (5.5) means that the polynomial $s_{(n^m)}(x/y|a)$ does not depend on the sequence a. It turns out that a similar phenomenon occurs for infinite families of variables. We shall regard the elements of the sequence a as independent variables to avoid convergence problems. Let us consider three families of variables $x = (x_i)$, $y = (y_i)$, $a = (a_i)$, $i \in \mathbb{Z}$. We define the functions $s_{\lambda/\mu}(x/y|a)$ in x, y, and a by formula (1.5), where we allow the primed and unprimed entries of the tableaux to run through the set of all integers. This definition can be shown to be equivalent to the following formula, where the factorial Schur functions $s_{\lambda/\mu}(x|a)$ are defined by (1.1) with T running over semistandard λ/μ-tableaux with entries from $\mathbb{Z}$ (see [9] and [21]).

PROPOSITION 6.1.

$$s_{\lambda/\mu}(x/y|a) = \sum_{\mu \subset \nu \subset \lambda} s_{\lambda/\nu}(x|a) s_{\nu'/\mu'}(y|-a). \tag{6.1}$$

PROOF. Repeating the arguments of the proof of Proposition 1.2, we see that the assertion follows from the formula

$$\sum_T \prod_{\alpha \in \nu/\mu} (y_{T(\alpha)} + a_{T(\alpha)+c(\alpha)}) = s_{\nu'/\mu'}(y|-a), \tag{6.2}$$

with summation over ν/μ-tableaux T with entries from $\mathbb{Z}$ whose rows strictly decrease and columns weakly decrease.

It was shown in [**9**] and [**21**] that in the case of infinite number of variables (parametrized by $\mathbb{Z}$) the factorial and supersymmetric Schur functions coincide with each other:

(6.3) $$s_{\lambda/\mu}(x|a) = s_{\lambda/\mu}(x/-a),$$

where the supersymmetric Schur functions $s_{\lambda/\mu}(x/y)$ are still defined by (0.2). In particular, $s_{\lambda/\mu}(x|a)$ is symmetric in a (which is not true in the finite case). Hence, we have
$$s_{\nu'/\mu'}(y|-a) = s_{\nu'/\mu'}(\tilde{y}|-\tilde{a}),$$
where $\tilde{y} = (y_{-i})$ and $\tilde{a} = (a_{-i})$. So, by (1.1),

(6.4) $$s_{\nu'/\mu'}(y|-a) = \sum_{T'} \prod_{\alpha' \in \nu'/\mu'} (y_{-T'(\alpha')} + a_{-T'(\alpha')-c(\alpha')}),$$

with summation over semistandard ν'/μ'-tableaux T' with entries from $\mathbb{Z}$. Note that the map
$$T'(\alpha') \to T(\alpha) = -T'(\alpha'),$$
where $\alpha = (i,j) \in \nu/\mu$ and $\alpha' = (j,i) \in \nu'/\mu'$, is a bijection between the set of semistandard ν'/μ'-tableaux and the set of ν/μ-tableaux whose rows strictly decrease and columns weakly decrease. Obviously, $c(\alpha) = -c(\alpha')$, and hence, (6.4) coincides with the left-hand side of (6.2), which completes the proof.

For $a = (0)$ formula (6.1) turns into the definition of the supersymmetric Schur functions $s_{\lambda/\mu}(x/y)$. It turns out that the right-hand side of (6.1) does not depend on the variables a_i, and thus the Macdonald–Goulden–Greene formula (see [**9**] and [**21**]) holds for the functions $s_{\lambda/\mu}(x/y|a)$ as well.

THEOREM 6.2. *One has the formula*

(6.5) $$s_{\lambda/\mu}(x/y|a) = \sum_T \prod_{\alpha \in \lambda/\mu} (x_{T(\alpha)} + y_{T(\alpha)+c(\alpha)}),$$

where T runs over all semistandard λ/μ-tableaux with entries from $\mathbb{Z}$.

PROOF. For $a = (0)$ formula (6.5) was proved in [**9**] and [**21**]. So, it suffices to verify that $s_{\lambda/\mu}(x/y|a) = s_{\lambda/\mu}(x/y)$. Using (6.1) and (6.3), we obtain

(6.6) $$s_{\lambda/\mu}(x/y|a) = \sum_{\nu} s_{\lambda/\nu}(x/-a) s_{\nu'/\mu'}(y/a).$$

By the symmetry property (0.3), we have $s_{\nu'/\mu'}(y/a) = s_{\nu/\mu}(a/y)$. Hence, using the definition of the supersymmetric Schur functions, we can rewrite (6.6) as follows:

(6.7) $$\begin{aligned}s_{\lambda/\mu}(x/y|a) &= \sum_{\nu,\rho,\sigma} s_{\lambda/\rho}(x) s_{\rho'/\nu'}(-a) s_{\nu/\sigma}(a) s_{\sigma'/\mu'}(y) \\ &= \sum_{\rho,\sigma} s_{\lambda/\rho}(x) s_{\sigma'/\mu'}(y) s_{\rho'/\sigma'}(-a/a).\end{aligned}$$

Note that $s_{\rho'/\sigma'}(-a/a) = 0$ unless $\rho = \sigma$. Indeed, assume that there exists a cell $\alpha_0 \in \rho'/\sigma'$. Then, applying (6.5) with $a = (0)$, we obtain
$$s_{\rho'/\sigma'}(-a/a) = s_{\rho'/\sigma'}(-a/\tau^{-c(\alpha_0)}a) = \sum_T \prod_{\alpha \in \rho'/\sigma'} (-a_{T(\alpha)} + a_{T(\alpha)+c(\alpha)-c(\alpha_0)}) = 0.$$

Thus, (6.7) coincides with $s_{\lambda/\mu}(x/y)$, which completes the proof.

§7. Shifted supersymmetric Schur polynomials and a basis in the center of $U(\mathfrak{gl}(m|n))$

We need to introduce super-analogs of the shifted symmetric polynomials (cf. [**29, 31, 32**]). A polynomial in two families of variables $u = (u_1, \ldots, u_m)$ and $v = (v_1, \ldots, v_n)$ will be called *shifted supersymmetric* if it is supersymmetric in the variables

$$(u_1 + m - 1, u_2 + m - 2, \ldots, u_m) \quad \text{and} \quad (v_1, v_2 - 1, \ldots, v_n - n + 1).$$

We denote the algebra of shifted supersymmetric polynomials by $\Lambda^*(m|n)$. It follows from the definition that $\Lambda^*(m|n)$ is isomorphic to the algebra of supersymmetric polynomials in x and y.

Let us consider the polynomials $s_{\lambda/\mu}(x/y|a)$ with the sequence a defined by $a_i = -m + i$ and the following values of the variables in $s_{\lambda/\mu}(x/y|a)$:

$$x_i = u_{m-i+1} - m + i \quad \text{for } i = 1, \ldots, m,$$
$$y_j = v_j + m - j \quad \text{for } j = 1, \ldots, n.$$

Then we obtain a shifted supersymmetric polynomial in u and v, which is denoted by $s^*_{\lambda/\mu}(u/v)$ and will be called *shifted supersymmetric Schur polynomial*. In the case $n = 0$, it coincides with the shifted Schur polynomial $s^*_{\lambda/\mu}(u)$ (see [**29, 31, 32**]), which can be defined by the formula

$$(7.1) \qquad s^*_{\lambda/\mu}(u) = \sum_T \prod_{\alpha \in \lambda/\mu} (u_{T(\alpha)} - c(\alpha)),$$

where the sum is taken over λ/μ-tableaux T with entries in $\{1, \ldots, m\}$ whose rows weakly decrease and columns strictly decrease. The highest component of $s^*_{\lambda/\mu}(u)$ is the usual skew Schur polynomial $s_{\lambda/\mu}(u)$.

Now we formulate some properties of the polynomials $s^*_{\lambda/\mu}(u/v)$ which can be easily derived from the corresponding properties of the polynomials $s_{\lambda/\mu}(x/y|a)$.

First we give a combinatorial interpretation of $s^*_{\lambda/\mu}(u/v)$.

To distinguish the indices of u and v, identify the indices of v with the symbols $1', \ldots, n'$. Consider the diagram of shape λ/μ and fill it with the indices $1', \ldots, n', 1, \ldots, m$ so that:

(a) in each row (resp., column) each primed index is to the left (resp., above) of each unprimed index;

(b) primed indices are strictly decreasing along rows and weakly decrease down columns;

(c) unprimed indices are weakly decreasing along rows and strictly decrease down columns.

Denote the resulting tableau by T.

PROPOSITION 7.1. *One has the formula*

$$(7.2) \qquad s^*_{\lambda/\mu}(u/v) = \sum_T \prod_{\substack{\alpha \in \lambda/\mu \\ T(\alpha) \text{ unprimed}}} (u_{T(\alpha)} - c(\alpha)) \prod_{\substack{\alpha \in \lambda/\mu \\ T(\alpha) \text{ primed}}} (v_{T(\alpha)} + c(\alpha)).$$

PROOF. This follows immediately from Proposition 1.2. It suffices to use (1.5) with x replaced by $\widetilde{x} = (x_m, \ldots, x_1)$.

In particular, for the *elementary* and *complete shifted supersymmetric polynomials* $e_k^*(u/v) := s_{(1^k)}^*(u/v)$ and $h_k^*(u/v) := s_{(k)}^*(u/v)$ we have

$$e_k^*(u/v) = \sum_{p+q=k} \sum_{\substack{i_1 > \cdots > i_p \\ j_1 \geqslant \cdots \geqslant j_q}} v_{j_1}(v_{j_2} - 1) \cdots (v_{j_q} - q + 1)(u_{i_1} + q) \cdots (u_{i_p} + k - 1),$$

$$h_k^*(u/v) = \sum_{p+q=k} \sum_{\substack{i_1 \geqslant \cdots \geqslant i_p \\ j_1 > \cdots > j_q}} v_{j_1}(v_{j_2} + 1) \cdots (v_{j_q} + q - 1)(u_{i_1} - q) \cdots (u_{i_p} - k + 1).$$

Using (7.1), we can rewrite (7.2) in the following equivalent form.

COROLLARY 7.2. *One has*

(7.3) $$s_{\lambda/\mu}^*(u/v) = \sum_{\mu \subset \nu \subset \lambda} s_{\lambda/\nu}^*(u) s_{\nu'/\mu'}^*(v).$$

This implies that the highest component of $s_{\lambda/\mu}^*(u/v)$ is the supersymmetric Schur polynomial $s_{\lambda/\mu}(u/v)$. So, the polynomials $s_\lambda^*(u/v)$ with $\lambda \subset (m,n)$-hook form a basis in $\Lambda^*(m|n)$.

The following results are reformulations of Theorems 4.4 and 4.5' for the shifted supersymmetric polynomials; cf. [**29, 31**] (we use the notation from §4).

THEOREM 7.3. *Let λ, ζ be partitions which are contained in the (m,n)-hook.*
(i) *If $\lambda \not\subset \zeta$, then*

(7.4) $$s_\lambda^*(\xi/\eta) = 0.$$

(ii) *If $\lambda = \zeta$, then*

(7.5) $$s_\lambda^*(\xi/\eta) = H(\lambda),$$

where $H(\lambda)$ is the product of the hook lengths of all cells of λ.

THEOREM 7.4. *Let $f(u/v)$ be a shifted supersymmetric polynomial such that*

$$f(u/v) = s_\lambda(u/v) + \text{lower terms}$$

for some partition $\lambda \subset (m,n)$-hook, and $f(\xi/\eta) = 0$ for any partition $\zeta \subset (m,n)$-hook with $|\zeta| < |\lambda|$. Then $f(u/v) = s_\lambda^*(u/v)$.

A distinguished linear basis in the center of the universal enveloping algebra $U(\mathfrak{gl}(m))$ was constructed in [**29**]. The eigenvalue of a basis element in a highest weight representation is a shifted Schur polynomial $s_\lambda^*(u)$. It turns out that this construction can be easily carried over to the case of the Lie superalgebra $\mathfrak{gl}(m|n)$. Below we formulate the corresponding theorem and briefly outline its proof. Another approach to this construction, based on the super-analogs of the higher Capelli identities, is presented in §8.

We denote by E_{ij}, $i,j = 1, \ldots, m+n$, the standard basis of the Lie superalgebra $\mathfrak{gl}(m|n)$. The $\mathbb{Z}_2$-grading on $\mathfrak{gl}(m|n)$ is defined by $E_{ij} \mapsto p(i) + p(j)$, where $p(i) = 0$

or 1 depending on whether $i \leqslant m$ or $i > m$. The commutation relations in this basis are given by

(7.6) $$[E_{ij}, E_{kl}] = \delta_{kj} E_{il} - \delta_{il} E_{kj}(-1)^{(p(i)+p(j))(p(k)+p(l))}.$$

Given $w = (u_1, \ldots, u_m, v_1, \ldots, v_n) \in \mathbb{C}^{m+n}$, we consider an arbitrary highest weight $\mathfrak{gl}(m|n)$-module $L(w)$ with the highest weight w. That is, $L(w)$ is generated by a nonzero vector ψ such that

$$E_{ii}\psi = u_i\psi \quad \text{for } i = 1, \ldots, m,$$
$$E_{m+j,m+j}\psi = v_j\psi \quad \text{for } j = 1, \ldots, n,$$
$$E_{ij}\psi = 0 \quad \text{for } 1 \leqslant i < j \leqslant m+n.$$

Every element z from the center $\mathrm{Z}(\mathfrak{gl}(m|n))$ of the universal enveloping algebra $\mathrm{U}(\mathfrak{gl}(m|n))$ acts in $L(w)$ as a scalar $\chi(z)$. For a fixed z, the scalar $\chi(z)$ is a shifted supersymmetric polynomial in u and v, and the map $z \mapsto \chi(z)$ defines an algebra isomorphism

(7.7) $$\chi\colon \mathrm{Z}(\mathfrak{gl}(m|n)) \to \Lambda^*(m|n),$$

which is called the *Harish–Chandra isomorphism* (see [16, 38, 40]).

Our present goal is to give an explicit description of the basis of the algebra $\mathrm{Z}(\mathfrak{gl}(m|n))$ formed by the preimages $\chi^{-1}(s^*_\lambda(u/v))$ of the basis elements of $\Lambda^*(m|n)$. Let us introduce some more notation.

We need to consider matrices with entries from superalgebras. All our matrices will be even. That is, if $B = (B_{ia})$ is an $(m,n) \times (m',n')$-matrix whose entries are homogeneous elements of a superalgebra $\mathcal{B}$, we always have $p(B_{ia}) = p(i) + p(a)$, where $p(a) = 0$ or 1 depending on whether $a \leqslant m'$ or $a > m'$. A matrix B will be identified with an element of the tensor product

$$B = \sum_{i,a} e_{ia} \otimes B_{ia}(-1)^{p(a)(p(i)+1)} \in \mathrm{Mat}_{(m,n) \times (m',n')} \otimes \mathcal{B},$$

where the e_{ia} are the standard matrix units.

More generally, given k matrices $B^{(1)}, \ldots, B^{(k)}$ of size $(m,n) \times (m',n')$, we define their tensor product $B^{(1)} \otimes \cdots \otimes B^{(k)}$ as an element

$$\sum e_{i_1 a_1} \otimes \cdots \otimes e_{i_k a_k} \otimes B^{(1)}_{i_1 a_1} \cdots B^{(k)}_{i_k a_k}(-1)^{\gamma(I,A)} \in (\mathrm{Mat}_{(m,n) \times (m',n')})^{\otimes k} \otimes \mathcal{B},$$

where

$$\gamma(I,A) = \sum_{r=1}^k p(a_r)(p(i_r)+1) + \sum_{1 \leqslant r < s \leqslant k}(p(i_r)+p(a_r))(p(i_s)+p(a_s)).$$

The supertrace of an element

$$B = \sum e_{i_1 j_1} \otimes \cdots \otimes e_{i_k j_k} \otimes B_{i_1,\ldots,i_k;j_1,\ldots,j_k} \in (\mathrm{Mat}_{(m,n) \times (m,n)})^{\otimes k} \otimes \mathcal{B}$$

is defined by

$$\mathrm{str}\, B = \sum_{i_1,\ldots,i_k} B_{i_1,\ldots,i_k;i_1,\ldots,i_k}(-1)^{p(i_1)+\cdots+p(i_k)}.$$

Using the natural action of the symmetric group S_k in the space $(\mathbb{C}^{m|n})^{\otimes k}$, we represent each element of S_k as a linear combination of tensor products of matrices.

In particular, the transposition $(i,j) \in S_k$, $i < j$, corresponds to the element

$$P_{ij} = \sum_{a,b} 1 \otimes \cdots \otimes 1 \otimes e_{ab} \otimes 1 \otimes \cdots \otimes 1 \otimes e_{ba} \otimes 1 \otimes \cdots \otimes 1 (-1)^{p(b)},$$

where the tensor factors e_{ab} and e_{ba} are in the ith and jth places, respectively.

Now we can describe the construction of a basis in $Z(\mathfrak{gl}(m|n))$.

Set $\widehat{E}_{ij} = E_{ij}(-1)^{p(j)}$ and denote by $\widehat{E}$ the $(m,n) \times (m,n)$-matrix whose ijth entry is $\widehat{E}_{ij}$.

Following [29], for a partition λ and a standard λ-tableau T, we denote by v_T the corresponding vector of the Young orthonormal basis with respect to an invariant inner product $(\,,\,)$ in the irreducible representation V^λ of the symmetric group S_k, $k = |\lambda|$. We let $c_T(r) = j - i$ if the cell $(i,j) \in \lambda$ is occupied by the entry r of the tableau T. Given two standard λ-tableaux T and T', introduce the matrix element

(7.8) $$\Psi_{TT'} = \sum_{s \in S_k} (s \cdot v_T, v_{T'}) \cdot s^{-1} \in \mathbb{C}[S_k].$$

THEOREM 7.5. *The element*

(7.9) $$\mathbb{S}_\lambda = \frac{1}{H(\lambda)} \operatorname{str}(\widehat{E} - c_T(1)) \otimes \cdots \otimes (\widehat{E} - c_T(k)) \cdot \Psi_{TT}$$

is independent of the λ-tableau T. The set of elements $\mathbb{S}_\lambda$ with $\lambda \subset (m,n)$-hook forms a basis in $Z(\mathfrak{gl}(m|n))$. Moreover, the image of $\mathbb{S}_\lambda$ under the Harish–Chandra isomorphism is $s_\lambda^*(u/v)$.

OUTLINE OF THE PROOF. In the case $n = 0$ this theorem was proved in [29]. It constitutes the "difficult part" of the proof of the higher Capelli identities. A straightforward generalization of those arguments proves Theorem 7.5. For this, one uses the following R-matrix form of the defining relations in $U(\mathfrak{gl}(m|n))$ (cf. [27]):

$$R(u - v) \cdot \widehat{E}(u) \otimes \widehat{E}(v) = \widehat{E}(v) \otimes \widehat{E}(u) \cdot R(u - v),$$

where $R(u) = 1 + P_{12}u$.

EXAMPLES. For the partitions of weight $\leqslant 2$ we have

$$\mathbb{S}_{(1)} = \operatorname{str} \widehat{E} = \sum_i E_{ii},$$

$$\mathbb{S}_{(2)} = \frac{1}{2} \operatorname{str}(\widehat{E} \otimes (\widehat{E} - 1) \cdot (1 + P_1))$$

$$= \frac{1}{2} \sum_{i,j} (E_{ii}(E_{jj} - (-1)^{p(j)}) + E_{ij}(E_{ji} - \delta_{ji}(-1)^{p(i)})(-1)^{p(j)}),$$

$$\mathbb{S}_{(1^2)} = \frac{1}{2} \operatorname{str}(\widehat{E} \otimes (\widehat{E} + 1) \cdot (1 - P_1))$$

$$= \frac{1}{2} \sum_{i,j} (E_{ii}(E_{jj} + (-1)^{p(j)}) - E_{ij}(E_{ji} + \delta_{ji}(-1)^{p(i)})(-1)^{p(j)}).$$

§8. Super Capelli identities

Here we formulate a super-analog of the higher Capelli identities obtained in [28–30].

Let us consider the supercommutative algebra $\mathcal{Z}$ with the generators z_{ia}, where $i = 1, \ldots, m + n$ and $a = 1, \ldots, m' + n'$, and the $\mathbb{Z}_2$-grading given by $z_{ia} \mapsto p(i) + p(a)$. Define the representation π of the Lie superalgebra $\mathfrak{gl}(m|n)$ in $\mathcal{Z}$ by

$$\pi(E_{ij}) = \sum_{a=1}^{m'+n'} z_{ia} \partial_{ja},$$

where $\partial_{ja} = \partial/\partial z_{ja}$ is the left derivation. This definition can be rewritten in matrix form as follows:

(8.1) $$\pi(\widehat{E}) = ZD',$$

where Z is the $(m, n) \times (m', n')$-matrix (z_{ia}) and D' is the $(m', n') \times (m, n)$-matrix (∂'_{ai}) with $\partial'_{ai} = \partial_{ia}(-1)^{p(i)}$.

For a partition λ with $|\lambda| = k$, denote by χ^λ the irreducible character of S_k. We identify χ^λ with an element of the group algebra $\mathbb{C}[S_k]$:

$$\chi^\lambda = \sum_{s \in S_k} \chi^\lambda(s) \cdot s.$$

Define the differential operator Δ_λ by

$$\Delta_\lambda = \frac{1}{k!} \operatorname{str}(Z^{\otimes k} \cdot D'^{\otimes k} \cdot \chi^\lambda).$$

The following is a super-analog of the higher Capelli identities (cf. [28–30]).

THEOREM 8.1. *One has*

(8.2) $$\pi(\mathbb{S}_\lambda) = \Delta_\lambda.$$

We outline two proofs of this identity. The first proof is based on the properties of the shifted supersymmetric polynomials. The second one uses a super-analog of a more general identity obtained in [28] and [30] (see Theorem 8.2 below).

FIRST PROOF. Again we follow the corresponding arguments from [29]. First, one verifies that the operator Δ_λ commutes with both actions of the Lie superalgebras $\mathfrak{gl}(m|n)$ and $\mathfrak{gl}(m'|n')$ in $\mathcal{Z}$; the latter is given by

$$\pi'(E'_{ab}) = \sum_{i=1}^{m+n} z_{ia} \partial_{ib}(-1)^{(p(a)+p(b))p(i)},$$

where the E'_{ab} are the standard generators of $\mathfrak{gl}(m'|n')$. This implies that Δ_λ is the image of a certain element $\mathbb{S}'_\lambda \in Z(\mathfrak{gl}(m|n))$ under π (cf. [12, 27]). To prove that $\mathbb{S}_\lambda = \mathbb{S}'_\lambda$, we compare their images under the Harish–Chandra isomorphism. Set $s'_\lambda(u/v) = \chi(\mathbb{S}'_\lambda)$. By Theorem 7.5, $\chi(\mathbb{S}_\lambda) = s^*_\lambda(u/v)$. We use Theorem 7.4 to prove that $s'_\lambda(u/v) = s^*_\lambda(u/v)$. Using (8.1), we check that both sides of (8.2) agree modulo lower terms (cf. [29, 31]). This proves that the polynomials $s'_\lambda(u/v)$ and $s^*_\lambda(u/v)$ have the same highest component, which coincides with the supersymmetric Schur polynomial $s_\lambda(u/v)$.

By Theorem 7.3, to complete the proof we must verify that $s'_\lambda(\xi/\eta)$ is zero for any partition $\zeta \subset (m,n)$-hook such that $|\zeta| < |\lambda|$. Let us consider the superalgebra $\mathcal{Z}$ with the parameters m' and n' chosen sufficiently large, so that $m' \geqslant \max\{m, |\zeta|\}$ and $n' \geqslant \max\{n, |\zeta|\}$.

Introduce the following element of $\mathcal{Z}$:

$$\psi_\zeta = \Delta_1^{\eta_1-\eta_2} \Delta_2^{\eta_2-\eta_3} \cdots \Delta_n^{\eta_n} \prod_{(i,j) \in \xi} z_{i,m'+j},$$

where $\Delta_r = \det[z_{m+i,m'+j}]_{1 \leqslant i,j \leqslant r}$ and the product is taken in any fixed order. It can be easily checked that ψ_ζ satisfies the relations

$$\pi(E_{ii})\psi_\zeta = \xi_i \psi_\zeta \quad \text{for } i = 1,\ldots,m,$$
$$\pi(E_{m+j,m+j})\psi_\zeta = \eta_j \psi_\zeta \quad \text{for } j = 1,\ldots,n,$$
$$\pi(E_{ij})\psi_\zeta = 0 \quad \text{for } 1 \leqslant i < j \leqslant m+n.$$

This means that ψ_ζ generates a $\mathfrak{gl}(m|n)$-module with the highest weight (ξ,η). Hence, ψ_ζ is an eigenvector for the operator Δ_λ with the eigenvalue $s'_\lambda(\xi/\eta)$. However, the degree of ψ_ζ equals $|\zeta|$ and so, if $|\zeta| < |\lambda| = k$, then ψ_ζ is annihilated by Δ_λ, that is, $s'_\lambda(\xi/\eta) = 0$, completing the proof.

THEOREM 8.2. *Let T and T' be two standard tableaux of the shape λ. Then*

(8.3) $\quad \pi((\widehat{E} - c_T(1)) \otimes \cdots \otimes (\widehat{E} - c_T(k)) \cdot \Psi_{TT'}) = Z^{\otimes k} \cdot (D')^{\otimes k} \cdot \Psi_{TT'}.$

PROOF. Following [30], we use some properties of the Jucys–Murphy elements in the group algebra for the symmetric group. However, the arguments from [30] can be modified to avoid using the Wick formula and the Olshanskiĭ special symmetrization map.

We use induction on k. Denote by U the tableau obtained from T by removing the cell with the entry k. From the branching property of the Young basis $\{v_T\}$, one can easily derive that

$$\Psi_{TT'} = \text{const} \cdot \Psi_{UU} \Psi_{TT'},$$

where "const" is a nonzero constant (more precisely, const $= \dim \mu/(k-1)!$, where μ is the shape of U and $\dim \mu = \dim V^\mu$).

So, we can rewrite the left-hand side of (8.3) as follows:

$$\text{const} \cdot (ZD' - c_T(1)) \otimes \cdots \otimes (ZD' - c_T(k-1)) \cdot \Psi_{UU} \otimes (ZD' - c_T(k)) \cdot \Psi_{TT'}.$$

By the induction hypothesis, this equals

$$\text{const} \cdot Z^{\otimes k-1} \cdot (D')^{\otimes k-1} \cdot \Psi_{UU} \otimes (ZD' - c_T(k)) \cdot \Psi_{TT'}$$
$$= Z^{\otimes k-1} \cdot (D')^{\otimes k-1} \otimes (ZD' - c_T(k)) \cdot \Psi_{TT'}$$
$$= \Big(\sum e_{i_1 j_1} \otimes \cdots \otimes e_{i_k j_k} \otimes z_{i_1 a_1} \cdots z_{i_{k-1} a_{k-1}} \partial'_{a_1 j_1} \cdots \partial'_{a_{k-1} j_{k-1}}$$
$$\times \Big(\sum z_{i_k a_k} \partial'_{a_k j_k} - \delta_{i_k j_k} c_T(k)\Big)\Big)(-1)^{\alpha(I,J,A)} \cdot \Psi_{TT'},$$

where

$$\alpha(I, J, A) = \sum_{r=1}^{k} p(j_r)(p(i_r) + 1) + \sum_{1 \leqslant r < s \leqslant k-1} (p(a_r) + p(j_r))(p(i_s) + p(a_s))$$
$$+ \sum_{1 \leqslant r < s \leqslant k} (p(i_r) + p(j_r))(p(i_s) + p(j_s)).$$

Now we transform this expression using the relations

$$\partial'_{bj} z_{ia} = z_{ia} \partial'_{bj} (-1)^{(p(i)+p(a))(p(j)+p(b))} + \delta_{ab}\delta_{ij}(-1)^{p(j)}$$

to obtain
(8.4)
$$\left(\sum e_{i_1 j_1} \otimes \cdots \otimes e_{i_k j_k} \otimes z_{i_1 a_1} \cdots z_{i_k a_k} \partial'_{a_1 j_1} \cdots \partial'_{a_k j_k} (-1)^{\beta_k(I,J,A)} \right) \cdot \Psi_{TT'}$$
$$+ \left(\sum e_{i_1 j_1} \otimes \cdots \otimes e_{i_{k-1} j_{k-1}} \otimes 1 \otimes z_{i_1 a_1} \cdots z_{i_{k-1} a_{k-1}} \partial'_{a_1 j_1} \cdots \partial'_{a_{k-1} j_{k-1}} \right.$$
$$\left. \times (-1)^{\beta_{k-1}(I,J,A)} \right) \times (P_{1k} + \cdots + P_{k-1,k} - c_T(k)) \cdot \Psi_{TT'},$$

where

$$\beta_k(I, J, A) = \sum_{r=1}^{k} p(j_r)(p(i_r) + 1) + \sum_{1 \leqslant r < s \leqslant k} (p(a_r) + p(j_r))(p(i_s) + p(a_s))$$
$$+ \sum_{1 \leqslant r < s \leqslant k} (p(i_r) + p(j_r))(p(i_s) + p(j_s)).$$

Note that $P_{1k} + \cdots + P_{k-1,k}$ is the image of the Jucys–Murphy element (see [14] and [26]) $(1k) + \cdots + (k-1,k) \in \mathbb{C}[S_k]$. It has the property

$$((1k) + \cdots + (k-1,k)) \cdot \Psi_{TT'} = c_T(k) \cdot \Psi_{TT'},$$

which was also used in [30] and can be easily derived from the following formula due to Jucys and Murphy:

$$((1k) + \cdots + (k-1,k)) \cdot v_T = c_T(k) \cdot v_T.$$

This proves that the second summand in (8.4) is zero, while the first one coincides with the right-hand side of (8.3). Theorem 8.2 is proved.

SECOND PROOF OF THEOREM 8.1. Put $T = T'$ in Theorem 8.2 and take the supertrace of both sides of (8.3). On the left-hand side we get $H(\lambda)\pi(\mathbb{S}_\lambda)$, while for the right-hand side we have

$$\operatorname{str} Z^{\otimes k} \cdot (D')^{\otimes k} \cdot \Psi_{TT} = \frac{1}{k!} \sum_{s \in S_k} \operatorname{str} s \cdot Z^{\otimes k} \cdot (D')^{\otimes k} \cdot \Psi_{TT} \cdot s^{-1}$$
$$= \frac{1}{\dim \lambda} \operatorname{str} Z^{\otimes k} \cdot (D')^{\otimes k} \cdot \chi^\lambda.$$

Here we have used the invariance of $Z^{\otimes k}$ and $(D')^{\otimes k}$ under conjugation by elements $s \in S_k$ and the following equality of elements of the group algebra of S_k:

$$\frac{1}{k!} \sum_{s \in S_k} s \cdot \Psi_{TT} \cdot s^{-1} = \frac{1}{\dim \lambda} \chi^\lambda.$$

So, on the right-hand side we get $H(\lambda)\Delta_\lambda$, which completes the proof.

EXAMPLE. In the case of $\lambda = (1^k)$, the identity (8.2) was obtained by Nazarov [**27**] in another form. Namely, a formal series $B(t)$ whose coefficients are generators of $Z(\mathfrak{gl}(m|n))$ was constructed in [**27**] with the use of some properties of the Yangian for the Lie superalgebra $\mathfrak{gl}(m|n)$. The explicit expression for $B(t)$ has the form of a "quantum" analog of the Berezinian:

(8.5)
$$B(t) = \sum_{\sigma \in S_m} \text{sgn}(\sigma) \left(1 + \frac{\widehat{E}}{t}\right)_{\sigma(1),1} \cdots \left(1 + \frac{\widehat{E}}{t-m+1}\right)_{\sigma(m),m}$$
$$\times \sum_{\tau \in S_n} \text{sgn}(\tau) \left(1 + \frac{\widehat{E}}{t-m+1}\right)^*_{m+\tau(1),m+1} \cdots \left(1 + \frac{\widehat{E}}{t-m+n}\right)^*_{m+\tau(n),m+n},$$

where $A^* = (A^{-1})^{st}$ and st is the following matrix supertransposition: $(B^{st})_{ij} = B_{ji}(-1)^{p(i)(p(j)+1)}$. The image of $B(t)$ under the Harish–Chandra isomorphism coincides with its eigenvalue on the highest vector ψ of the highest weight $\mathfrak{gl}(m|n)$-module $L(w)$, $w = (u,v) \in \mathbb{C}^{m|n}$. The eigenvalue of the first determinant in (8.5) on ψ is
$$\frac{(t+u_1)\cdots(t+u_m-m+1)}{t(t-1)\cdots(t-m+1)}.$$

To find the eigenvalue of the second determinant, we can replace the matrix $\widehat{E}$ with its submatrix $\widetilde{E} = (\widehat{E}_{ij})_{m+1 \leqslant i,j \leqslant m+n}$. Now, the determinant

$$\sum_{\tau \in S_n} \text{sgn}(\tau) \left(1 + \frac{\widetilde{E}}{t-m+1}\right)^*_{m+\tau(1),m+1} \cdots \left(1 + \frac{\widetilde{E}}{t-m+n}\right)^*_{m+\tau(n),m+n}$$

equals

$$\left(\sum_{\tau \in S_n} \text{sgn}(\tau) \left(1 + \frac{\widetilde{E}}{t-m+1}\right)_{m+\tau(1),m+1} \cdots \left(1 + \frac{\widetilde{E}}{t-m+n}\right)_{m+\tau(n),m+n}\right)^{-1},$$

which follows from [**27**, Proposition 3] and can be also proved directly by using the R-matrix form of the defining relations in $U(\mathfrak{gl}(n))$ (see, e.g., [**25**]). So, its eigenvalue on ψ is
$$\frac{(t-m+1)\cdots(t-m+n)}{(t-v_1-m+1)\cdots(t-v_n-m+n)}.$$

Thus,
$$\chi(B(t)) = \frac{(t+u_1)\cdots(t+u_m-m+1)(t-m+1)\cdots(t-m+n)}{t(t-1)\cdots(t-m+1)(t-v_1-m+1)\cdots(t-v_n-m+n)}.$$

Relation (2.5) implies that
$$\chi(B(t)) = 1 + \sum_{k=1}^{\infty} \frac{e_k^*(u/v)}{t(t-1)\cdots(t-k+1)}.$$

By Theorem 7.5, $\chi(\mathbb{S}_{(1^k)}) = e_k^*(u/v)$; hence,
$$B(t) = 1 + \sum_{k=1}^{\infty} \frac{\mathbb{S}_{(1^k)}}{t(t-1)\cdots(t-k+1)}.$$

Using (8.2), we get the following identity (see [**27**]):

$$\pi(B(t)) = 1 + \sum_{k=1}^{\infty} \frac{\Delta_{(1^k)}}{t(t-1)\cdots(t-k+1)}.$$

For $n = n' = 0$ this turns into the classical Capelli identity (see, e.g., [**11, 12**]).

References

1. A. Abderrezzak, *Généralisation d'identités de Carlitz, Howard et Lehmer*, Aequationes Math. **49** (1995), 36–46.
2. A. Berele and A. Regev, *Hook Young diagrams with applications to combinatorics and to representations of Lie superalgebras*, Adv. Math. **64** (1987), 118–175.
3. N. Bergeron and A. M. Garsia, *Sergeev's formula and the Littlewood–Richardson rule*, Linear and Multilinear Algebra **27** (1990), 79–100.
4. L. C. Biedenharn and J. D. Louck, *A new class of symmetric polynomials defined in terms of tableaux*, Adv. in Appl. Math. **10** (1989), 396–438.
5. _____, *Inhomogeneous basis set of symmetric polynomials defined by tableaux*, Proc. Nat. Acad. Sci. U.S.A. **87** (1990), 1441–1445.
6. W. Y. C. Chen and J. D. Louck, *The factorial Schur function*, J. Math. Phys. **34** (1993), 4144–4160.
7. P. H. Dondi and P. D. Jarvis, *Diagram and superfield techniques in the classical superalgebras*, J. Phys. A **14** (1981), 547–563.
8. I. M. Gessel and G. X. Viennot, *Binomial determinants, paths, and hook length formulas*, Adv. Math. **58** (1985), 300–321.
9. I. Goulden and C. Greene, *A new tableau representation for supersymmetric Schur functions*, J. Algebra **170** (1994), 687–703.
10. I. P. Goulden and A. M. Hamel, *Shift operators and factorial symmetric functions*, J. Combin. Theory Ser. A **69** (1995), 51–60.
11. R. Howe, *Remarks on classical invariant theory*, Trans. Amer. Math. Soc. **313** (1989), 539–570.
12. R. Howe and T. Umeda, *The Capelli identity, the double commutant theorem, and multiplicity-free actions*, Math. Ann. **290** (1991), 569–619.
13. J. van der Jeugt, J. W. B. Hughes, R. C. King, and J. Thierry-Mieg, *Character formulae for irreducible modules of the Lie superalgebra $sl(m|n)$*, J. Math. Phys. **31** (1990), 2278–2304.
14. A.-A. A. Jucys, *Symmetric polynomials and the center of the symmetric group ring*, Rep. Math. Phys. **5** (1974), 107–112.
15. V. G. Kac, *Lie superalgebras*, Adv. Math. **26** (1977), 8–96.
16. _____, *Representations of classical Lie superalgebras*, Differential Geometry Methods in Mathematical Physics II (K. Bleuer, H. R. Petry, and A. Reetz, eds.), Lecture Notes in Math., vol. 676, Springer-Verlag, Berlin–Heidelberg–New York, 1978, pp. 597–626.
17. A. Lascoux, Letter to the author.
18. _____, *Classes de Chern des variétés de drapeaux*, C. R. Acad. Sci. Paris Sér. I **295** (1982), 393–398.
19. D. E. Littlewood, *The theory of group characters and matrix representations of groups*, 2nd edition, Clarendon Press, Oxford, 1950.
20. I. G. Macdonald, *Symmetric functions and Hall polynomials*, 2nd edition, Oxford University Press, Oxford, 1995.
21. _____, *Schur functions: theme and variations*, Actes 28-e Séminaire Lotharingien, Publ. I.R.M.A. Strasbourg, 498/S-27, 1992, pp. 5–39.
22. _____, *Notes on Schubert polynomials*, Publ. LACIM, Univ. du Québec à Montréal, 1991.
23. N. Metropolis, G. Nicoletti, and G. C. Rota, *A new class of symmetric functions*, Mathematical Analysis and Applications, Advances in Mathematics Supplementary Studies, vol. 7B, Academic Press, New York, 1981, pp. 563–575.
24. A. Molev and M. Nazarov, *Capelli identities for classical Lie algebras*, Preprint CMA MRR 003-97, Australian National University (1997), Canberra.

25. A. I. Molev, M. L. Nazarov, and G. I. Olshanskiĭ, *Yangians and classical Lie algebras*, Uspekhi Mat. Nauk **51** (1996), no. 2, 27–104; English transl. in Russian Math. Surveys **51** (1996), no. 2, 205–282.
26. G. E. Murphy, *A new construction of Young's seminormal representation of the symmetric group*, J. Algebra **69** (1981), 287–291.
27. M. L. Nazarov, *Quantum Berezinian and the classical Capelli identity*, Lett. Math. Phys. **21** (1991), 123–131.
28. _____, *Yangians and Capelli identities*, In this volume.
29. A. Okounkov, *Quantum immanants and higher Capelli identities*, Transformation Groups **1** (1996), 99–126.
30. _____, *Young basis, Wick formula and higher Capelli identities*, Int. Math. Research Notes (1996), 817–839.
31. A. Okounkov and G. Olshanskiĭ, *Shifted Schur functions*, Algebra i Analiz **9** (1997), no. 2, 79–147; English transl. in St. Petersburg Math. J. **9** (1998).
32. G. I. Olshanskiĭ, *Quasi-symmetric functions and factorial Schur functions*, Preprint (1995).
33. P. Pragacz, *Algebro-geometric applications of Schur S- and Q-polynomials*, Topics in Invariant Theory, Seminaire d'Algebre Paul Dubriel et Marie-Paule Malliavin, Proceedings, Lecture Notes in Math., vol. 1478, Springer-Verlag, New York–Berlin, 1991, pp. 130–191.
34. P. Pragacz and A. Thorup, *On a Jacobi–Trudi identity for supersymmetric polynomials*, Adv. Math. **95** (1992), 8–17.
35. J. B. Remmel, *The combinatorics of (k,l)-hook Schur functions*, Contemp. Math. **34** (1984), 253–287.
36. B. Sagan, *The symmetric group: representations, combinatorial algorithms, and symmetric functions*, Wadsworth & Brooks, Pacific Grove, CA, 1991.
37. S. Sahi, *The spectrum of certain invariant differential operators associated to a Hermitian symmetric space*, Lie Theory and Geometry, Progr. Math., vol. 123, Birkhauser, Boston, 1994, pp. 569–576.
38. M. Scheunert, *Casimir elements of Lie superalgebras*, Differential Geometry Methods in Mathematical Physics, Reidel, Dordrecht, 1984, pp. 115–124.
39. R. P. Stanley, *Unimodality and Lie superalgebras*, Stud. Appl. Math. **72** (1985), 263–281.
40. J. R. Stembridge, *A characterization of supersymmetric polynomials*, J. Algebra **95** (1985), 439–444.

CENTRE FOR MATHEMATICS AND ITS APPLICATIONS, AUSTRALIAN NATIONAL UNIVERSITY, CANBERRA, ACT 0200, AUSTRALIA

E-mail address: molev@pell.anu.edu.au

Yangians and Capelli Identities

Maxim Nazarov

To Professor Alexander Kirillov on the occasion of his sixtieth birthday

ABSTRACT. We study the image of the universal R-matrix for the Yangian $Y(\mathfrak{gl}_N)$ with respect to the evaluation homomorphism of $Y(\mathfrak{gl}_N)$ to the enveloping algebra $U(\mathfrak{gl}_N)$. We use the fusion procedure as defined by I. Cherednik. As a corollary we get a generalization of the classical Capelli identity.

§1. Introduction

In this article we apply the representation theory of Yangians to classical invariant theory. Let us consider the action of the Lie algebra $\mathfrak{gl}_N \times \mathfrak{gl}_M$ in the space $\mathcal{P}$ of polynomial functions on $\mathbb{C}^N \otimes \mathbb{C}^M$. This action is multiplicity-free, and its irreducible components are parametrized by Young diagrams λ with at most $\min(M, N)$ rows. As a result, the space $\mathcal{I}$ of $\mathfrak{gl}_N \times \mathfrak{gl}_M$-invariant differential operators on $\mathbb{C}^N \otimes \mathbb{C}^M$ with polynomial coefficients splits into the direct sum of one-dimensional subspaces parametrized by the diagrams λ. It is easy to describe these subspaces.

Let x_{ia}, with $i = 1, \ldots, N$ and $a = 1, \ldots, M$, be the standard coordinates on the vector space $\mathbb{C}^N \otimes \mathbb{C}^M$. Let ∂_{ia} be the partial derivation with respect to the coordinate x_{ia}. Suppose that the diagram λ consists of n boxes. Let χ_λ be the irreducible character of the symmetric group S_n parametrized by λ. Then the one-dimensional subspace in $\mathcal{I}$ corresponding to λ is spanned by the operator

$$(1.1) \qquad \sum_{\sigma \in S_n} \sum_{i_1,\ldots,i_n} \sum_{a_1,\ldots,a_n} \frac{\chi_\lambda(\sigma)}{n!} \, x_{i_1 a_1} \cdots x_{i_n a_n} \, \partial_{i_{\sigma(1)} a_1} \cdots \partial_{i_{\sigma(n)} a_n},$$

where the indices $i_1, \ldots, i_n$ and $a_1, \ldots, a_n$ run through $1, \ldots, N$ and $1, \ldots, M$.

On the other hand, the action of the Lie algebra $\mathfrak{gl}_N$ in the space $\mathcal{P}$ extends to the action of the universal enveloping algebra $U(\mathfrak{gl}_N)$ by differential operators with polynomial coefficients. The space $\mathcal{I}$ is then the image of the center of $U(\mathfrak{gl}_N)$; see for instance [**HU**]. In this article we give an explicit formula for the central element of $U(\mathfrak{gl}_N)$ corresponding to the operator (1.1). For the case when the diagram λ

1991 *Mathematics Subject Classification.* Primary 17B35, 22E46; Secondary 15A72, 81R50.

©1998 American Mathematical Society

has only one column, this formula was discovered by Capelli [C]. For the other particular case in which λ consists of only one row, this formula was found in [N1].

To derive an explicit formula for the general diagram λ, we employ the representation theory of the Yangian $Y(\mathfrak{gl}_N)$ of the Lie algebra $\mathfrak{gl}_N$. The Yangian $Y(\mathfrak{gl}_N)$ is a canonical deformation of the universal enveloping algebra $U(\mathfrak{gl}_N[z])$ in the class of Hopf algebras [D1]. Moreover, it contains $U(\mathfrak{gl}_N)$ as a subalgebra and admits a homomorphism $\pi: Y(\mathfrak{gl}_N) \to U(\mathfrak{gl}_N)$ identical on $U(\mathfrak{gl}_N)$. Thus the irreducible representation π_λ of the Lie algebra $\mathfrak{gl}_N$ corresponding to λ can be regarded as a representation of the algebra $Y(\mathfrak{gl}_N)$.

We use the notion of the universal R-matrix for the Hopf algebra $Y(\mathfrak{gl}_N)$, cf. [D1]. Let $\Delta: Y(\mathfrak{gl}_N) \to Y(\mathfrak{gl}_N) \otimes Y(\mathfrak{gl}_N)$ be the comultiplication of $Y(\mathfrak{gl}_N)$. Denote by Δ' the composition of Δ with permutation of tensor factors in $Y(\mathfrak{gl}_N) \otimes Y(\mathfrak{gl}_N)$. The algebra $Y(\mathfrak{gl}_N)$ has a canonical family of automorphisms τ_z parametrized by $z \in \mathbb{C}$. The universal R-matrix for $Y(\mathfrak{gl}_N)$ is a formal power series $\mathcal{R}(z)$ in z^{-1} with coefficients from $Y(\mathfrak{gl}_N) \otimes Y(\mathfrak{gl}_N)$ and leading term 1 such that

$$(1.2) \qquad \mathcal{R}(z) \cdot \mathrm{id} \otimes \tau_z(\Delta'(Y)) = \mathrm{id} \otimes \tau_z(\Delta(Y)) \cdot \mathcal{R}(z), \qquad Y \in Y(\mathfrak{gl}_N).$$

For the description of the series $\mathcal{R}(z)$ see §3. The image $\pi_\lambda \otimes \pi(\mathcal{R}(z))$ is a rational function in z with at most n poles. These poles are contained in the collection of the contents $c_1, \ldots, c_n$ of the diagram λ (see §2). Denote by ψ_λ the trace of the representation π_λ of $Y(\mathfrak{gl}_N)$ and consider the polynomial in z

$$(1.3) \qquad \psi_\lambda \otimes \pi(\mathcal{R}(z)) \cdot (z - c_1) \cdots (z - c_n)$$

with values in the algebra $U(\mathfrak{gl}_N)$; cf. [D2, RS]. We prove that the value of polynomial (1.3) at $z = 0$ is the central element corresponding to the operator (1.1).

There is an explicit formula for the polynomial (1.3). Denote by Λ^c the Young tableau obtained by filling the boxes of the diagram λ with the numbers $1, \ldots, n$ by columns. Suppose that $c_1, \ldots, c_n$ are the contents of the respective boxes of λ. Denote by S_λ and T_λ the subgroups in S_n preserving the collections of numbers appearing respectively in every row and column of the tableau Λ^c. Consider the element of the group ring $\mathbb{C} \cdot S_n$

$$y \cdot \sum_{\sigma \in S_\lambda} \sum_{\rho, \rho' \in T_\lambda} \rho \sigma \rho' \, \mathrm{sgn}(\rho \rho') = \sum_{\sigma \in S_n} y_\sigma \sigma, \qquad y_\sigma \in \mathbb{C},$$

where the nonzero factor $y \in \mathbb{C}$ is chosen to make this element of $\mathbb{C} \cdot S_n$ an idempotent.

The Lie algebra $\mathfrak{gl}_N$ acts in space $\mathbb{C}^N \otimes \mathbb{C}^M$ by linear combinations of the operators

$$\sum_{1 \leqslant a \leqslant M} x_{ia} \partial_{ja}, \qquad i, j = 1, \ldots, N.$$

Then the differential operator corresponding to the value of (1.3) at $z = 0$ is

$$(1.4) \qquad \sum_{\sigma \in S_n} \sum_{i_1, \ldots, i_n} \sum_{a_1, \ldots, a_n} y_\sigma \cdot \overrightarrow{\prod_k} (x_{i_k a_k} \partial_{i_{\sigma(k)} a_k} - c_k \cdot \delta_{i_k i_{\sigma(k)}}),$$

where the index k runs through $1, \ldots, n$ and factors in the ordered product are arranged from the left to right while k increases. Here δ_{ij} is the Kronecker delta.

The equality of the differential operators (1.1) and (1.4) was proved in [**O1**]. It can be also verified by using the Jucys–Murphy elements in the group ring of S_n; see the forthcoming article [**O2**]. The present article contains a proof of that equality going back to the origin (1.3) of the formula (1.4). This proof is based on the estimation of the order of the pole at $z = 0$ of the rational function $\pi_\lambda \otimes \pi_\mu(\mathcal{R}(z))$ for diagrams μ with less than n boxes; cf. [**OO, S**]. That estimation is performed in §4. We employ the notion of fusion procedure for the symmetric group S_n introduced by Cherednik [**C1**]. In §2 we give a concise account of the fusion procedure; see [**JKMO**] for another exposition of relevant results from [**C1**].

I am very grateful to I. Cherednik for numerous illuminating conversations. I am also grateful to F. Knop, M. Noumi, and V. Tolstoy for stimulating discussions and valuable remarks. This work is a part of the project on representation theory of Yangians which started at A. Kirillov's seminar in Moscow. I thank all the participants of that seminar for providing a favorable atmosphere for the start. I am especially indebted to A. Okounkov and G. Olshanski. Their results from [**O1, OO**] have inspired the present work.

§2. Fusion procedure for the symmetric group

We start by recalling several classical facts about irreducible modules of the symmetric group S_n over the complex field. They are parametrized by the Young diagrams λ with exactly n boxes. We shall denote by U_λ the irreducible module of S_n corresponding to the diagram λ. Consider the chain of subgroups

$$S_1 \subset S_2 \subset \cdots \subset S_n$$

with respect to the standard embeddings. There is a canonical decomposition of the space U_λ into the direct sum of one-dimensional subspaces associated with this chain. These subspaces are parametrized by the *Young tableaux* of shape λ. Each of these tableaux is a bijective filling of the boxes of λ with numbers $1, \ldots, n$ such that in every row and column the numbers increase from left to right and from top to bottom, respectively. Denote by $\mathcal{T}_\lambda$ the set of these tableaux.

For every tableau $\Lambda \in \mathcal{T}_\lambda$ denote by U_Λ the corresponding one-dimensional subspace in U_λ. For any $m \in \{1, \ldots, n\}$ consider the tableau obtained from Λ by removing each of the numbers $m+1, \ldots, n$. Let the Young diagram μ be its shape. Then U_Λ is contained in an irreducible S_m-submodule of U_λ corresponding to μ. Any basis of U_λ formed by vectors $u_\Lambda \in U_\Lambda$ is called a *Young basis*. Let us fix an S_n-invariant inner product $\langle \ , \ \rangle$ on U_λ. Then the subspaces U_Λ are pairwise orthogonal. We shall assume that $\langle u_\Lambda, u_\Lambda \rangle = 1$ for each tableau $\Lambda \in \mathcal{T}_\lambda$.

Consider the *column tableau* of the shape λ obtained by filling the boxes of λ with $1, \ldots, n$ by columns from left to right, downward in each column. We denote this tableau by Λ^c. Consider the diagonal matrix element of the S_n-module U_λ corresponding to the vector u_{Λ^c},

$$(2.1) \qquad \Phi_\lambda = \sum_{\sigma \in S_n} \langle \sigma u_{\Lambda^c}, u_{\Lambda^c} \rangle \sigma^{-1} \in \mathbb{C} \cdot S_n.$$

We use the explicit formula for this matrix element contained in [**Y1**]. Denote by S_λ and T_λ the subgroups in S_n preserving the collections of numbers appearing respectively in every row and column of the tableau Λ^c. Put

$$P_\lambda = \sum_{\sigma \in S_\lambda} \sigma, \qquad Q_\lambda = \sum_{\sigma \in T_\lambda} \operatorname{sgn}(\sigma)\,\sigma\,.$$

As usual, let $\lambda'_1, \lambda'_2, \ldots$ be the numbers of boxes in the columns of the diagram λ. Then

$$\Phi_\lambda = \frac{Q_\lambda P_\lambda Q_\lambda}{\lambda'_1!\,\lambda'_2!\,\cdots}\,.$$

We shall also need an expression of a different kind for the matrix element Φ_λ. For any distinct $i,j = 1, \ldots, n$ let $(i\,j)$ be the transposition in the symmetric group S_n. Consider the rational function of two complex variables u, v with values in the group ring $\mathbb{C} \cdot S_n$

$$\varphi_{ij}(u,v) = 1 - \frac{(i\,j)}{u - v}\,.$$

As a direct calculation shows, this rational function satisfies the equations

(2.2) $\qquad \varphi_{ij}(u,v)\,\varphi_{ik}(u,w)\,\varphi_{jk}(v,w) = \varphi_{jk}(v,w)\,\varphi_{ik}(u,w)\,\varphi_{ij}(u,v)$

for all pairwise distinct i, j, k. Evidently, for all pairwise distinct i, j, k, l, we have

(2.3) $\qquad \varphi_{ij}(u,v)\,\varphi_{kl}(z,w) = \varphi_{kl}(z,w)\,\varphi_{ij}(u,v)\,.$

Consider the rational function of u, v, w appearing at either side of (2.2). Denote this function by $\varphi_{ijk}(u,v,w)$. The factor $\varphi_{ik}(u,w)$ in (2.2) has a pole at $u = w$. However, we have the following lemma.

LEMMA 2.1. *The restriction of $\varphi_{ijk}(u,v,w)$ to the set of (u,v,w) such that $u = v \pm 1$, is regular at $u = w$.*

PROOF. As a direct calculation shows, for the restriction we have

$$\varphi_{ijk}(v \pm 1, v, w) = (1 \mp (i\,j)) \cdot \left(1 - \frac{(i\,k) + (j\,k)}{v - w}\right).$$

The latter rational function of v, w is manifestly regular at $w = v \pm 1$. $\square$

For each $i \in \{1, \ldots, n\}$ we set $c_i = s - t$ if the number i appears in the sth column and the tth row of the tableau Λ^c. The difference $s - t$ is called the *content* of the box of the diagram λ occupied by the number i. For each i let z_i be a complex parameter. Equip the set of all pairs (i,j), where $1 \leqslant i < j \leqslant n$, with the lexicographical ordering. Introduce the ordered product over this set

(2.4) $\qquad \displaystyle\overrightarrow{\prod_{(i,j)}} \varphi_{ij}(c_i + z_i, c_j + z_j)\,.$

Consider this product as a rational function of the parameters $z_1, \ldots, z_n$ with values in $\mathbb{C} \cdot S_n$. Denote by $\Phi_\lambda(z_1, \ldots, z_n)$ this rational function. Denote by $\mathcal{Z}$ the set of all tuples $(z_1, \ldots, z_n)$ such that $z_i = z_j$ whenever the numbers i and j appear in the same row of the tableau Λ^c. The following theorem goes back to [**C1**] and [**J**].

THEOREM 2.2. *The restriction of $\Phi_\lambda(z_1,\ldots,z_n)$ to $\mathcal{Z}$ is regular at $z_1 = \cdots = z_n$. The value of this restriction at $z_1 = \cdots = z_n$ coincides with the matrix element Φ_λ.*

We shall present the main steps of the proof as separate propositions. This proof follows [**N2**] and is based on Lemma 2.1. Another proof is contained in [**JKMO**].

PROPOSITION 2.3. *The restriction of $\Phi_\lambda(z_1,\ldots,z_n)$ to $\mathcal{Z}$ is regular at $z_1 = \cdots = z_n$.*

PROOF. We shall provide an expression for the restriction of the function (2.4) to $\mathcal{Z}$ which is manifestly regular at $z_1 = \cdots = z_n$. Let us reorder the pairs (i,j) in the product (2.4) as follows. (This reordering will not affect the value of the product due to relations (2.2) and (2.3).) Let $\mathcal{C}$ be the sequence of numbers obtained by reading the tableau Λ^c in the usual way, that is by rows from top to bottom, eastward in every row. For each $j \in \{1,\ldots,n\}$ denote by $\mathcal{B}_j$ and $\mathcal{A}_j$ the subsequences of $\mathcal{C}$ consisting of all numbers $i < j$ that appear respectively before and after j in that sequence. Now set $(i,j) \prec (k,l)$ if one of the following conditions is satisfied:
  (i) the number i appears in $\mathcal{B}_j$ while k appears in $\mathcal{A}_l$;
  (ii) the numbers i and k appear respectively in $\mathcal{B}_j$ and $\mathcal{B}_l$ where $j < l$;
  (iii) the numbers i and k appear respectively in $\mathcal{A}_j$ and $\mathcal{A}_l$ where $j > l$;
  (iv) we have the equality $j = l$ and i appears before k in $\mathcal{B}_j$ or $\mathcal{A}_j$.

From now on we assume that the factors in (2.4) corresponding to the pairs (i,j) are arranged with respect to this new ordering. The factor $\varphi_{ij}(c_i + z_i, c_j + z_j)$ has a pole at $z_i = z_j$ if and only if the numbers i and j stand on the same diagonal of the tableau Λ^c. Such a pair (i,j) will be called *singular*. Note that the number i occurs in the subsequence $\mathcal{A}_j$ exactly when i stands to the left and below j in the tableau Λ^c. In this case $c_j - c_i > 1$ and the pair (i,j) cannot be singular.

Let a singular pair (i,j) be fixed. Suppose that the number i appears in the sth column and the tth row of the tableau Λ^c. In our new ordering, the next pair after (i,j) is (h,j), where the number h appears in the $(s+1)$th column and the tth row of Λ^c. In particular, we have $c_i = c_j = c_h - 1$. Moreover, $(i,h) \prec (i,j)$. Due to relations (2.2), (2.3), the product

$$\overrightarrow{\prod_{(k,l)\prec(i,j)}} \varphi_{kl}(c_k + z_k, c_l + z_l)$$

is divisible on the right by $\varphi_{ih}(c_i + z_i, c_h + z_h)$. The restriction of the latter function to $z_i = z_h$ is just $1 + (i\,h)$. Note that the element $(1 + (i\,h))/2$ is an idempotent.

Now for each singular pair (i,j) let us replace the two adjacent factors in (2.4)

$$\varphi_{ij}(c_i + z_i, c_j + z_j)\varphi_{hj}(c_h + z_h, c_j + z_j)$$

by

$$\varphi_{ih}(c_i + z_i, c_h + z_h)\varphi_{ij}(c_i + z_i, c_j + z_j)\varphi_{hj}(c_h + z_h, c_j + z_j)/2$$
$$= \varphi_{ihj}(c_i + z_i, c_h + z_h, c_j + z_j)/2.$$

This replacement does not affect the value of the restriction of the function (2.4) to $\mathcal{Z}$. But the restriction of the function $\varphi_{ihj}(c_i + z_i, c_h + z_h, c_j + z_j)$ to $z_i = z_h$ is regular at $z_i = z_j$ by Lemma 2.1. □

In our new ordering we have the decomposition

$$\Phi_\lambda(z_1,\ldots,z_n) = \Upsilon_\lambda(z_1,\ldots,z_n)\Theta_\lambda(z_1,\ldots,z_n),$$

where $\Upsilon_\lambda(z_1,\ldots,z_n)$ and $\Theta_\lambda(z_1,\ldots,z_n)$ are products of the factors in (2.4) that correspond to the pairs (i,j) with i appearing in $\mathcal{B}_j$ and $\mathcal{A}_j$, respectively. The function $\Theta_\lambda(z_1,\ldots,z_n)$ is regular at $z_1 = \cdots = z_n$. Moreover, the value Θ of this function at $z_1 = \cdots = z_n$ is invertible in $\mathbb{C}\cdot S_n$. Denote by Φ and Υ the values at $z_1 = \cdots = z_n$ of the restrictions to $\mathcal{Z}$ of $\Phi_\lambda(z_1,\ldots,z_n)$ and $\Upsilon_\lambda(z_1,\ldots,z_n)$, respectively. Thus we have the equality $\Phi = \Upsilon\Theta$. We shall prove that $\Phi = \Phi_\lambda$.

PROPOSITION 2.4. *Let the numbers k and l stand next to each other in a row of the tableau Λ^c. Then the element Υ in $\mathbb{C}\cdot S_n$ is divisible on the right by $1+(k\,l)$.*

PROOF. We can assume that $k < l$. Then relations (2.2) and (2.3) imply that the product $\Upsilon_\lambda(z_1,\ldots,z_n)$ is divisible on the right by $\varphi_{kl}(c_k + z_k, c_l + z_l)$. The restriction of the latter function to $z_k = z_l$ is exactly the element $1 + (k\,l)$. □

Let α be the involutive antiautomorphism of the group ring $\mathbb{C}\cdot S_n$ defined by $\alpha(g) = g^{-1}$ for $g \in S_n$. By relations (2.2) and (2.3), the product (2.4) is invariant with respect to this antiautomorphism. So is the value Φ of its restriction to $\mathcal{Z}$.

COROLLARY 2.5. *Let the numbers $k < l$ stand next to each other in the first row of the tableau Λ^c. Then the element Φ in $\mathbb{C}\cdot S_n$ is divisible on the right by*

$$(2.5) \qquad \varphi_{kl}(c_k, c_l) \cdot \prod_{k<m<l}^{\rightarrow} \varphi_{ml}(c_m, c_l).$$

The element Φ is then also divisible on the left by

$$(2.6) \qquad \prod_{k<m<l}^{\leftarrow} \varphi_{ml}(c_m, c_l) \cdot 0\varphi_{kl}(c_k, c_l).$$

PROOF. By Proposition 2.4, Υ is divisible on the right by $\varphi_{kl}(c_k, c_l)$. But by relations (2.2) and (2.3), the product $\varphi_{kl}(c_k,c_l)\cdot\Theta$ is divisible on the right by (2.5). Since $\Phi = \Upsilon\Theta$, we obtain the first statement of Corollary 2.5. Note that the image of (2.5) with respect to the antiautomorphism α is (2.6). Since Φ is invariant with respect to α, we get the second statement of Corollary 2.5. □

Again, consider the rational function $\varphi_{ijk}(u,v,w)$ appearing at either side of (2.2). The value at $u = w$ of its restriction to $u = v - 1$ is not divisible on the right by $\varphi_{jk}(v, v-1) = 1 - (j\,k)$, see the proof of Lemma 2.1. In the proof of Proposition 2.7, we use the following obvious observation.

LEMMA 2.6. *The value of the function $\varphi_{ijk}(v-1,v,w)\varphi_{kj}(w,v)$ at $w = v - 1$ coincides with that of the function $-2(i\,k)\cdot\varphi_{kj}(w,v)$.*

The next proposition is the central part of the proof of Theorem 2.2; cf. [**JKMO**].

PROPOSITION 2.7. *Let the numbers k and $k+1$ stand in the same column of the tableau Λ^c. Then the element Υ in $\mathbb{C}\cdot S_n$ is divisible on the left by $1-(k,k+1)$.*

PROOF. Observe first that Proposition 2.7 follows from its particular case $k+1=n$. Indeed, let ν be the shape of the tableau obtained from Λ^c by removing each of the numbers $k+2,\ldots,n$. Then

$$\Upsilon_\lambda(z_1,\ldots,z_n) = \Upsilon_\nu(z_1,\ldots,z_{k+1}) \cdot \overrightarrow{\prod_{(i,j)}} \varphi_{ij}(c_i+z_i, c_j+z_j),$$

where $j = k+2,\ldots,n$ and i runs through the sequence $\mathcal{B}_j$. Denote by Υ' the value at $z_1 = \cdots = z_{k+1}$ of the restriction of $\Upsilon_\nu(z_1,\ldots,z_{k+1})$ to $\mathcal{Z}$. According to our proof of Proposition 2.3, we get $\Upsilon = \Upsilon'\Upsilon''$ for a certain element Υ'' in $\mathbb{C} \cdot S_n$.

From now on we assume that $k+1 = n$. Since the element Θ is invertible, it suffices to prove that $\Upsilon\Theta = \Phi$ is divisible by $1-(n-1,n)$ on the left. But the element Φ is invariant with respect to the antiautomorphism α. We prove that Φ is divisible by $1-(n-1,n)$ on the right. This is equivalent to the formula

(2.7) $$\Phi \cdot \varphi_{n,n-1}(c_n, c_{n-1}) = \Phi \cdot (1+(n-1,n)) = 0.$$

Suppose that the number n appears in the sth column and the tth row of the tableau Λ^c. Let $i_1,\ldots,i_s$ be all the numbers in the tth row. So we have $i_s = n$. Then due to (2.2) and (2.3), for a certain element Θ' in $\mathbb{C} \cdot S_n$ we have

$$\Theta \cdot \varphi_{n,n-1}(c_n, c_{n-1}) = \overrightarrow{\prod_{p<s}} \varphi_{i_p, n-1}(c_{i_p}, c_{n-1}) \cdot \varphi_{n,n-1}(c_n, c_{n-1}) \cdot \Theta'.$$

Therefore to get (2.7) we must prove that

(2.8) $$\Upsilon \cdot \overrightarrow{\prod_{p<s}} \varphi_{i_p, n-1}(c_{i_p}, c_{n-1}) \cdot \varphi_{n,n-1}(c_n, c_{n-1}) = 0.$$

We prove this by induction on s. If $s=1$, then $\Upsilon_\lambda(z_1,\ldots,z_n) = \Phi_\lambda(z_1,\ldots,z_n)$, so that $\Upsilon = \Phi$. Moreover, then none of the pairs (i,j) in (2.4) is singular. Then Φ has the form

$$\Phi' \cdot \varphi_{n-1,n}(c_{n-1}, c_n) = \Phi' \cdot (1-(n-1,n))$$

for a certain element Φ' in $\mathbb{C} \cdot S_n$. So we get the equality $\Upsilon \cdot \varphi_{n,n-1}(c_n, c_{n-1}) = 0$.

Now suppose that $s > 1$. We must prove that the restriction of the product

(2.9) $$\Upsilon_\lambda(z_1,\ldots,z_n) \cdot \overrightarrow{\prod_{p \leqslant s}} \varphi_{i_p, n-1}(c_{i_p}+z_{i_p}, c_{n-1}+z_{n-1})$$

to $\mathcal{Z}$ vanishes at $z_1 = \cdots = z_n$. Denote $i_{s-1} = m$. The number $m-1$ appears in the $(s-1)$th column and the $(t-1)$th row of Λ^c. So we have $c_{m-1} = c_n$. Let μ be the shape of the tableau obtained from Λ^c by removing each of the numbers $m+1,\ldots,n$. Then the function $\Upsilon_\lambda(z_1,\ldots,z_n)$ has the form

$$\Upsilon_\mu(z_1,\ldots,z_m) \Psi(z_1,\ldots,z_{n-1}) \varphi_{m-1,n-1}(c_{m-1}+z_{m-1}, c_{n-1}+z_{n-1})$$
$$\times \mathrm{X}(z_1,\ldots,z_n) \varphi_{m-1,n}(c_{m-1}+z_{m-1}, c_n+z_n)$$
$$\times \varphi_{n-1,n}(c_{n-1}+z_{n-1}, c_n+z_n) \cdot \overrightarrow{\prod_{p<s}} \varphi_{i_p n}(c_{i_p}+z_{i_p}, c_n+z_n).$$

Here we have denoted by $\Psi(z_1,\ldots,z_{n-1})$ the product

(2.10) $$\prod_{(i,j)}^{\rightarrow} \varphi_{ij}(c_i + z_i, c_j + z_j), \qquad j = m+1,\ldots,n-1,$$

where i runs through $\mathcal{B}_j$, but $(i,j) \neq (m-1, n-1)$. Here we have also denoted

(2.11) $$\mathrm{X}(z_1,\ldots,z_n) = \prod_{(i,n)}^{\rightarrow} \varphi_{in}(c_i + z_i, c_n + z_n),$$

where i runs over the sequence $\mathcal{B}_n$ but $i \notin \{m-1, n-1,\ldots,m\}$. In particular, any factor in the product (2.11) commutes with

$$\varphi_{m-1,n-1}(c_{m-1} + z_{m-1}, c_{n-1} + z_{n-1})$$

because of (2.3). Therefore the product (2.9) takes the form

(2.12)
$$\Upsilon_\mu(z_1,\ldots,z_m)\, \Psi(z_1,\ldots,z_{n-1})\, \mathrm{X}(z_1,\ldots,z_n)$$
$$\times \varphi_{m-1,n-1,n}(c_{m-1} + z_{m-1}, c_{n-1} + z_{n-1}, c_n + z_n)$$
$$\times \prod_{p<s}^{\rightarrow} \varphi_{i_p n}(c_{i_p} + z_{i_p}, c_n + z_n) \cdot \prod_{p<s}^{\rightarrow} \varphi_{i_p, n-1}(c_{i_p} + z_{i_p}, c_{n-1} + z_{n-1})$$
$$\times \varphi_{n,n-1}(c_n + z_n, c_{n-1} + z_{n-1})$$
$$= \Upsilon_\mu(z_1,\ldots,z_m)\, \Psi(z_1,\ldots,z_{n-1})\, \mathrm{X}(z_1,\ldots,z_n)$$
$$\times \varphi_{m-1,n-1,n}(c_{m-1} + z_{m-1}, c_{n-1} + z_{n-1}, c_n + z_n)$$
$$\times \varphi_{n,n-1}(c_n + z_n, c_{n-1} + z_{n-1}) \cdot \prod_{p<s}^{\rightarrow} \varphi_{i_p, n-1}(c_{i_p} + z_{i_p}, c_{n-1} + z_{n-1})$$
$$\times \prod_{p<s}^{\rightarrow} \varphi_{i_p n}(c_{i_p} + z_{i_p}, c_n + z_n).$$

To get the latter equality, we used relations (2.2) and (2.3). The restriction to $\mathcal{Z}$ of the product of factors in the first line of (2.12) is regular at $z_1 = \cdots = z_n$ according to our proof of Proposition 2.3. Each of the factors in the third line is also regular at $z_1 = \cdots = z_n$. Therefore by Lemma 2.6 the restriction of (2.12) to $\mathcal{Z}$ has the same value at $z_1 = \cdots = z_n$ as the restriction to $\mathcal{Z}$ of

(2.13)
$$-2\Upsilon_\mu(z_1,\ldots,z_m)\, \Psi(z_1,\ldots,z_{n-1})\, \mathrm{X}(z_1,\ldots,z_n)$$
$$\times (m-1,n) \cdot \varphi_{n,n-1}(c_n + z_n, c_{n-1} + z_{n-1})$$
$$\times \prod_{p<s}^{\rightarrow} \varphi_{i_p, n-1}(c_{i_p} + z_{i_p}, c_{n-1} + z_{n-1}) \cdot \prod_{p<s}^{\rightarrow} \varphi_{i_p n}(c_{i_p} + z_{i_p}, c_n + z_n)$$
$$= -2\Upsilon_\mu(z_1,\ldots,z_m)\, \Psi(z_1,\ldots,z_{n-1})\, \mathrm{X}(z_1,\ldots,z_n)$$
$$\times \prod_{p<s}^{\rightarrow} \varphi_{i_p, m-1}(c_{i_p} + z_{i_p}, c_n + z_n) \cdot \prod_{p<s}^{\rightarrow} \varphi_{i_p, n-1}(c_{i_p} + z_{i_p}, c_{n-1} + z_{n-1})$$
$$\times (m-1,n) \cdot \varphi_{n,n-1}(c_n + z_n, c_{n-1} + z_{n-1}).$$

Here each factor $\varphi_{i_p, m-1}(c_{i_p} + z_{i_p}, c_n + z_n)$ commutes with $\mathrm{X}(z_1,\ldots,z_n)$ due to (2.3). In each of these factors we can replace $c_n + z_n$ by $c_{m-1} + z_{m-1}$ without affecting the value at $z_1 = \cdots = z_n$ of the restriction of (2.13) to $\mathcal{Z}$. Denote

$$\Gamma(z_1, \ldots, z_m) = \overrightarrow{\prod_{p<s}} \varphi_{i_p, m-1}(c_{i_p} + z_{i_p}, c_{m-1} + z_{m-1}).$$

According to our proof of Proposition 2.3, it now suffices to demonstrate the vanishing at $z_1 = \cdots = z_{n-1}$ of the restriction to $\mathcal{Z}$ of the product

(2.14) $$\Upsilon_\mu(z_1, \ldots, z_m) \Psi(z_1, \ldots, z_{n-1}) \Gamma(z_1, \ldots, z_m).$$

Consider the product (2.10). Here the factors corresponding to the pairs (i, j) are arranged with respect to ordering chosen in the proof of Proposition 2.3. Let us now reorder the pairs (i, j) in (2.10) as follows. For every $j = m+1, \ldots, m+\lambda'_{s-1}-t$ change the sequence

$$(m-1, j), (i_1, j), \ldots, (i_{s-1}, j) \quad \text{to} \quad (i_1, j), \ldots, (i_{s-1}, j), (m-1, j).$$

Denote by $\Psi'(z_1, \ldots, z_{n-1})$ the resulting ordered product. Then by (2.2) and (2.3)

(2.15) $$\Psi(z_1, \ldots, z_{n-1}) \Gamma(z_1, \ldots, z_m) = \Gamma(z_1, \ldots, z_m) \Psi'(z_1, \ldots, z_{n-1}).$$

Now let (i, j) be any singular pair in (2.10). Let (h, j) be the pair following (i, j) in the ordering from the proof of Proposition 2.3. Then (h, j) follows (i, j) in our new ordering as well. Furthermore, by relations (2.2) and (2.3) the product

(2.16) $$\Upsilon_\mu(z_1, \ldots, z_m) \Gamma(z_1, \ldots, z_m)$$

is divisible on the right by $\varphi_{ih}(c_i + z_i, c_h + z_h)$. Therefore

(2.17) $$\begin{aligned} &\Upsilon_\mu(z_1, \ldots, z_m) \Gamma(z_1, \ldots, z_m) \Psi'(z_1, \ldots, z_{n-1}) \\ &= \Upsilon_\mu(z_1, \ldots, z_m) \Gamma(z_1, \ldots, z_m) \Psi''(z_1, \ldots, z_{n-1}), \end{aligned}$$

where $\Psi''(z_1, \ldots, z_{n-1})$ is a rational function whose restriction to $\mathcal{Z}$ is regular at $z_1 = \cdots = z_{n-1}$. But, by the induction assumption, the restriction of (2.16) to $\mathcal{Z}$ vanishes at $z_1 = \cdots = z_m$. Thus by (2.15) and (2.17) the restriction of (2.14) to $\mathcal{Z}$ vanishes at $z_1 = \cdots = z_{n-1}$. □

We have shown that the element $\Phi = \Upsilon \Theta$ of $\mathbb{C} \cdot S_n$ is invariant with respect to the antiautomorphism α.

COROLLARY 2.8. *Let the numbers k and $k+1$ stand in the same column of the tableau Λ^c. Then the element Φ in $\mathbb{C} \cdot S_n$ is divisible on the left and on the right by $\varphi_{k,k+1}(c_k, c_{k+1})$.*

The next proposition completes the proof of Theorem 2.2.

PROPOSITION 2.9. *We have the equality $\Phi = \Phi_\lambda$.*

PROOF. Due to Propositions 2.4 and 2.7, we have the equality $\Upsilon = Q_\lambda X_\lambda P_\lambda$ for some element $X_\lambda \in \mathbb{C} \cdot S_n$. It is a classical fact that here one can assume $X_\lambda \in \mathbb{C}$; see for instance [**GM**]. By Corollary 2.8, the decomposition $\Phi = \Upsilon \Theta$ now implies that Φ equals $Q_\lambda P_\lambda Q_\lambda$ up to a factor from $\mathbb{C}$. We show that in the expansion of $\Phi \in \mathbb{C} \cdot S_n$ with respect to the basis of $g \in S_n$ the coefficient at the identity is 1.

Let g_0 be the element of maximal length in S_n. Consider the product (2.4) with the initial lexicographical ordering of the pairs (i,j). For each $k = 1,\ldots,n-1$ denote

$$\varphi_k(u,v) = \varphi_{k,k+1}(u,v) \cdot (k,k+1) = (k,k+1) - \frac{1}{u-v}.$$

Then

$$\Phi_\lambda(z_1,\ldots,z_n)\, g_0 = \prod_{(i,j)}^{\rightarrow} \varphi_{j-i}(c_i + z_i, c_j + z_j).$$

But

$$g_0 = \prod_{(i,j)}^{\rightarrow} (j-i, j-i+1)$$

is a reduced decomposition in S_n. Therefore in the expansion of the element

$$\Phi_\lambda(z_1,\ldots,z_n)\, g_0 \in \mathbb{C}(z_1,\ldots,z_n) \cdot S_n$$

with respect to the basis of $g \in S_n$, the coefficient at g_0 is 1. So is the coefficient at g_0 in the expansion of the element $\Phi g_0 \in \mathbb{C} \cdot S_n$. □

The continuation of (2.4) to $z_1 = \cdots = z_n$ along the set $\mathcal{Z}$ is called the *fusion procedure*. In the course of the proof of Proposition 2.9 we have established the following fact.

COROLLARY 2.10. *We have the equality* $\Upsilon = Q_\lambda P_\lambda$.

We shall now make a concluding remark about the element $\Upsilon \in \mathbb{C} \cdot S_n$. Consider the *row tableau* of shape λ obtained by filling the boxes of λ with $1,\ldots,n$ by rows downward, from left to right in every row. Denote this tableau by Λ^r. It is obtained from the tableau Λ^c by a certain permutation of the numbers $1,\ldots,n$. Denote this permutation by σ_λ. For the vectors u_{Λ^c}, u_{Λ^r} of the Young basis in U_λ, we have $\langle \sigma_\lambda u_{\Lambda^c}, u_{\Lambda^r} \rangle \neq 0$ due to [Y2]. The element

$$\Upsilon_\lambda = \sum_{\sigma \in S_n} \frac{\langle \sigma u_{\Lambda^c}, u_{\Lambda^r} \rangle}{\langle \sigma_\lambda u_{\Lambda^c}, u_{\Lambda^r} \rangle} \sigma^{-1} \sigma_\lambda$$

of the group ring $\mathbb{C} \cdot S_n$ does not depend on the choice of the vectors u_{Λ^c}, u_{Λ^r}. In fact one has $\Upsilon_\lambda = Q_\lambda P_\lambda$; see for instance [GM]. Thus $\Upsilon = \Upsilon_\lambda$ by Corollary 2.10.

LEMMA 2.11. *The restriction of $\varphi_{ijk}(u,v,w)$ to the set of all (u,v,w) with $w = v \pm 1$ is regular at $u = w$.*

PROOF. As a direct calculation shows, for the restriction we have

$$\varphi_{ijk}(u, v, v \pm 1) = \left(1 - \frac{(i\,j) + (i\,k)}{u-v}\right) \cdot (1 \pm (j\,k)).$$

The latter rational function of u, v is manifestly regular at $u = v \pm 1$. □

We conclude this section with one more proposition. It will be used in the next section. Let the superscript $^\vee$ denote the embedding of the group S_n to S_{n+1} determined by assigning to the transposition of i and j that of $i+1$ and $j+1$.

PROPOSITION 2.12. *We have the following equality of rational functions in u with values in $\mathbb{C} \cdot S_{n+1}$:*

$$(2.18) \qquad \prod_{1 \leqslant i \leqslant n}^{\rightarrow} \varphi_{1,i+1}(u, c_i) \cdot \Phi_\lambda^\vee = \left(1 - \sum_{1 \leqslant i \leqslant n} \frac{(1, i+1)}{u}\right) \cdot \Phi_\lambda^\vee.$$

PROOF. Denote by $Z(u)$ the rational function in the left-hand side of (2.18). The value of this function at $u = \infty$ is $\Phi_\lambda^\vee$. Moreover, the residue of $Z(u)$ at $u = 0$ is

$$-\sum_{1 \leqslant i \leqslant n} (1, i+1) \Phi_\lambda^\vee.$$

It remains to prove that $Z(u)$ has a pole only at $u = 0$ and this pole is simple. Let an index $i \in \{2, \ldots, n\}$ be fixed. The factor $\varphi_{1,i+1}(u, c_i)$ in (2.18) has a pole at $u = c_i$. We prove that when we estimate the order of the pole of $Z(u)$ at $u = c_i$ from above, that factor does not count.

Suppose that the number i does not appear in the first row of the tableau Λ^c. Then the number $i - 1$ appears in Λ^c directly above i. In particular, $c_{i-1} = c_i + 1$. By Corollary 2.8, the element $\Phi_\lambda^\vee$ is divisible on the left by $\varphi_{i,i+1}(c_{i-1}, c_i)$. But the product

$$\varphi_{1i}(u, c_{i-1}) \varphi_{1,i+1}(u, c_i) \varphi_{i,i+1}(c_{i-1}, c_i)$$

is regular at $u = c_i$ due to Lemma 2.11.

Now suppose that the number i appears in the first row of Λ^c. Let the number k be next to the left of i in the first row of Λ^c. Then $c_k = c_i - 1$. By Corollary 2.5, the element $\Phi_\lambda^\vee$ is divisible on the left by

$$(2.19) \qquad \prod_{k < j < i}^{\leftarrow} \varphi_{j+1,i+1}(c_j, c_i) \cdot \varphi_{k+1,i+1}(c_k, c_i).$$

Consider the product of factors in (2.18)

$$(2.20) \qquad \varphi_{1,k+1}(u, c_k) \cdot \prod_{k < j < i}^{\rightarrow} \varphi_{1,j+1}(u, c_j) \cdot \varphi_{1,i+1}(u, c_i).$$

Multiplying the product (2.20) on the right by (2.19) and using (2.2), (2.3), we get

$$\prod_{k < j < i}^{\leftarrow} \varphi_{j+1,i+1}(c_j, c_i) \cdot \varphi_{1,k+1,i+1}(u, c_k, c_i) \cdot \prod_{k < j < i}^{\rightarrow} \varphi_{1,j+1}(u, c_j).$$

The latter product is regular at $u = c_i$ by Lemma 2.11. The proof is complete. □

§3. The Yangian of the general linear Lie algebra

In this section we shall collect several known facts from [**C2, D1**] about the *Yangian* of the Lie algebra $\mathfrak{gl}_N$. This is a complex associative unital algebra $Y(\mathfrak{gl}_N)$ with the countable set of generators $T_{ij}^{(s)}$, where $s = 1, 2, \ldots$ and $i, j = 1, \ldots, N$. The defining relations in the algebra $Y(\mathfrak{gl}_N)$ are

$$(3.1) \quad [T_{ij}^{(r+1)}, T_{kl}^{(s)}] - [T_{ij}^{(r)}, T_{kl}^{(s+1)}] = T_{kj}^{(r)} T_{il}^{(s)} - T_{kj}^{(s)} T_{il}^{(r)}, \qquad r, s = 0, 1, 2, \ldots,$$

where $T_{ij}^{(0)} = \delta_{ij} \cdot 1$. We shall also use the following matrix form of these relations.

Let $E_{ij} \in \mathrm{End}\,(\mathbb{C}^N)$ be the standard matrix units. We introduce the element

$$P = \sum_{i,j} E_{ij} \otimes E_{ji} \in \mathrm{End}\,(\mathbb{C}^N) \otimes \mathrm{End}\,(\mathbb{C}^N),$$

where the indices i, j run through $1, \ldots, N$. Consider the *Yang R-matrix*, which is the rational function of two complex variables u, v with values in $\mathrm{End}\,(\mathbb{C}^N) \otimes \mathrm{End}\,(\mathbb{C}^N)$

$$R(u,v) = \mathrm{id} - \frac{P}{u-v}.$$

We also introduce the formal power series in u^{-1}

$$T_{ij}(u) = T_{ij}^{(0)} + T_{ij}^{(1)} u^{-1} + T_{ij}^{(2)} u^{-2} + \cdots$$

and combine all these series into the single element of $\mathrm{End}\,(\mathbb{C}^N) \otimes \mathrm{Y}(\mathfrak{gl}_N)[[u^{-1}]]$

$$T(u) = \sum_{i,j} E_{ij} \otimes T_{ij}(u).$$

For any associative unital algebra A, denote by ι_s its embedding into the tensor product $\mathrm{A}^{\otimes n}$ as the sth tensor factor:

$$\iota_s(X) = 1^{\otimes(s-1)} \otimes X \otimes 1^{\otimes(n-s)}, \qquad X \in \mathrm{A},\ s = 1, \ldots, n.$$

We shall also use various embeddings of the algebra $\mathrm{A}^{\otimes m}$ into $\mathrm{A}^{\otimes n}$ for any $m \leqslant n$. For $s_1, \ldots, s_m \in \{1, \ldots, n\}$ pairwise distinct and $X \in \mathrm{A}^{\otimes m}$, we put

$$X_{s_1 \ldots s_m} = \iota_{s_1} \otimes \cdots \otimes \iota_{s_m}(X) \in \mathrm{A}^{\otimes n}.$$

Denote

$$T_s(u) = \iota_s \otimes \mathrm{id}\,(T(u)) \in \mathrm{End}\,(\mathbb{C}^N)^{\otimes n} \otimes \mathrm{Y}(\mathfrak{gl}_N)[[u^{-1}]].$$

In this notation, the defining relations (3.1) can be rewritten as a single equation

(3.2) $$R(u,v) \otimes 1 \cdot T_1(u) T_2(v) = T_2(v) T_1(u) \cdot R(u,v) \otimes 1.$$

After multiplying each side of (3.2) by $u - v$, it becomes a relation in the algebra

$$\mathrm{End}\,(\mathbb{C}^N) \otimes \mathrm{End}\,(\mathbb{C}^N) \otimes \mathrm{Y}(\mathfrak{gl}_N)((u^{-1}, v^{-1})).$$

For further comments on the definition of the algebra $\mathrm{Y}(\mathfrak{gl}_N)$, see [**MNO**].

The relation (3.2) implies that for any $z \in \mathbb{C}$ the assignment $T_{ij}(u) \mapsto T_{ij}(u-z)$ determines an automorphism of the algebra $\mathrm{Y}(\mathfrak{gl}_N)$. Here the formal series in $(u-z)^{-1}$ should be reexpanded in u^{-1}. We denote this automorphism by τ_z.

We also regard E_{ij} as generators of the universal enveloping algebra $\mathrm{U}(\mathfrak{gl}_N)$. The algebra $\mathrm{Y}(\mathfrak{gl}_N)$ contains $\mathrm{U}(\mathfrak{gl}_N)$ as a subalgebra: due to (3.1), the assignment $E_{ij} \mapsto -T_{ji}^{(1)}$ defines the embedding. Moreover, there is a homomorphism

(3.3) $$\pi: \mathrm{Y}(\mathfrak{gl}_N) \to \mathrm{U}(\mathfrak{gl}_N): T_{ij}(u) \mapsto \delta_{ij} - E_{ji} u^{-1}.$$

By definition, this homomorphism is identical on the subalgebra $\mathrm{U}(\mathfrak{gl}_N)$. It is called the *evaluation homomorphism* for the algebra $\mathrm{Y}(\mathfrak{gl}_N)$. We regard $\mathbb{C}^N$ as a $\mathrm{Y}(\mathfrak{gl}_N)$-module using this homomorphism. Denote by $V(z)$ the $\mathrm{Y}(\mathfrak{gl}_N)$-module obtained from $\mathbb{C}^N$ by applying the automorphism τ_z. The action of the generators $T_{ij}^{(s)}$ in $V(z)$ can be determined by the assignment

$$\mathrm{End}\,(\mathbb{C}^N) \otimes \mathrm{Y}(\mathfrak{gl}_N)[[u^{-1}]] \to \mathrm{End}\,(\mathbb{C}^N) \otimes \mathrm{End}\,(\mathbb{C}^N)[[u^{-1}]]: T(u) \mapsto R(u,z).$$

Furthermore, there is a natural Hopf algebra structure on $Y(\mathfrak{gl}_N)$. Again due to (3.2), the comultiplication $\Delta: Y(\mathfrak{gl}_N) \to Y(\mathfrak{gl}_N) \otimes Y(\mathfrak{gl}_N)$ can be defined by

$$T_{ij}(u) \mapsto \sum_h T_{ih}(u) \otimes T_{hj}(u). \tag{3.4}$$

Here the tensor product is taken over the subalgebra $\mathbb{C}[[u^{-1}]]$ in $Y(\mathfrak{gl}_N)[[u^{-1}]]$ and h runs through $1, \ldots, N$. For any $z_1, \ldots, z_n \in \mathbb{C}$, consider the $Y(\mathfrak{gl}_N)$-module $V(z_1) \otimes \cdots \otimes V(z_n)$. By definition (3.4), the action of the generators $T_{ij}^{(s)}$ in the space $(\mathbb{C}^N)^{\otimes n}$ of this module can be determined by the assignment

$$\begin{aligned}\operatorname{End}(\mathbb{C}^N) \otimes Y(\mathfrak{gl}_N)[[u^{-1}]] &\to \operatorname{End}(\mathbb{C}^N) \otimes \operatorname{End}(\mathbb{C}^N)^{\otimes n}[[u^{-1}]]: \\ T(u) &\mapsto R_{12}(u, z_1) \cdots R_{1,n+1}(u, z_n).\end{aligned} \tag{3.5}$$

The symmetric group S_n acts in the space $(\mathbb{C}^N)^{\otimes n}$ by permutations of the tensor factors:

$$(k,l) \mapsto P_{kl} \in \operatorname{End}(\mathbb{C}^N)^{\otimes n}, \qquad k \neq l.$$

The function $R_{kl}(u,v)$ with values in $\operatorname{End}(\mathbb{C}^N)^{\otimes n}$ corresponds to the function $\varphi_{kl}(u,v)$ with values in the group ring $\mathbb{C} \cdot S_n$. Let λ be a Young diagram with n boxes and not more than N rows. Denote by F_λ the element of $\operatorname{End}(\mathbb{C}^N)^{\otimes n}$ corresponding to

$$\frac{\dim U_\lambda}{n!} \Phi_\lambda \in \mathbb{C} \cdot S_n.$$

So $F_\lambda^2 = F_\lambda$ and the image of F_λ in $(\mathbb{C}^N)^{\otimes n}$ is an irreducible $\mathfrak{gl}_N$-submodule [**Y1**]. Denote this image by V_λ. We identify the algebra $\operatorname{End}(V_\lambda)$ with the subalgebra in $\operatorname{End}(\mathbb{C}^N)^{\otimes n}$ that consists of all elements of the form $F_\lambda X F_\lambda$.

By virtue of the homomorphism (3.3), we can regard V_λ as a $Y(\mathfrak{gl}_N)$-module. Denote the action o $Y(\mathfrak{gl}_N)$ in the module V_λ by π_λ. Just as in the previous section, let c_k be the content of the box with the number k in the column tableaux Λ^c. The next proposition is contained in [**C2**].

PROPOSITION 3.1. *The action π_λ of $Y(\mathfrak{gl}_N)$ can be described as the restriction to the vector subspace V_λ in $(\mathbb{C}^N)^{\otimes n}$ of the action (3.5) with $z_1 = c_1, \ldots, z_n = c_n$.*

PROOF. Consider the vector space $(\mathbb{C}^N)^{\otimes n}$ as a $Y(\mathfrak{gl}_N)$-module by virtue of (3.3). The action of $Y(\mathfrak{gl}_N)$ in the latter module can be determined by the assignment

$$\begin{aligned}\operatorname{End}(\mathbb{C}^N) \otimes Y(\mathfrak{gl}_N)[[u^{-1}]] &\to \operatorname{End}(\mathbb{C}^N) \otimes \operatorname{End}(\mathbb{C}^N)^{\otimes n}[[u^{-1}]]: \\ T(u) &\mapsto 1 - \sum_{1 \leqslant k \leqslant n} \frac{P_{1,k+1}}{u}.\end{aligned} \tag{3.6}$$

The action of $Y(\mathfrak{gl}_N)$ in V_λ is by definition the restriction of (3.6) to the subspace V_λ in $(\mathbb{C}^N)^{\otimes n}$. So it suffices to prove the equality of rational functions in u

$$\left(1 - \sum_{1 \leqslant k \leqslant n} \frac{P_{1,k+1}}{u}\right) \cdot (\operatorname{id} \otimes F_\lambda) = \prod_{1 \leqslant k \leqslant n}^{\rightarrow} R_{1,k+1}(u, c_k) \cdot (\operatorname{id} \otimes F_\lambda)$$

with values in $\operatorname{End}(\mathbb{C}^N)^{\otimes(n+1)}$. But this equality follows from Proposition 2.12. $\square$

Denote by F_N the antisymmetrization map in $\mathrm{End}\,(\mathbb{C}^N)^{\otimes N}$. Thus $F_N = F_\lambda$ for $n = N$ and the diagram λ consisting of one column only. Consider the element

$$(3.7) \qquad F_N \otimes 1 \cdot T_1(u) \ldots T_N(u - N + 1)$$

of the algebra $\mathrm{End}\,(\mathbb{C}^N)^{\otimes N} \otimes \mathrm{Y}(\mathfrak{gl}_N)[[u^{-1}]]$. Due to Theorem 2.2 and to relations (3.2), this element is divisible by $F_N \otimes 1$ also on the right. Since the map F_N is one-dimensional, the element (3.7) has the form $F_N \otimes D(u)$ for a certain series $D(u) \in \mathrm{Y}(\mathfrak{gl}_N)[[u^{-1}]]$. This series is called the *quantum determinant* for $\mathrm{Y}(\mathfrak{gl}_N)$. The coefficients $D_1, D_2, \ldots$ of this series at $u^{-1}, u^{-2} \ldots$ are free generators of the center of the algebra $\mathrm{Y}(\mathfrak{gl}_N)$ (see [**MNO**] for the proof). The definitions (3.4) and (3.7) imply

$$\Delta(D(u)) = D(u) \otimes D(u)\,.$$

Hence the center of the algebra $\mathrm{Y}(\mathfrak{gl}_N)$ is a Hopf subalgebra.

The relations (3.2) imply that for any formal series $f(u) \in 1 + u^{-1}\mathbb{C}[[u^{-1}]]$ the assignment $T_{ij}(u) \mapsto f(u)T_{ij}(u)$ determines an automorphism of the algebra $\mathrm{Y}(\mathfrak{gl}_N)$. We shall denote this automorphism by ω_f. Consider the fixed point subalgebra in $\mathrm{Y}(\mathfrak{gl}_N)$ with respect to all the automorphisms ω_f. This subalgebra is called the *Yangian* of the simple Lie algebra $\mathfrak{sl}_N$ and denoted by $\mathrm{Y}(\mathfrak{sl}_N)$; cf. [**D1**].

PROPOSITION 3.2. *The algebra $\mathrm{Y}(\mathfrak{sl}_N)$ is a Hopf subalgebra in $\mathrm{Y}(\mathfrak{gl}_N)$. The algebra $\mathrm{Y}(\mathfrak{gl}_N)$ is isomorphic to the tensor product of its center and its subalgebra $\mathrm{Y}(\mathfrak{sl}_N)$.*

PROOF. We shall follow the arguments from [**MNO**]. Observe that by our definition

$$\omega_f(D(u)) = D(u)\,f(u)\cdots f(u - N + 1)\,.$$

Determine the series $A(u) \in \mathrm{Y}(\mathfrak{gl}_N)[[u^{-1}]]$ by the equation

$$D(u) = A(u) \cdots A(u - N + 1)\,.$$

Then we have $\omega_f(A(u)) = f(u)\,A(u)$. The coefficients $A_1, A_2 \ldots$ of the series $A(u)$ at $u^{-1}, u^{-2} \ldots$ are free generators of the center of $\mathrm{Y}(\mathfrak{gl}_N)$. Moreover, we have

$$(3.8) \qquad \Delta(A(u)) = A(u) \otimes A(u)\,.$$

Every coefficient of the series $T_{ij}(u)\,A(u)^{-1}$ belongs to the subalgebra $\mathrm{Y}(\mathfrak{sl}_N)$. Therefore the algebra $\mathrm{Y}(\mathfrak{gl}_N)$ is generated by its center and its subalgebra $\mathrm{Y}(\mathfrak{sl}_N)$. Now suppose that for some positive integer s there exists a nonzero polynomial $\mathcal{Q}$ in s variables with coefficients from $\mathrm{Y}(\mathfrak{sl}_N)$ such that $\mathcal{Q}(A_1, \ldots, A_s) = 0$. Choose the number s to be minimal. But for $f(u) = 1 + au^{-s}$ with any $a \in \mathbb{C}$ we have

$$\omega_f : \mathcal{Q}(A_1, \ldots, A_s) \mapsto \mathcal{Q}(A_1, \ldots, A_s + a)\,.$$

Thus $\mathcal{Q}(A_1, \ldots, A_s + a) = 0$ for all $a \in \mathbb{C}$. So we may assume that the polynomial $\mathcal{Q}$ does not depend on the last variable, which contradicts the choice of s. This contradiction completes the proof of the second statement of Proposition 3.2.

In particular, the coefficients of all the series $T_{ij}(u)A(u)^{-1}$ generate the algebra $\mathrm{Y}(\mathfrak{sl}_N)$. Now the first statement of Proposition 3.2 follows from (3.4) and (3.8). $\square$

COROLLARY 3.3. *The center of the algebra* $Y(\mathfrak{sl}_N)$ *is trivial.*

Note that the subalgebra $Y(\mathfrak{sl}_N)$ in $Y(\mathfrak{gl}_N)$ is preserved by the automorphism τ_z for any $z \in \mathbb{C}$. Denote by Δ' the composition of Δ with the permutation of tensor factors in $Y(\mathfrak{gl}_N) \otimes Y(\mathfrak{gl}_N)$. According to [D1], there is a unique formal series
$$\mathcal{S}(z) \in 1 + Y(\mathfrak{sl}_N) \otimes Y(\mathfrak{sl}_N)[[z^{-1}]] \cdot z^{-1}$$
such that in the algebra $Y(\mathfrak{sl}_N)^{\otimes 3}[[z^{-1}]]$ we have

(3.9) $\quad \mathrm{id} \otimes \Delta(\mathcal{S}(z)) = \mathcal{S}_{12}(z)\,\mathcal{S}_{13}(z), \quad \Delta \otimes \mathrm{id}\,(\mathcal{S}(z)) = \mathcal{S}_{13}(z)\,\mathcal{S}_{23}(z)$

and for any element $Y \in Y(\mathfrak{sl}_N)$ we also have

(3.10) $\quad \mathcal{S}(z) \cdot \mathrm{id} \otimes \tau_z(\Delta'(Y)) = \mathrm{id} \otimes \tau_z(\Delta(Y)) \cdot \mathcal{S}(z)\,.$

The series $\mathcal{S}(z)$ is called the *universal R-matrix* for the Hopf algebra $Y(\mathfrak{sl}_N)$. For any $v \in \mathbb{C}$ denote by π_v the action of the algebra $Y(\mathfrak{gl}_N)$ in the module $V(v)$.

LEMMA 3.4. *We have the equality* $\pi_v \otimes \mathrm{id}\,(\mathcal{S}(z)) = T(v-z) \cdot \mathrm{id} \otimes A(v-z)^{-1}$.

PROOF. Let us keep the parameter $v \in \mathbb{C}$ fixed. Denote by $T'(z)$ and $T''(z)$ the formal series in z^{-1} in the left- and the right-hand side, respectively, of the equality to be proved. Due to (3.10) and Proposition 3.2, the series $T'(z)$ satisfies the equation

(3.11) $\quad T'(z) \cdot \pi_v \otimes \tau_z(\Delta'(Y)) = \pi_v \otimes \tau_z(\Delta(Y)) \cdot T'(z)$

for any element $Y \in Y(\mathfrak{gl}_N)$. We assume that here Y is one of the generators $T_{ij}^{(s)}$. By (3.4) we have the equalities in $\mathrm{End}\,(\mathbb{C}^N) \otimes \mathrm{End}\,(\mathbb{C}^N) \otimes Y(\mathfrak{gl}_N)[[u^{-1}]]$

$$\mathrm{id} \otimes (\pi_v \otimes \tau_z \circ \Delta)(T(u)) = R_{12}(u,v) \otimes 1 \cdot T_1(u-z),$$
$$\mathrm{id} \otimes (\pi_v \otimes \tau_z \circ \Delta')(T(u)) = T_1(u-z) \cdot R_{12}(u,v) \otimes 1\,.$$

So the collection of all equations (3.11) is equivalent to the single equation for $T'(z)$

(3.12) $\quad R_{12}(u+z,v) \otimes 1 \cdot T_1(u)\,T'_2(z) = T'_2(z)\,T_1(u) \cdot R_{12}(u+z,v) \otimes 1,$

which, after being multiplied by $u+z-v$, becomes an equation in the algebra
$$\mathrm{End}\,(\mathbb{C}^N) \otimes \mathrm{End}\,(\mathbb{C}^N) \otimes Y(\mathfrak{gl}_N)((u^{-1},z^{-1}))\,.$$

But the element $T''(z)$ satisfies the same equation. Indeed, since the coefficients of the series $A(v-z)$ in z^{-1} are central in $Y(\mathfrak{gl}_N)$ we have (by (3.2))

(3.13) $\quad R_{12}(u+z,v) \otimes 1 \cdot T_1(u)\,T''_2(z) = T''_2(z)\,T_1(u) \cdot R_{12}(u+z,v) \otimes 1\,.$

Consider the series
$$X(z) = T'(z+v)\,T''(z+v)^{-1}$$
in z^{-1} with the coefficients in $\mathrm{End}\,(\mathbb{C}^N) \otimes Y(\mathfrak{sl}_N)$. Comparing (3.12), (3.13) we obtain

(3.14) $\quad [R_{12}(u+z+v,v) \otimes 1 \cdot T_1(u),\, X_2(z)] = 0\,.$

We shall write
$$X(z) = \sum_{i,j} E_{ij} \otimes X_{ij}(z),$$

where
$$X_{ij}(z) = X_{ij}^{(0)} + X_{ij}^{(1)}z^{-1} + X_{ij}^{(2)}z^{-2} + \cdots$$
for some $X_{ij}^{(s)} \in Y(\mathfrak{sl}_N)$. Let us multiply equation (3.14) by $u+z$. Considering the coefficient at $u^0 z^{-s}$, we obtain

(3.15) $\qquad [T_{ij}^{(1)}, X_{kl}^{(s)}] = \delta_{kj} X_{il}^{(s)} - \delta_{il} X_{kj}^{(s)}, \qquad s = 1, 2, \ldots.$

Further, by considering the coefficient at $u^{-1} z^{-s}$ we obtain

(3.16) $\qquad [T_{ij}^{(2)}, X_{kl}^{(s)}] + [T_{ij}^{(1)}, X_{kl}^{(s+1)}] = T_{kj}^{(1)} X_{il}^{(s)} - \delta_{il} \sum_h X_{kh}^{(s)} T_{hj}^{(1)},$

where the index h runs through $1, \ldots, N$.

Let us now prove by induction on $s = 0, 1, 2, \ldots$ that $X_{kl}^{(s)} = \delta_{kl} X_s$ for some element $X_s \in Y(\mathfrak{sl}_N)$ commuting with each $T_{ij}^{(1)}$. The condition
$$X(z) \in 1 + \mathrm{End}(\mathbb{C}^N) \otimes Y(\mathfrak{sl}_N)[[z^{-1}]] \cdot z^{-1}$$
provides the base for induction. Let us consider equation (3.16). Due to (3.15) and to the induction assumption, it takes the form
$$\delta_{kl}[T_{ij}^{(2)}, X_s] + \delta_{kj} X_{il}^{(s+1)} - \delta_{il} X_{kj}^{(s+1)} = 0.$$
The latter equation shows that for $k \neq l$ we have $X_{kk}^{(s+1)} = X_{ll}^{(s+1)}$ and $X_{kl}^{(s+1)} = 0$. Thus $X_{kl}^{(s+1)} = \delta_{kl} X_{s+1}$ for some $X_{s+1} \in Y(\mathfrak{sl}_N)$. Now by using (3.15) with $s+1$ instead of s, we get the equality $[T_{ij}^{(1)}, X_{s+1}] = 0$. Therefore
$$X(z) = \mathrm{id} \otimes (1 + X_1 z^{-1} + X_2 z^{-2} + \cdots)$$
for some elements $X_1, X_2, \cdots \in Y(\mathfrak{sl}_N)$. These elements are central due to (3.14). But the center of the algebra $Y(\mathfrak{sl}_N)$ is $\mathbb{C}$ by Corollary 3.3. So $X_1, X_2, \cdots \in \mathbb{C}$.

Thus the series $T'(z)$ and $T''(z)$ coincide up to a factor from $\mathbb{C}[[z^{-1}]]$. That factor must be 1, since the equation
$$\pi_v \otimes \Delta(\mathcal{S}(z)) = \pi_v \otimes \mathrm{id} \otimes \mathrm{id}(\mathcal{S}_{12}(z)\mathcal{S}_{13}(z))$$
for $\pi_v \otimes \mathrm{id}(\mathcal{S}(z)) = T'(z)$ is also satisfied for $T''(z)$. The latter fact follows from definition (3.4) and from (3.8). $\qquad \square$

Now we introduce the formal series in z^{-1} with the coefficients in the center of $Y(\mathfrak{gl}_N)$

(3.17) $\qquad D_\lambda(z) = A(c_1 - z) \cdots A(c_n - z).$

We also introduce the following element of the algebra $\mathrm{End}(\mathbb{C}^N)^{\otimes n} \otimes Y(\mathfrak{gl}_N)[[z^{-1}]]$:

(3.18) $\qquad T_\lambda(z) = F_\lambda \otimes 1 \cdot T_1(c_1 - z) \cdots T_n(c_n - z).$

By Theorem 2.2 and relations (3.2), the latter element is divisible by $F_\lambda \otimes 1$ also on the right. Thus we can regard $T_\lambda(z)$ as an element of $\mathrm{End}(V_\lambda) \otimes Y(\mathfrak{gl}_N)[[z^{-1}]]$. By (3.9) and by Lemma 3.4, we have the following corollary to Proposition 3.1.

COROLLARY 3.5. *We have the equality* $\pi_\lambda \otimes \mathrm{id}(\mathcal{S}(z)) = T_\lambda(z) \cdot \mathrm{id} \otimes D_\lambda(z)^{-1}.$

The next theorem is the main result of this section.

THEOREM 3.6. *There is a series* $\mathcal{R}(z) \in 1 + Y(\mathfrak{gl}_N) \otimes Y(\mathfrak{gl}_N)[[z^{-1}]] \cdot z^{-1}$ *satisfying* (1.2) *such that the equality* $\pi_\lambda \otimes \mathrm{id}\,(\mathcal{R}(z)) = T_\lambda(z)$ *holds for any Young diagram* λ.

PROOF. We prove that there is a formal power series $\mathcal{D}(z)$ in z^{-1} with coefficients in the tensor square of the center of $Y(\mathfrak{gl}_N)$ and leading term 1 such that $\pi_\lambda \otimes \mathrm{id}\,(\mathcal{D}(z)) = \mathrm{id} \otimes D_\lambda(z)$ for any Young diagram λ. Then by Proposition 3.2 and Corollary 3.5 it suffices to put $\mathcal{R}(z) = \mathcal{S}(z)\mathcal{D}(z)$.

Let each of the indices $d_1, d_2, \ldots$ run over $0, 1, 2, \ldots$ but only a finite number of them differ from zero. Put $d_1 + d_2 + \cdots = d$. Introduce the formal series in z^{-1}

$$(3.19) \qquad f_{d_1,d_2,\ldots}(z) = \sum_{s_1,\ldots,s_n} (c_1 - z)^{-s_1} \cdots (c_n - z)^{-s_n},$$

where the sum is taken over all the sequences $s_1, \ldots, s_n$ of the numbers $0, 1, 2, \ldots$ such that the multiplicities of $1, 2, \ldots$ are $d_1, d_2, \ldots$, respectively. Then by (3.17)

$$(3.20) \qquad D_\lambda(z) = \sum_{d_1,d_2,\ldots} f_{d_1,d_2,\ldots}(z) A_1^{d_1} A_2^{d_2} \cdots,$$

where the sum is taken over all sequences $d_1, d_2, \ldots$ such that $d \leqslant n$.

Now let the indices $d_1, d_2, \ldots$ be fixed but the number n and the diagram λ vary. For each $i = 1, \ldots, N$ denote by λ_i the length of ith row of the diagram λ. Let $f_\lambda^{(s)}$ be the coefficient of the series (3.19) at z^{-s}. It depends on λ as a fixed polynomial of

$$c_1^r + \cdots + c_n^r, \qquad r = 0, 1, 2, \ldots.$$

So $f_\lambda^{(s)}$ can be expressed as a fixed symmetric polynomial in $\lambda_1 - 1, \ldots, \lambda_N - N$. Therefore it is the eigenvalue in the $\mathfrak{gl}_N$-module V_λ of a certain element F_s of the center $Z(\mathfrak{gl}_N)$ of the algebra $U(\mathfrak{gl}_N)$ that does not depend on the diagram λ.

Note that by (3.19) we have $f_\lambda^{(s)} = 0$ if $s < d_1 + 2d_2 + \ldots$ for any λ. Hence for any $s < d_1 + 2d_2 + \ldots$ we have $F_s = 0$. For $d_1 = d_2 = \cdots = 0$ we have $F_0 = 1$.

The image of the center of the algebra $Y(\mathfrak{gl}_N)$ with respect to the evaluation homomorphism π coincides with $Z(\mathfrak{gl}_N)$. Indeed, the images $\pi(D_1), \ldots, \pi(D_N)$ of the coefficients of the quantum determinant $D(u)$ generate the center $Z(\mathfrak{gl}_N)$ of the algebra $U(\mathfrak{gl}_N)$ (see [**C, HU**]). So we can choose a central element of $Y(\mathfrak{gl}_N)$

$$B_{d_1,d_2,\ldots}^{(s)} \in \pi^{-1}(F_s).$$

Moreover, we can assume that for any $s < d_1 + 2d_2 + \ldots$ this element is zero. If $d_1 = d_2 = \cdots = 0$ and $s = 0$, we can assume that this element is 1. Now set

$$\mathcal{D}(z) = \sum_{s \geqslant 0} z^{-s} \sum_{d_1,d_2,\ldots} B_{d_1,d_2,\ldots}^{(s)} \otimes A_1^{d_1} A_2^{d_2} \cdots,$$

where the inner sum is taken over all sequences $d_1, d_2, \ldots$ with $d_1 + 2d_2 + \ldots \leqslant s$.

Consider $f_\lambda^{(s)}$ with fixed $d_1, d_2, \ldots$ as a polynomial in $\lambda_1, \ldots, \lambda_N$. To prove that $\pi_\lambda \otimes \mathrm{id}\,(\mathcal{D}(z)) = \mathrm{id} \otimes D_\lambda(z)$ for any diagram λ, it remains to show that this polynomial vanishes when the diagram λ satisfies the condition $n < d$ (see (3.20)).

Observe that the coefficient $f_\lambda^{(s)}$ of the series (3.19) has the form of the sum

$$\sum_{0 \leqslant k \leqslant d} (n-k) \cdots (n-d) g_\lambda^{(k)},$$

where $g_\lambda^{(k)}$ depends on λ as a certain symmetric polynomial in $c_1,\ldots,c_n$ and belongs to the ideal generated by the kth *elementary* symmetric polynomial

$$\sum_{1\leqslant i_1<\cdots<i_k\leqslant n} c_{i_1}\cdots c_{i_k}.$$

Let us express the latter sum as a symmetric polynomial in $\lambda_1-1,\ldots,\lambda_N-N$. We denote this polynomial by $b_k(\lambda_1,\ldots,\lambda_N)$. It suffices to prove that $b_k(\lambda_1,\ldots,\lambda_N)=0$ when the diagram λ satisfies the condition $n<k$. We shall do that by induction on the difference $k-n=1,2,\ldots$.

If λ is a diagram with $n\geqslant 1$ boxes, then $b_n(\lambda_1,\ldots,\lambda_N)=c_1\cdots c_n=0$. Indeed, then there is at least one box on the main diagonal of the diagram λ. The content of this box is zero. Further, suppose that λ is a diagram such that $\lambda_1>\cdots>\lambda_N$. Then due to the equality

$$\sum_{0\leqslant k\leqslant n} b_k(\lambda_1,\ldots,\lambda_N)u^{n-k} = \prod_{1\leqslant k\leqslant n}(u+c_k),$$

for any $i=1,\ldots,N$ and $k=1,\ldots,n$ we have the relation

$$b_k(\lambda_1,\ldots,\lambda_N) = b_k(\lambda_1,\ldots,\lambda_i+1,\ldots,\lambda_N) - (\lambda_i-i+1)b_{k-1}(\lambda_1,\ldots,\lambda_N).$$

With a fixed index k, this relation must be valid for any $\lambda_1,\ldots,\lambda_N\in\mathbb{C}$. In particular, if λ is the empty diagram, then $b_1(\lambda_1,\ldots,\lambda_N)=0$. The latter fact provides the base for induction. Furthermore, we have already established that $b_k(\lambda_1,\ldots,\lambda_N)=0$ when $n=k\geqslant 1$. Now for each $k\geqslant 2$ the above relation along with the induction assumption implies that $b_k(\lambda_1,\ldots,\lambda_N)=0$ for any Young diagram λ with less than k boxes. $\square$

Now let m be an arbitrary positive integer and μ a Young diagram with m boxes. Equip the set of all pairs (k,l), where $1\leqslant k\leqslant n$ and $1\leqslant l\leqslant m$, with the lexicographical ordering. Let d_l be the content of the box with l in the column tableaux M^c of shape μ. Consider the rational function of z

$$R_{\lambda\mu}(z) = (F_\lambda\otimes\mathrm{id})\cdot \overrightarrow{\prod_{(k,l)}} R_{k,l+n}(c_k,d_l+z)\cdot(\mathrm{id}\otimes F_\mu)$$

with values in $\mathrm{End}\,(\mathbb{C}^N)^{\otimes(n+m)}$, where the factors corresponding to the pairs (k,l) are arranged with respect to the ordering above. By Theorem 2.2 and by relations (2.2) and (2.3), any value of $R_{\lambda\mu}(z)$ is divisible by $F_\lambda\otimes\mathrm{id}$ on the right and by $\mathrm{id}\otimes F_\mu$ on the left. Thus we can regard $R_{\lambda\mu}(z)$ as a function with values in $\mathrm{End}\,(V_\lambda)\otimes\mathrm{End}\,(V_\mu)$. We have the following corollary to Proposition 3.1 and Theorem 3.6.

COROLLARY 3.7. *We have the equality* $\pi_\lambda\otimes\pi_\mu(\mathcal{R}(z))=R_{\lambda\mu}(z)$.

In the next section we will study the behavior of the function $R_{\lambda\mu}(z)$ at $z=0$.

§4. The estimation theorem

Let any two Young diagrams λ and μ consisting of n and m boxes, respectively, be fixed. As before, we denote by c_k and d_k the contents of the boxes with k in the column tableaux Λ^c and M^c, respectively. We shall employ the elements $\Phi_\lambda\in\mathbb{C}\cdot S_n$

and $\Phi_\mu \in \mathbb{C} \cdot S_m$ determined by (2.1). Let us equip the set of all pairs (i,j), where $1 \leqslant i \leqslant n$ and $1 \leqslant j \leqslant m$, with the lexicographical ordering:

(4.1) $$(1,1) \prec \cdots \prec (1,m) \prec \cdots \prec (n,1) \prec \cdots \prec (n,m).$$

We shall regard S_n as a subgroup in S_{n+m} with respect to the standard embedding. The superscript $^\vee$ denotes the embedding of the group S_m into S_{n+m} determined by assigning to the transposition of j and k the transposition of $j+n$ and $k+n$. Let z be a complex parameter. Consider the following rational function of z with values in the group ring $\mathbb{C} \cdot S_{n+m}$:

(4.2) $$\Phi_{\lambda\mu}(z) = \Phi_\lambda \cdot \overrightarrow{\prod_{(i,j)}} \varphi_{i,j+n}(c_i, d_j + z) \cdot \Phi_\mu^\vee,$$

where the factors corresponding to the pairs (i,j) are arranged with respect to the ordering (4.1). Let r be the number of boxes on the main diagonal of the diagram λ. This number is called the *rank* of the diagram λ.

THEOREM 4.1. *The order of the pole of $\Phi_{\lambda\mu}(z)$ at $z = 0$ does not exceed the rank r of λ. This order does not exceed $r - 1$ if the diagram λ is not contained in μ.*

PROOF. When estimating the order of the pole of $\Phi_{\lambda\mu}(z)$ at $z = 0$, we use arguments similar to those already used in the proof of Proposition 2.12, but more elaborate.

The factor $\varphi_{i,n+j}(c_i, d_j+z)$ in (4.2) has a pole at $z = 0$ if and only if the numbers i and j stand on the same diagonals of the tableaux Λ^c and M^c, respectively. This pole is simple. We call such a pair (i,j) *singular*. Let us confine every such pair to a certain segment of the sequence (4.1) as follows.
  (i) If $j = 1$, then the pair (i,j) itself makes a segment.
  (ii) If $j \neq 1$ and j is not in the first row of M^c, then the two pairs $(i, j-1)$, (i,j) form a segment. Note that $j - 1$ stands directly above j in the tableau M^c.
  (iii) If $j \neq 1$ appears in the first row of M^c, then the corresponding segment is

(4.3) $$(i,k), (i, k+1), \ldots, (i, j-1), (i,j),$$

where k is the number next to the left of the number j in the first row of M^c.

Within each of these segments, (i,j) is the only singular pair, and all segments are disjoint. Note that the number of all segments of type (i) is exactly r.

Consider the segment (ii), where the singular pair (i,j) is fixed. Here we have $c_i = d_j = d_{j-1} - 1$. By Corollary 2.8, the element Φ_μ of the group ring $\mathbb{C} \cdot S_m$ is divisible on the left by $\varphi_{j-1,j}(d_{j-1}, d_j)$. Therefore, by relations (2.2) and (2.3) in $\mathbb{C} \cdot S_{n+m}$, the rational function of z

(4.4) $$\overrightarrow{\prod_{(i,j)\prec(p,q)}} \varphi_{p,q+n}(c_p, d_q + z) \cdot \Phi_\mu^\vee$$

is divisible on the left by $\varphi_{j-1+n, j+n}(d_{j-1}, d_j)$. But by Lemma 2.11 the product

$$\varphi_{i,j-1+n}(c_i, d_{j-1} + z)\, \varphi_{i,j+n}(c_i, d_j + z)\, \varphi_{j-1+n, j+n}(d_{j-1}, d_j)$$

is regular at the point $z = 0$. Thus when estimating the order of the pole of (4.2) at $z = 0$ from above, factors of type (ii) do not count.

Now consider the segment (iii), where the singular pair (i,j) is again fixed. Here we have $c_i = d_j = d_k + 1$. By Corollary 2.5, the element Φ_μ of $\mathbb{C} \cdot S_m$ is divisible on the left by

$$\overleftarrow{\prod_{k<q<j}} \varphi_{qj}(d_q, d_j) \cdot \varphi_{kj}(d_k, d_j).$$

Therefore due to relations (2.2) and (2.3), the rational function determined by (4.4) is divisible on the left by

$$\overleftarrow{\prod_{k<q<j}} \varphi_{q+n,j+n}(d_q, d_j) \cdot \varphi_{k+n,j+n}(d_k, d_j).$$

Consider the product of the factors in (4.2) corresponding to the pairs from the segment (iii). Multiplying this product on the right by (4.5), we obtain

(4.5)
$$\varphi_{i,k+n}(c_i, d_k + z) \cdot \overrightarrow{\prod_{k<q<j}} \varphi_{i,q+n}(c_i, d_q + z) \cdot \varphi_{i,j+n}(c_i, d_j + z)$$
$$\times \overleftarrow{\prod_{k<q<j}} \varphi_{q+n,j+n}(d_q, d_j) \cdot \varphi_{k+n,j+n}(d_k, d_j)$$
$$= \overleftarrow{\prod_{k<q<j}} \varphi_{q+n,j+n}(d_q, d_j) \cdot \varphi_{i,k+n}(c_i, d_k + z) \varphi_{i,j+n}(c_i, d_j + z)$$
$$\times \varphi_{k+n,j+n}(d_k, d_j) \cdot \overrightarrow{\prod_{k<q<j}} \varphi_{i,q+n}(c_i, d_q + z),$$

where we again used relations (2.2), (2.3). The rational function in the right-hand side of the latter equality is regular at $z = 0$ by Lemma 2.11. Thus when estimating the order of the pole of (4.2) at $z = 0$, the factors of type (iii) do not count either. This completes the proof of the first statement in Theorem 4.1.

The proof of the second statement is similar but more involved. As before, denote by λ'_s and μ'_s the lengths of the sth columns of the diagrams λ and μ, respectively. Assume that the diagram λ is not contained in μ. We prove that when estimating the order of the pole of (4.2) at $z = 0$, we may exclude from our count the factor corresponding to the singular pair $(1,1)$, as well as those corresponding to the singular pairs (i,j) with $i \neq 1$.

Fix the minimal number s such that $\lambda'_s > \mu'_s$. Let $i_1, \ldots, i_s$ and $j_1, \ldots, j_s$ be the first s numbers in the first rows of the tableaux Λ^c and M^c, respectively. Of course, here $i_1 = j_1 = 1$. Denote $\mu'_s - 1 = t$, so that the numbers in the sth column of M^c are $j_s, j_s + 1, \ldots, j_s + t$. The singular pairs

$$(i_1, j_1), \ldots, (i_s, j_s) \quad \text{and} \quad (i_s + 1, j_s + 1), \ldots, (i_s + t, j_s + t)$$

will be called *special*. We shall now proceed in several steps.

I. Let us reorder the pairs (i, j) in the product (4.2) as described below; this reordering will not affect the value of the product due to the relations (2.3). For each $i = i_s, i_s + 1, \ldots, i_s + t$, change the sequence of pairs

$$(i, j), (i, j+1), \ldots, (i, m), (i+1, 1), \ldots, (i+1, j-1), (i+1, j),$$

where $j = j_s, j_s+1, \ldots, j_s+t$ respectively, to the sequence
$$(i+1,1), \ldots, (i+1, j-1), (i,j), (i+1, j), (i, j+1), \ldots, (i,m).$$
The latter reordering preserves all the segments (i)–(iii) except for the segment with (i_s, j_s) and the t segments of type (ii)
$$(i_s+q, j_s+q-1), (i_s+q, j_s+q), \qquad q = 1, \ldots, t.$$
Instead of these $t+1$ segments, we introduce the segments in our new ordering

(4.6) $\qquad (i_s+q, j_s+q), (i_s+q+1, j_s+q), \qquad q = 0, 1, \ldots, t.$

Observe that here
$$d_{j_s+q} = c_{i_s+q} = c_{i_s+q+1} + 1.$$
Moreover, by Corollary 2.8, the element Φ_λ of $\mathbb{C} \cdot S_n$ is divisible on the right by
$$\varphi_{i_s+q, i_s+q+1}(c_{i_s+q}, c_{i_s+q+1}).$$
Just as in the proof of the first statement of Theorem 4.1, but using Lemma 2.1, we can now exclude from our count the factors in (4.2) corresponding to each of the special singular pairs
$$(i_s+q, j_s+q), \qquad q = 0, 1, \ldots, t.$$
If $\mu'_s = 0$ this step is empty. If $s = 1$, then the proof of Theorem 4.1 is complete.

II. Suppose that $s > 1$. Note that by our assumption $\lambda'_{s-1} \leqslant \mu'_{s-1}$. Denote
$$i_s = i, \quad j_s = j, \quad i_{s-1} = h, \quad j_{s-1} = k, \quad \text{and} \quad j_{s-1} + \lambda'_{s-1} - 1 = l.$$

On the left of the next picture we show the $(s-1)$th and sth columns of the tableau Λ^c, while on the right we have the same columns of the tableau M^c.

h	i
·	·
·	·
·	·
·	

k	j
·	·
·	·
·	l
·	

Observe that in this notation the segment of (i, j) in (4.1) was exactly (4.3).

Let us perform further reordering of the factors in (4.2). Consider the subsequence in (4.1) consisting of all the pairs (p, q) such that
$$(h, k) \preccurlyeq (p, q) \preccurlyeq (i, l);$$
this subsequence has been not affected by the reordering at the previous step. Let us now consecutively take the pairs (p, q) with $h \leqslant p < i$ and $n < q \leqslant m$ to the left of that subsequence. Then take the pairs (p, q) with $h < p \leqslant i$ and $1 \leqslant q < k$ to the right. Between them the following sequence will remain:
$$(h, k), \ldots, (h, l), (h+1, k), \ldots, (h+1, l), \ldots, \ldots, \ldots, (i, k), \ldots, (i, l).$$

This reordering does not change the value of (4.2), again due to relations (2.3). It does not break any of the new segments (4.6). It also preserves all the segments (i)–(iii) except for those containing the singular pairs (i,j) and (h,k). The factor in (4.2) corresponding to (i,j) has been excluded from our count at the previous step. By Corollary 2.5, the element Φ_λ of $\mathbb{C} \cdot S_n$ is divisible on the right by

$$\varphi_{hi}(c_h, c_i) \cdot \overrightarrow{\prod_{h<p<i}} \varphi_{pi}(c_p, c_i).$$

Note that here $d_k = c_h = c_i - 1$. Therefore by using Lemma 2.1 along with relations (2.2) and (2.3), we can exclude from our count the factor in (4.2) corresponding to the pair (h,k) as well. This ends the second step.

Let us now repeat the last step for $s-1, \ldots, 2$ instead of s. Then all the factors in (4.2) corresponding to special singular pairs will be excluded from our count. In particular, the factor corresponding to $(1,1)$ will be excluded. The segments (i)–(iii) with nonspecial singular pairs will be preserved by every reordering. So we shall exclude from count the factors in (4.2) corresponding to all nonspecial singular pairs (i,j) with $j \neq 1$. This can be demonstrated by arguments used in the proof of the first statement of Theorem 4.1. The proof of the second statement is now also completed. $\square$

§5. Higher Capelli identities

Let two positive integers N and M be fixed. In this section we consider invariants of the action of Lie algebra $\mathfrak{gl}_N \times \mathfrak{gl}_M$ in the space $\mathcal{PD}$ of differential operators on $\mathbb{C}^N \otimes \mathbb{C}^M$ with polynomial coefficients. Let the indices i, j and a, b run over $1, \ldots, N$ and $1, \ldots, M$, respectively. Let x_{ia} be the standard coordinates on the vector space $\mathbb{C}^N \otimes \mathbb{C}^M$. We shall define the actions of $\mathfrak{gl}_N$ and $\mathfrak{gl}_M$ in the space $\mathcal{P}$ of polynomial functions on $\mathbb{C}^N \otimes \mathbb{C}^M$ by

(5.1) $$E_{ij} \mapsto \sum_b x_{ib} \partial_{jb} \quad \text{and} \quad E_{ab} \mapsto \sum_j x_{ja} \partial_{jb},$$

respectively. Here ∂_{jb} is the partial derivation with respect to the coordinate x_{jb}. We have the irreducible decomposition [**HU**] of the $\mathfrak{gl}_N \times \mathfrak{gl}_M$-module

(5.2) $$\mathcal{P} = \bigoplus_\lambda V_\lambda \otimes W_\lambda,$$

where λ runs through the set of all Young diagrams with $0, 1, 2, \ldots$ boxes and not more than M, N rows. Here W_λ is the irreducible $\mathfrak{gl}_M$-module corresponding to the diagram λ. Thus the action of $\mathfrak{gl}_N \times \mathfrak{gl}_M$ in $\mathcal{P}$ is multiplicity free.

The space $\mathcal{D}$ of differential operators on $\mathbb{C}^N \otimes \mathbb{C}^M$ with constant coefficients can be regarded as the $\mathfrak{gl}_N \times \mathfrak{gl}_M$-module dual to $\mathcal{P}$. Hence the space $\mathcal{I}$ of all the $\mathfrak{gl}_N \times \mathfrak{gl}_M$-invariants in

(5.3) $$\mathcal{PD} = \mathcal{P} \otimes \mathcal{D}$$

is the direct sum of one-dimensional subspaces corresponding to the diagrams λ with not more than M, N rows. Denote by $\mathcal{I}_\lambda$ the one-dimensional subspace corresponding to the partition λ. Let us give an explicit formula for a nonzero vector from $\mathcal{I}_\lambda$. Suppose that the diagram λ consists of n boxes. Let χ_λ be the character of the irreducible S_n-module U_λ. Consider the element of $\mathcal{PD}$ defined by (1.1). We denote this element by c_λ.

PROPOSITION 5.1. *We have $c_\lambda \in \mathcal{I}_\lambda$.*

PROOF. Consider the $\mathfrak{gl}_N \times \mathfrak{gl}_M$-submodule in $\mathcal{P}$ formed by polynomials of degree n. This submodule can be identified with the subspace of invariants in $(\mathbb{C}^N)^{\otimes n} \otimes (\mathbb{C}^M)^{\otimes n}$ with respect to simultaneous permutations of the first n and the last n tensor factors. The permutation σ of the first n tensor factors corresponds to the map

$$x_{i_1 a_1} \cdots x_{i_n a_n} \mapsto x_{i_{\sigma(1)} a_1} \cdots x_{i_{\sigma(n)} a_n}$$

in $\mathcal{P}$. We have the irreducible decomposition of the $S_n \times \mathfrak{gl}_N$-module

$$(\mathbb{C}^N)^{\otimes n} = \bigoplus_\lambda U_\lambda \otimes V_\lambda,$$

where λ runs over the set of all diagrams with n boxes and not more than N rows. So the image of the map

$$\eta: x_{i_1 a_1} \cdots x_{i_n a_n} \mapsto \sum_{\sigma \in S_n} \frac{\chi_\lambda(\sigma)}{n!} x_{i_{\sigma^{-1}(1)} a_1} \cdots x_{i_{\sigma^{-1}(n)} a_n}$$

in $\mathcal{P}$ is the subspace $V_\lambda \otimes W_\lambda$, if the diagram λ has no more than M, N boxes. The element c_λ in (5.3) is the image with respect to the map $\eta \otimes \mathrm{id}$ of

$$\sum_{i_1, \ldots, i_n} \sum_{a_1, \ldots, a_n} x_{i_1 a_1} \cdots x_{i_n a_n} \partial_{i_1 a_1} \cdots \partial_{i_n a_n}.$$

The latter element of $\mathcal{PD}$ is manifestly $\mathfrak{gl}_N \times \mathfrak{gl}_M$-invariant. □

Extend the action (5.1) of the Lie algebra $\mathfrak{gl}_N$ in $\mathcal{P}$ to the action of the universal enveloping algebra $\mathrm{U}(\mathfrak{gl}_N)$. The subspace of $\mathfrak{gl}_N \times \mathfrak{gl}_M$-invariants in $\mathcal{PD}$ coincides with the image of the center $\mathrm{Z}(\mathfrak{gl}_N)$ of the algebra $\mathrm{U}(\mathfrak{gl}_N)$ (see [**HU**]). We shall specify an element $e_\lambda \in \mathrm{Z}(\mathfrak{gl}_N)$ whose image is c_λ. Let $\psi_\lambda: \mathrm{Y}(\mathfrak{gl}_N) \to \mathbb{C}$ be the trace of the representation π_λ. Consider the element

$$\psi_\lambda \otimes \pi(\mathcal{R}(z)) \in \mathrm{U}(\mathfrak{gl}_N)[[z^{-1}]],$$

where π is the evaluation homomorphism (3.3). Then by Theorem 3.6, the product (1.3) is a polynomial in z. Denote this polynomial by $e_\lambda(z)$. The following general observation goes back to [**D2**].

PROPOSITION 5.2. *The coefficients of the polynomial $e_\lambda(z)$ belong to $\mathrm{Z}(\mathfrak{gl}_N)$.*

PROOF. Consider the enveloping algebra $\mathrm{U}(\mathfrak{gl}_N)$ as a subalgebra in $\mathrm{Y}(\mathfrak{gl}_N)$. Every element of this subalgebra is invariant with respect to the automorphism τ_z for any $z \in \mathbb{C}$. Furthermore, the restriction of the comultiplication (3.4) to $\mathrm{U}(\mathfrak{gl}_N)$ is cocommutative:

$$\Delta(Y) = Y \otimes 1 + 1 \otimes Y, \qquad Y \in \mathfrak{gl}_N.$$

Therefore, by the definition of the series $\mathcal{R}(z)$, for any $Y \in \mathfrak{gl}_N$ we have

$$[\mathcal{R}(z), Y \otimes 1 + 1 \otimes Y] = 0.$$

By applying the map $\psi_\lambda \otimes \pi$ to the latter equality, we obtain

$$[\psi_\lambda \otimes \pi(\mathcal{R}(z)), Y] = 0. \square$$

The next theorem is the main result of this section. Denote the value $e_\lambda(0)$ by e_λ.

THEOREM 5.3. *The image in $\mathcal{PD}$ of the element $e_\lambda \in \mathrm{Z}(\mathfrak{gl}_N)$ is c_λ.*

We shall present the main steps of the proof as separate propositions. First let us give an explicit formula for the polynomial in z with values in $\mathrm{End}\,(V_\lambda) \otimes \mathrm{U}(\mathfrak{gl}_N)$

(5.4) $$E_\lambda(z) = \pi_\lambda \otimes \pi(\mathcal{R}(z))(z - c_1) \cdots (z - c_n).$$

Just as before, here we regard $\mathrm{End}\,(V_\lambda)$ as a subalgebra in $\mathrm{End}\,(\mathbb{C}^N)^{\otimes n}$. We introduce the element

$$E = \sum_{i,j} E_{ji} \otimes E_{ij} \in \mathrm{End}\,(\mathbb{C}^N) \otimes \mathrm{U}(\mathfrak{gl}_N),$$

where the indices i, j run through $1, \ldots, N$. For $s = 1, \ldots, n$, we shall write

$$E_s = \iota_s \otimes \mathrm{id}\,(E) \in \mathrm{End}\,(\mathbb{C}^N)^{\otimes n} \otimes \mathrm{U}(\mathfrak{gl}_N).$$

Then Theorem 3.6 and definition (3.3) imply

(5.5) $$\begin{aligned} E_\lambda(z) &= \mathrm{id} \otimes \pi(T_\lambda(z))(z - c_1) \cdots (z - c_n) \\ &= F_\lambda \otimes 1 \cdot (E_1 + z - c_1) \cdots (E_n + z - c_n). \end{aligned}$$

Let tr denote the standard matrix trace on $\mathrm{End}\,(\mathbb{C}^N)^{\otimes n}$. By definition we have

(5.6) $$e_\lambda(z) = \mathrm{tr} \otimes \mathrm{id}\,(E_\lambda(z)).$$

We denote by D the following element of the algebra $\mathrm{End}\,(\mathbb{C}^N)^{\otimes n} \otimes \mathcal{PD}$

$$\sum_{i_1,\ldots,j_n} \sum_{j_1,\ldots,j_n} \sum_{a_1,\ldots,a_n} E_{j_1 i_1} \otimes \cdots \otimes E_{j_n i_n} \otimes x_{i_1 a_1} \cdots x_{i_n a_n} \partial_{j_1 a_1} \cdots \partial_{j_n a_n}.$$

Consider the product $C_\lambda = F_\lambda \otimes 1 \cdot D$. We shall need the following observation.

PROPOSITION 5.4. *We have the equality $c_\lambda = \mathrm{tr} \otimes \mathrm{id}(C_\lambda)$.*

PROOF. For any F from the image of the symmetric group S_n in $\mathrm{End}\,(\mathbb{C}^N)^{\otimes n}$, the elements $F \otimes 1$ and D of the algebra $\mathrm{End}\,(\mathbb{C}^N)^{\otimes n} \otimes \mathcal{PD}$ commute. Thus Proposition 5.4 follows from the definition of c_λ and the identity in $\mathbb{C} \cdot S_n$

$$\frac{1}{n!} \sum_{\sigma \in S_n} \sigma \Phi_\lambda \sigma^{-1} = \frac{1}{\dim U_\lambda} \sum_{\sigma \in S_n} \chi_\lambda(\sigma) \sigma. \quad \square$$

By Proposition 5.4 and by (5.6), the next proposition implies Theorem 5.3.

PROPOSITION 5.5. *The image of $E_\lambda(0)$ in $\mathrm{End}\,(\mathbb{C}^N)^{\otimes n} \otimes \mathcal{PD}$ is C_λ.*

PROOF. Consider the standard grading on the vector space $\mathcal{PD}$ by degree of differential operators. Extend this grading to the vector space $\mathrm{End}\,(\mathbb{C}^N)^{\otimes n} \otimes \mathcal{PD}$. Denote the image of $E_\lambda(0)$ in that space by C'_λ. By (5.5), the leading term of C'_λ coincides with the homogeneous element C_λ. Thus it remains to show that the element C'_λ is homogeneous as well. By (5.2), it suffices to check that for any Young diagram μ with less than n boxes we have

(5.7) $$\mathrm{id} \otimes \pi_\mu(E_\lambda(0)) = 0.$$

In this case the diagram λ is not contained in μ. The number of zeros among the contents $c_1, \ldots, c_n$ of the boxes of the diagram λ is exactly its rank r. So by the definition (5.4), relation (5.7) follows from Corollary 3.7 and Theorem 4.1. $\quad \square$

Acknowledgements. The author was supported by EPSRC Advanced Research Fellowship. He was also supported by the Erwin Schrödinger Institute in Vienna.

References

[C]   A. Capelli, *Sur les opérations dans la théorie des formes algébriques*, Math. Ann. **37** (1890), 1–37.

[C1]   I. Cherednik, *Special bases of irreducible finite-dimensional representations of the degenerate affine Hecke algebra*, Funktsional. Anal. i Prilozhen. **20** (1986), no. 1, 87–88; English transl. in Functional Anal. Appl. **20** (1986).

[C2]   _____, *A new interpretation of Gelfand–Zetlin bases*, Duke Math. J. **54** (1987), 563–577.

[D1]   V. Drinfeld, *Hopf algebras and the quantum Yang–Baxter equation*, Dokl. Akad. Nauk SSSR **283** (1985), no. 5, 1060–1064; English transl., Soviet Math. Dokl. **32** (1985), 254–258.

[D2]   _____, *On almost cocommutative Hopf algebras*, Algebra i Analiz **1** (1989), no. 2, 30–46; English transl., Leningrad Math. J. **1** (1990), 321–342.

[GM]   A. Garsia and T. Maclarnan, *Relations between Young's natural and the Kazhdan–Lusztig representations of S_n*, Adv. Math. **69** (1988), 32–92.

[HU]   R. Howe and T. Umeda, *The Capelli identity, the double commutant theorem, and multiplicity-free actions*, Math. Ann. **290** (1991), 569–619.

[J]   A. Jucys, *On the Young operators of the symmetric groups*, Lietuvos Fizikos Rinkinys **6** (1966), 163–180.

[JKMO] M. Jimbo, A. Kuniba, T. Miwa, and M. Okado, *The $A_n^{(1)}$ face models*, Comm. Math. Phys. **119** (1988), 543–565.

[KRS]   P. Kulish, N. Reshetikhin, and E. Sklyanin, *Yang–Baxter equation and the representation theory I*, Lett. Math. Phys. **5** (1981), 393–403.

[MNO] A. Molev, M. Nazarov, and G. Olshanski, *Yangians and classical Lie algebras*, Uspekhi Mat. Nauk **51** (1996), no. 2, 27–104; English transl., Russian Math. Surveys **51** (1996), 205–282.

[N1]   M. Nazarov, *Quantum Berezinian and the classical Capelli identity*, Lett. Math. Phys. **21** (1991), 123–131.

[N2]   _____, *Young's symmetrizers for projective representations of the symmetric group*, Adv. Math. **127** (1997), 190–257.

[O]   G. Olshanski, *Representations of infinite-dimensional classical groups, limits of enveloping algebras, and Yangians*, Topics in Representation Theory, Advances in Soviet Math., vol. 2, Amer. Math. Soc., Providence, 1991, pp. 1–66.

[O1]   A. Okounkov, *Quantum immanants and higher Capelli identities*, Transformation Groups **1** (1996), 99–126.

[O2]   _____, *Young basis, Wick formula, and higher Capelli identities*, Int. Math. Res. Notes (1996), no. 17, 817–839.

[OO]   A. Okounkov and G. Olshanski, *Shifted Schur functions*, Algebra i Analiz **9** (1997), no. 2, 79–147; English transl. in St.-Petersburg Math. J. **9** (1998) (to appear).

[RS]   N. Reshetikhin and M. Semenov-Tian-Shansky, *Central extensions of quantum current groups*, Lett. Math. Phys. **19** (1990), 133–142.

[S]   S. Sahi, *The spectrum of certain invariant differential operators associated to a Hermitian symmetric space*, in: Lie Theory and Geometry, Progress in Math., vol. 123, Birkhäuser, Boston, 1994, pp. 569–576.

[Y1]   A. Young, *On quantitative substitutional analysis*. I, Proc. London Math. Soc. **33** (1901), 97–146; II **34** (1902), 361–397.

[Y2]   _____, *On quantitative substitutional analysis*. VI, Proc. London Math. Soc. **31** (1931), 253–289.

DEPARTMENT OF MATHEMATICS, UNIVERSITY OF YORK, YORK YO1 5DD, ENGLAND
E-mail address: `mln1@york.ac.uk`

Hinges and the Study–Semple–Satake–Furstenberg–De Concini–Procesi–Oshima Boundary

Yurii A. Neretin

To A. A. Kirillov on the occasion of his 60th anniversary

ABSTRACT. The aim of the present paper is twofold. One of the objectives is to present an elementary geometric description of the boundaries of symmetric spaces. These boundaries arise in mathematics independently and not simultaneously in quite different connections: enumerative algebraic geometry, harmonic analysis on symmetric spaces, the theory of automorphic forms. For spaces of rank one, e.g., for the Lobachevsky space $SO(n,1)/SO(n-1,1)$, the structure of the boundary is very simple. For symmetric spaces of rank > 1, there are many different boundaries and their structure is rather complicated. The second objective of the paper is to present the author's results announced in [**Ner4, Ner5, Ner7**]. We do not discuss applications of the constructions under consideration (except in the very simple §0) and also do not survey everything known about such objects. In Chapter I we describe in detail the variety of complete collineations defined by Semple in 1951 and show in Chapters II and III how to extract explicit constructions of other objects of the same kind from this description. Moreover, we give elementary constructions for the Satake-Furstenberg, Martin, and Karpelevich boundaries of symmetric spaces and construct some "new" boundaries.

Contents

§0. The problem of five conics and complete conics

Chapter I. Hinges and the canonical completion of the group $PGL_n(\mathbb{C})$

§1. Exterior algebras and the category GA

1991 *Mathematics Subject Classification.* Primary 22E15, 32M15, 53C35; Secondary 31C35.

Key words and phrases. Hausdorff distance, symmetric space, complete symmetric varieties, linear relation, Satake–Furstenberg boundary, Karpelevich boundary, Martin boundary, Tits building, compactifications.

Acknowledgements. I am indebted to C. De Concini, S. L. Tregub, M. M. Kapranov, W. Ballmann, M. A. Olshanetsky, B. Kostant, and È. B. Vinberg for discussion of this subject, useful remarks and references. I thank the administration of the Max-Planck-Institut für Mathematik (Bonn), the Mittag-Leffler Institute (Stockholm), and the Erwin Schrödinger Institut (Vienna) for their hospitality.

Supported by Russian Foundation for Basic Research under grant No. 95-01-00814.

©1998 American Mathematical Society

§2. Construction of a Hausdorff quotient space
§3. Hinges
§4. Projective embedding
§5. Semigroup of hinges
§6. Representations

Chapter II. Supplementary remarks
§7. Action of the semigroup $\widetilde{\mathrm{Hinge}}_n$ on the flag space
§8. Complete symmetric varieties
§9. Real forms of complete symmetric varieties. The Satake-Furstenberg boundary
§10. Category of hinges

Chapter III. Other boundaries
§11. Velocity compactifications of symmetric spaces
§12. Tits building on the matrix sky
§13. Hybridization: The Dynkin–Olshanetsky and Karpelevich boundaries
§14. The space of geodesics and sea urchins
§15. The boundary of a Bruhat–Tits building
§16. Remarks

§0. The problem of five conics and complete conics

We recall an old problem of enumerative algebraic geometry, which was widely discussed in the second half of the nineteenth century (see remarks in §16). Now this problem is mainly of historical interest, but still is not trivial.

On the complex projective plane $\mathbb{C}P^2$, let five conics $L_1, \ldots, L_5$ in general position be given. The question is: how many different conics are tangent to all five conics $L_1, \ldots, L_5$?

0.1. The space of conics. A conic on the plane $\mathbb{C}P^2$ is given by the equation

$$(0.1) \qquad \sum_{i,j=1}^{3} a_{ij} x_i x_j = 0,$$

where $a_{ij} \in \mathbb{C}$, $a_{ij} = a_{ji}$, and some a_{ij} are nonzero. Since the set of solutions of (0.1) is preserved under multiplication of all a_{ij} by a number, the space Q of all conics is identified with the 5-dimensional projective space $\mathbb{C}P^5$ (with coordinates $a_{11} : a_{12} : a_{13} : a_{22} : a_{23} : a_{33}$).

The group $PGL_3(\mathbb{C})$ acts on $\mathbb{C}P^2$ by projective transformations; therefore, this group acts also on the space $Q = \mathbb{C}P^5$.

As is known, the group $PGL_3(\mathbb{C})$ has the following three orbits in $Q = \mathbb{C}P^5$:

1° Nondegenerate conics. The stabilizer of the conic $x_1^2 + x_2^2 + x_3^2 = 0$ is the complex orthogonal group $O_3(\mathbb{C})$. Therefore, the space of all nondegenerate conics is the homogeneous (symmetric) space $PGL_3(\mathbb{C})/SO_3(\mathbb{C})$.

2° Pairs of intersecting lines.

3° Pairs of coinciding lines (double lines).

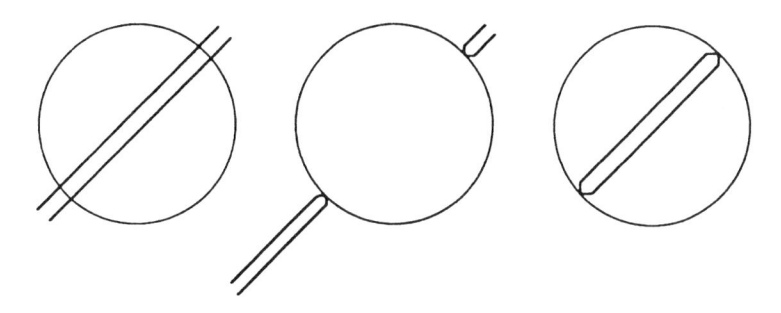

FIGURE 0.1

0.2. An attempt to solve the problem. Consider two conics given by the symmetric matrices $A = (a_{ij})$ and $P = (p_{ij})$.

The fact that the conics are tangent can be expressed by the following condition:
(∗)   the equation $f(\lambda) = \det(A - \lambda P) = 0$ has a multiple root.
Therefore, the discriminant $D(A, P)$ of this equation must be zero. Now the problem of five conics can be restated in the following form: given five matrices $A_1, \ldots, A_5$, find the number of solutions of the system of equations

(0.2) $$D(A_k, P) = 0, \quad k = 1, \ldots, 5.$$

If the surfaces of sixth order $D(A_k, P) = 0$ were in general position, then (by the Bézout theorem) the answer would be equal to 6^5.

However, these surfaces are not in general position. The point is that each conic is tangent (in the sense of condition (∗)) to every double line (see Figure 0.1). Therefore, the set of solutions to system (0.2) contains the two-dimensional surface of double lines, and we cannot apply the Bézout theorem.

0.3. Complete conics. We were interested in the set of solutions of system (0.2) in the space of nondegenerate quadrics $PGL_3(\mathbb{C})/SO_3(\mathbb{C})$. In fact we have set the problem in the completion $Q = \mathbb{C}P^5$ of the space $PGL_3(\mathbb{C})/SO_3(\mathbb{C})$.

An unpleasant phenomenon, namely, the appearance of a two-dimensional component in the space of solutions of system (0.2), occurred on the boundary of the space $PGL_3(\mathbb{C})/SO_3(\mathbb{C})$. It turns out that the situation will be modified if we take another boundary of the space $PGL_3(\mathbb{C})/SO_3(\mathbb{C})$.

Consider the space $\widehat{\mathbb{C}P^2}$ dual to $\mathbb{C}P^2$. Recall that, by definition, the points of $\widehat{\mathbb{C}P^2}$ are lines in $\mathbb{C}P^2$. Moreover, if $a \in \widehat{\mathbb{C}P^2}$, then the set $L(a)$ of all lines in $\mathbb{C}P^2$ (i.e., points of $\widehat{\mathbb{C}P^2}$) passing through a is a line in $\widehat{\mathbb{C}P^2}$.

To any *nondegenerate* conic $Q \subset \mathbb{C}P^2$ we assign the *dual conic* $\widehat{Q}$ that consists of all lines tangent to Q. If Q is defined by the matrix $A = (a_{ij})$ (see (0.1)), then $\widehat{Q}$ is defined by the matrix A^{-1}.

Consider the embedding

(0.3) $$PGL_3(\mathbb{C})/SO_3(\mathbb{C}) \to \mathbb{C}P^5 \times \mathbb{C}P^5$$

that assigns to each nondegenerate conic Q the pair of conics $(Q, \widehat{Q}) \in \mathbb{C}P^5 \times \mathbb{C}P^5$.

The variety C of "complete conics" is defined as the closure of the image of the embedding (0.3). This closure consists of four orbits of the group $PGL_3(\mathbb{C})$:

1. All pairs of the form $(Q, \widehat{Q})$, where Q is a nondegenerate quadric.

2. All pairs (R, T), where the quadric $R \subset \mathbb{C}P^2$ is a pair of lines ℓ_1 and ℓ_2 meeting at the point a and T is a double line $L(a)$.

3. All pairs (S, H), where $S \subset \mathbb{C}P^2$ is a double line and $H \subset \widehat{\mathbb{C}P^2}$ is a pair of lines $L(a)$ and $L(b)$, where $a, b \in S$.

4. All pairs (M, N), where $M \subset \mathbb{C}P^2$ is a double line and $N \subset \widehat{\mathbb{C}P^2}$ is a double line of the form $L(a)$, where $a \in M$.

The dimensions of these orbits are 5, 4, 4, and 3, respectively.

THEOREM 0.1. *C is a smooth algebraic variety.*

This assertion is rather simple (for details, see [**SR, DCP, GH**] and Theorem 4.6 below).

0.4. Solution of the problem. Now let X_Q be the set of all nondegenerate quadrics tangent to the given nondegenerate quadric Q. Let $\overline{X_Q}$ be the closure of X_Q in C. The following theorem is a pleasant exercise.

THEOREM 0.2. *"a)" The homology ring of C is generated by two cycles: the cycle λ that consists of all quadrics passing through a given point and the cycle μ that consists of all quadrics tangent to a given line.*

b) The cycles λ and μ satisfy the relations

(0.4) $\quad \lambda^5 = 1, \quad \lambda^4\mu = 2, \quad \lambda^3\mu^2 = 4, \quad \lambda^2\mu^3 = 4, \quad \lambda\mu^4 = 2, \quad \mu^5 = 1.$

c) Let ν be the homology class of the cycle $\overline{X_Q}$. Then we have $\nu = 2(\lambda + \mu)$.

REMARK 0.3. Equalities (0.4) have a clear geometric meaning. For example, the relation $\lambda^3\mu^2 = 4$ means that there exist four quadrics that pass through three points in general position and are tangent to two lines in general position.

Now it remains to compute

(0.5) $$\nu^5 = 2^5(\lambda + \mu)^5 = 3264.$$

Moreover, we must verify that the trouble we wish to avoid is now really absent. Namely, verify that there is no subvariety $K \subset C$ such that $K \subset \overline{X_Q}$ for all Q. Clearly, if such a variety K exists, then K must be $PGL_3(\mathbb{C})$-invariant; therefore, it can only coincide with the orbit of the fourth type from the list in subsection 0.3. It remains to show that the variety $\overline{X_Q}$ does not contain the orbit of the fourth type.

Chapter I. Hinges and the canonical completion of the group $PGL_n(\mathbb{C})$

§1. Exterior algebras and the category GA

1.1. Linear relations. Let V and W be finite-dimensional linear spaces. By a *linear relation* $P \colon V \rightrightarrows W$ we mean an arbitrary subspace $P \subset V \oplus W$.

Denote by $\mathrm{Gr}^k(H)$ the Grassmannian of all k-dimensional subspaces of the space H. The set of all linear relations $V \rightrightarrows W$ is the union of Grassmann varieties:

$$\bigcup_{k=0}^{\dim V + \dim W} \mathrm{Gr}^k(V \oplus W).$$

This set can be naturally endowed with the topology of disjoint union.

EXAMPLE 1.1. Let $A\colon V \to W$ be a linear operator. Consider its graph graph(A), that is, the set of pairs $(v, Av) \in V \oplus W$. Then graph(A) is a linear relation $V \rightrightarrows W$. Similarly, the graph of an operator $B\colon W \to V$ is also a linear relation $V \rightrightarrows W$. Below we do not distinguish between linear operators and their graphs.

Let $P\colon V \rightrightarrows W$ and $Q\colon W \rightrightarrows Y$ be linear relations. Then their product $QP\colon V \rightrightarrows Y$ is defined by the condition

$$(v, y) \in QP \iff \exists w \in W : (v, w) \in P, (w, y) \in Q.$$

Further, for every linear relation $P\colon V \rightrightarrows W$ we introduce the following objects:
a) by its *kernel* we mean $\operatorname{Ker} P = P \cap V$;
b) by its *image* $\operatorname{Im} P$ we mean the projection of P to W;
c) by its *domain* $\operatorname{Dom} P$ we mean the projection of P to V;
d) by its *indefiniteness* $\operatorname{Indef} P$ we mean the intersection of P with W.

REMARK 1.2. A linear relation P is the graph of an operator $A\colon V \to W$ if and only if we have $\operatorname{Dom} P = V$ and $\operatorname{Indef} P = \{0\}$. In this case we have $\operatorname{Ker} P = \operatorname{Ker} A$ and $\operatorname{Im} P = \operatorname{Im} A$.

For any linear relation $P\colon V \rightrightarrows W$ we introduce its *rank*

$$\operatorname{rk} P = \dim \operatorname{Dom} P - \dim \operatorname{Ker} P = \dim \operatorname{Im} P - \dim \operatorname{Indef} P$$
$$= \dim P - \dim \operatorname{Ker} P - \dim \operatorname{Indef} P.$$

Further, for $P\colon V \rightrightarrows W$ we define the *pseudoinverse* linear relation $P^{\square}\colon W \rightrightarrows V$; namely, we consider the same linear subspace $P \subset V \oplus W$ as a subspace of $W \oplus V$.

Let $P\colon V \rightrightarrows W$ be a linear relation and $R \subset V$ a subspace. Define the subspace $PR \subset W$ as the set of all $w \in W$ for which there exists a $v \in R$ such that $(v, w) \in P$.

Finally, let us define the *multiplication of a linear relation by a number* $\lambda \neq 0$. Suppose $P\colon V \rightrightarrows W$ is a linear relation. Then the linear relation λP consists of vectors of the form $(v, \lambda w)$, where $(v, w) \in P$.

1.2. **Category GA.** The objects of the category GA are finite-dimensional complex linear spaces. The set $\operatorname{Mor}(V, W) = \operatorname{Mor}_{GA}(V, W)$ of morphisms $V \to W$ consists of
a) all linear relations $V \rightrightarrows W$ defined up to a scalar multiplier,
b) a formal morphism $null = null_{V,W}$ that is not identified with any linear relation.

Let $P \in \operatorname{Mor}(V, W)$ and $Q \in \operatorname{Mor}(W, Y)$. Then the product $QP \in \operatorname{Mor}(V, Y)$ is defined by the following rule:
a) if at least one of the factors is $null$, then the product is $null$ as well;
b) if P and Q are linear relations and the following two conditions hold:

(1.1) $$\operatorname{Im} P + \operatorname{Dom} Q = W,$$
(1.2) $$\operatorname{Ker} Q \cap \operatorname{Indef} P = 0,$$

then the product QP can be calculated as the product of linear relations. For the case in which at least one of conditions (1.1) or (1.2) fails, the product is $null$.

We endow the set $\text{Mor}_{GA}(V,W)$ with a non-Hausdorff topology. We assume that a set R is closed if and only if the following two conditions hold:

a) *null* $\in R$;

b) $R \setminus \textit{null}$ is closed in the topology of the Grassmannian.

In particular, *null* is the only closed point, and the closure of any other singleton defined by the point $P \neq \textit{null}$ contains the point *null*.

THEOREM 1.3 (see [**Ner6**]). a) *The multiplication of morphisms of the category GA is associative.*

b) *If* $P \in \text{Mor}(V,W)$, $Q \in \text{Mor}(W,Y)$, *and if* P, Q, *and* QP *differ from null, then*

(1.3) $$\dim QP = \dim Q + \dim P - \dim W.$$

c) *The multiplication is continuous.*

PROOF. We restrict ourselves to the proof of b) (the same arguments prove c)). Thus, let assumptions (1.1) and (1.2) hold. Denote by Z the space $V \oplus W \oplus W \oplus Y$, and denote by X its subspace that consists of vectors of the form (v,w,w,y). Denote by T the subspace of X that consists of vectors of the form $(0,w,w,0)$. Then QP is the image of the subspace $(Q \oplus P) \cap X$ under the projection $X \to X/T = V \oplus Y$; by condition (1.1), we have $(Q \oplus P) + X = Z$, and therefore

$$\dim (Q \oplus P) \cap X = \dim (Q \oplus P) + \dim X - \dim Z = \dim Q + \dim P - \dim W.$$

By condition (1.2) we have $T \cap ((Q \oplus P) \cap X) = 0$; hence, the projection $X \to X/T$ is injective on T, and the assertion is proved. $\square$

REMARK 1.4. This theorem is one of the reasons (possibly not the most important) to introduce the element *null*; otherwise, assertions b) and c) of Theorem 1.4 would fail for the category of linear relations.

1.3. Exterior algebras. Let us introduce some standard notation. Let V be a complex linear space of dimension n with basis $e_1, \ldots, e_n$. Denote by $\Lambda^k V$ the kth exterior power of the space V, i.e., the set of all vectors of the form

$$\sum_{i_1 < \cdots < i_k} \alpha_{i_1 \ldots i_k} e_{i_1} \wedge \cdots \wedge e_{i_k}.$$

Denote by $\Lambda(V) = \bigoplus_{k=0}^{n} \Lambda^k V$ the exterior algebra over V.

Let $A \colon V \to W$ be a linear operator. Denote by $\Lambda^k A \colon \Lambda^k V \to \Lambda^k W$ the kth exterior power of A and denote by ΛA the operator

$$\Lambda A = \bigoplus_{k=0}^{n} \Lambda^k A \colon \Lambda V \to \Lambda W.$$

Let $v \in V$. Denote by $a(v)$ the *operator of exterior multiplication* (creation operator) $a(v)h = v \wedge h$ in ΛV. This operator maps $\Lambda^k V$ into $\Lambda^{k+1} V$.

Let V' be the dual space of V. Let $f \in V'$. Denote by $a^+(f)$ the *operator of interior multiplication* (annihilation operator) in $\Lambda^k V$. This operator is defined by the relation

$$a(f) v_1 \wedge \cdots \wedge v_k = \sum (-1)^{j+1} f(v_j) v_1 \wedge \cdots \wedge v_{j-1} \wedge v_{j+1} \wedge \cdots \wedge v_k.$$

The operators $a(v)$ and $a^+(f)$ satisfy the following identities (*canonical anticommutation relations*):

(1.4) $\quad a(v)a(w) + a(w)a(v) = 0, \qquad a^+(f)a^+(g) + a^+(g)a^+(f) = 0,$

(1.5) $\qquad\qquad\qquad a^+(f)a(v) + a(v)a^+(f) = f(v) \cdot E.$

1.4. Plücker embeddings. Let $S \subset V$ be a subspace of dimension k. In S we introduce a basis $s_1, \ldots, s_k$ and consider the vector $s_1 \wedge \cdots \wedge s_k \in \Lambda^k V$.

Let $t_j = \sum q_{ij} s_j$ be another basis in S. Let Q be the matrix with entries q_{ij}. Then we have
$$t_1 \wedge \cdots \wedge t_k = \det(Q) s_1 \wedge \cdots \wedge s_k.$$
Thus, the subspace S defines a vector in $\Lambda^k V$ which is determined up to a nonzero scalar factor.

Therefore, we have obtained the so-called *Plücker embedding*
$$\mathrm{Gr}^k(V) \to \mathbb{P}(\Lambda^k V)$$
(we denote by $\mathbb{P}W$ the projectivized space W).

1.5. Fundamental representation of the category GA. To any complex linear space V we assign the space ΛV. To any $P \in \mathrm{Mor}_{GA}(V, W)$ we assign an operator $\lambda(P) \colon \Lambda V \to \Lambda W$, which is defined up to a constant factor, so that for any V, W, Y and any $P \in \mathrm{Mor}(V, W)$ and $Q \in \mathrm{Mor}(W, Y)$ we have $\lambda(QP) = s(Q, P) \lambda(Q) \lambda(P)$, where $s(Q, P) \in \mathbb{C}$. In other words, λ is a *projective representation of the category GA*.

First we define the operator $\lambda(\cdot)$ for some special morphisms of the category GA.

1^+. We set $\lambda(null) = 0$ (this is another reason to introduce the morphism $null$).

2^+. Let X be a subspace of V. Let $T \colon V \rightrightarrows X$ be the graph of the embedding $X \to V$. Let $f_1, \ldots, f_\alpha$ be a basis in the space of linear functionals that annihilate the subspace X. Then the operator $\lambda(T) \colon \Lambda V \to \Lambda X$ is defined by the formula

(1.6) $\qquad\qquad\qquad \lambda(T) = a^+(f_1) \cdots a^+(f_\alpha).$

3^+. Let $Q \colon X \rightrightarrows Y$ be the graph of an operator $A \colon X \to Y$. Then we set $\lambda(Q) = \Lambda A = \bigoplus_j \Lambda^j A$.

4^+. Let Y be the quotient space $Y = W/L$. Let $R \colon Y \rightrightarrows W$ be the graph of the projection $W \to Y$. Choose a basis $e_1, \ldots, e_\beta$ in L. Then

(1.7) $\qquad\qquad\qquad \lambda(R) = a(e_1) a(e_2) \cdots a(e_\beta).$

It is important to note that the operator $\lambda(R)$ does not depend on the choice of a basis. Indeed, $\lambda(R) h = (e_1 \wedge \cdots \wedge e_\beta) \wedge h$, and, as we have seen in the previous subsection the product $e_1 \wedge \cdots \wedge e_\beta$ is defined by the subspace L uniquely up to a scalar factor. The same arguments show that the operator (1.6) is defined by the subspace X uniquely up to a scalar factor (and does not depend on the choice of the basis $f_1, \ldots, f_\alpha$). Indeed, by identities (1.4), operator (1.6) is defined by the element $f_1 \wedge \cdots \wedge f_\alpha \in \Lambda V^\circ$, where V° is the space dual to V.

Now let us take an arbitrary linear relation $P \colon V \rightrightarrows W$. We set $X = \mathrm{Dom}\, P$ and $Y = W/\mathrm{Indef}\, P$. Then P can be decomposed into the product $P = RQT$, where $T \colon V \rightrightarrows X$ has the form described in 2^+, $R \colon W/\mathrm{Indef}\, P \rightrightarrows W$ has the form

4^+, and Q is the graph of the operator $\operatorname{Dom} P \to W/\operatorname{Indef} P$, which is defined in an obvious way. Now we set $\lambda(P) = \lambda(R)\lambda(Q)\lambda(T)$.

THEOREM 1.5. *The functor $\lambda(\,\cdot\,)$ is a projective representation of the category GA, i.e., for any V, W, and Y and any $P \in \operatorname{Mor}(V,W)$ and $Q \in \operatorname{Mor}(W,Y)$ we have $\lambda(QP) = s(Q,P)\lambda(Q)\lambda(P)$, where $s(Q,P) \in \mathbb{C} \setminus \{0\}$.*

REMARK 1.6. Let $\dim V = n$, $\dim W = m$, and $\dim P = q$. Then it is clear that the operator $\lambda(P)\colon \Lambda V \to \Lambda W$ maps $\Lambda^j V$ into $\Lambda^{j+q-n}W$. In particular, if $V = W$ and $n = m = q$, then the operator $\lambda(P)$ leaves all exterior powers $\Lambda^j V$ invariant.

1.6. Proof of Theorem 1.5. Let V and W be linear spaces and let V° and W° be their dual spaces. Let $P\colon V \rightrightarrows W$ be a linear relation. We define the *dual linear relation* $P^\circ\colon V^\circ \rightrightarrows W^\circ$ as the set of all pairs $(f,g) \in V^\circ \oplus W^\circ$ such that for any $(v,w) \in P$ we have $f(v) = g(w)$.

We can readily verify that $(PQ)^\circ = P^\circ Q^\circ$.

REMARK 1.7. Let $P\colon V \rightrightarrows V$ be the graph of an invertible operator A. Then P° is the graph of the operator $(A^t)^{-1}$, where the operator $A^t\colon V^\circ \to V^\circ$ is adjoint to A.

THEOREM 1.8. *Let $P\colon V \rightrightarrows W$ be a linear relation.*

a) The operator $\lambda(P)\colon \Lambda V \to \Lambda W$ satisfies the relation $a(w)\lambda(P) = \lambda(P)a(v)$ for all pairs $(v,w) \in P$.

b) The operator $\lambda(P)$ satisfies the relation $a^+(g)\lambda(P) = \lambda(P)a^+(f)$ for all $(f,g) \in P^\circ$.

c) If a nonzero operator $\Delta\colon \Lambda(V) \to \Lambda(W)$ satisfies the relations

(1.8) $$a(w)\Delta = \Delta a(v),$$
(1.9) $$a^+(g)\Delta = \Delta a^+(f),$$

for all $(v,w) \in P$ and $(f,g) \in P^\circ$, then Δ coincides with $\lambda(P)$ up to a scalar factor.

PROOF. Choose bases $e_1, \ldots, e_n$; $f_1, \ldots, f_m$ such that the subspace $P \subset V \oplus W$ is a linear span of the vectors of the form

$$(e_1, f_1), \ldots, (e_\alpha, f_\alpha), \qquad (e_{\alpha+1}, 0), \ldots, (e_\beta, 0), \qquad (0, f_{\alpha+1}), \ldots, (0, f_\gamma).$$

Then all assertions become more or less evident.

Now let us pass directly to the proof of Theorem 1.5. Let $(v,w) \in P$ and $(w,y) \in Q$. Then

$$a(y)\lambda(Q)\lambda(P) = \lambda(Q)a(w)\lambda(P) = \lambda(Q)\lambda(P)a(v),$$

i.e., $\lambda(Q)\lambda(P)$ satisfies the same relations (1.8)–(1.9) as $\lambda(PQ)$. Therefore, by assertion c) of Theorem 1.8, $\lambda(QP)$ and $\lambda(Q)\lambda(P)$ coincide up to a scalar factor.

1.7. Categories of linear relations. The categories B, C, and GD defined below will be used in §8 only (for details, see [**Ner2, Ner6**]).

The objects of the category B are odd-dimensional complex linear spaces endowed with a nondegenerate *symmetric* bilinear form. Let V and W be objects of

B and let M_V and M_W be the corresponding bilinear forms. In $V \oplus W$ we introduce the symmetric bilinear form

$$M_{V \oplus W}((v,w),(v',w')) = M(v,v') - M(w,w').$$

The set $\mathrm{Mor}_B(V,W)$ of morphisms $V \to W$ consists of elements of two types:
 a) linear relations $P\colon V \rightrightarrows W$ such that $P \subset V \oplus W$ is a maximal isotropic (with respect to the form $M_{V \oplus W}$) subspace of $V \oplus W$;
 b) the element $null = null_{V,W}$.

The morphisms are multiplied in accordance with the rules of the category GA (note that conditions (1.1) and (1.2) are equivalent in this case).

REMARK 1.9. Recall that a subspace $H \subset Y$ is said to be *isotropic* with respect to the form M on Y if $M(h,h') = 0$ for all $h,h' \in H$. Note that in our case we have

$$\dim P = \frac{1}{2}(\dim V \oplus \dim W).$$

REMARK 1.10. Let $V = W$ and let A be an operator that preserves the form M_V, i.e.,

$$M(Av, Av') - M(v,v') = 0.$$

This is exactly equivalent to the condition that the graph P of the operator is isotropic in $V \oplus V$, that is, $P \in \mathrm{Mor}_B(V,V)$. We can readily show that the automorphism group of the object V is the orthogonal group of the space V.

The categories C and GD are defined in the same way; however, the objects of the category C are complex linear spaces endowed with a nondegenerate *skew-symmetric* bilinear form, and the objects of the category CD are complex *even-dimensional* spaces endowed with a nondegenerate *symmetric* bilinear form.

1.8. Category GA^*. Now we describe the category GA^*, which can be naturally regarded as a central extension of the category GA (we will first meet this category in §6).

The objects of the category GA^* are complex linear spaces. The morphisms from V to W are the operators $\Lambda(V) \to \Lambda(W)$ of the form $s \cdot \lambda(P)$, where $P \in \mathrm{Mor}_{GA}(V,W)$ and $s \in \mathbb{C}$.

The projection of GA^* onto GA is sufficiently evident (to the operator $s \cdot \lambda(P)$ we assign $P \in \mathrm{Mor}_{GA}(V,W)$ for $s \neq 0$ and $null$ for $s = 0$).

It would be of interest to obtain convenient explicit formulas for the product of operators $(s \cdot \lambda(P))(s' \cdot \lambda(P'))$. Clearly, this product has the form $s'' \cdot \lambda(PP')$, but the problem (possibly not very complicated) is to calculate s'' explicitly.

§2. Construction of a Hausdorff quotient space

2.1. Let M be a compact metric space and $M = \bigcup_{\alpha \in A} M_\alpha$ a partition of M into pairwise disjoint sets. Then the quotient set A is endowed with the *quotient topology*; namely, a subset $B \subset A$ is open if and only if the set $\bigcup_{\alpha \in B} M_\alpha$ is open in M.

Let $\alpha_j, \alpha \in A$. Then $\alpha = \lim_{j \to \infty} \alpha_j$ if and only if there exist $p \in M_\alpha$ and let $p_j \in M_{\alpha_j}$ be such that $p_j \to p$ in M.

In many interesting cases, the quotient topology on A is not Hausdorff (for instance, for the partition of $\mathbb{R}$ into positive numbers, negative numbers, and 0).

The following example is of main interest below.

2.2. Example: the Grassmannian.

Let $M = \mathrm{Gr}_{2n}^n$ be the Grassmannian of all n-dimensional linear subspaces of $\mathbb{C}^n \oplus \mathbb{C}^n$. Let the multiplicative group of the field of complex numbers $\mathbb{C}^* = \mathbb{C} \setminus 0$ act on $M = \mathrm{Gr}_{2n}^n$ by multiplication of linear relations by numbers. Consider the partition of the space $M = \mathrm{Gr}_{2n}^n$ into the orbits of the group $\mathbb{C}^*$.

If a linear relation $L \subset \mathbb{C}^n \oplus \mathbb{C}^n$ has the form $L = P \oplus Q$, where $P \subset \mathbb{C}^n \oplus 0$ and $Q \subset 0 \oplus \mathbb{C}^n$, then the point L is fixed with respect to $\mathbb{C}^*$. All other orbits, regarded as homogeneous spaces, are isomorphic to $\mathbb{C}^*$. Their closures in the Grassmannian (regarded as complex varieties) are isomorphic to the Riemann sphere $\overline{\mathbb{C}} = \mathbb{C} \cup \infty = \mathbb{C}^* \cup 0 \cup \infty$.

The one-point orbits correspond to the closed points of the quotient space $M = \mathrm{Gr}_{2n}^n / \mathbb{C}^*$ (i.e., one-point set is closed). If $S \in \mathrm{Gr}_{2n}^n$ is not a fixed point, then the closure of the corresponding one-point set in M contains two points:

$$\mathrm{Dom}\, S \oplus \mathrm{Indef}\, S \subset \mathbb{C}^n \oplus \mathbb{C}^n, \qquad \mathrm{Ker}\, S \oplus \mathrm{Im}\, S \subset \mathbb{C}^n \oplus \mathbb{C}^n.$$

2.3. Hausdorff metric: preliminaries.

Let M be a compact metric space with metric $\rho(\cdot, \cdot)$. Let $[M]$ be the set of all nonempty closed subsets of M. Recall the definition of the *Hausdorff metric* in M.

Let $m \in M$ and $N \in [M]$. Define the distance between m and N in the usual way by setting $\rho(m, N) = \min_{n \in N} \rho(m, n)$. Denote by N_ε the set of all points $m \in M$ at a distance $< \varepsilon$ from N, by $B_\varepsilon(m)$ the ball $\{x : \rho(x, m) < \varepsilon\}$, and by $\overline{Q}$ the closure of the set Q.

Let $N, L \in [M]$. Then the Hausdorff distance $h(N, L)$ is defined as the infimum of all $\varepsilon > 0$ such that $N \subset L_\varepsilon$ and $L \subset N_\varepsilon$.

Recall some simple facts about the Hausdorff metric.

LEMMA 2.1. *Let $N \in [M]$. Let $x_1, \ldots, x_p \in M$ be a collection of points such that $\rho(x_j, N) < \varepsilon$ and the balls $B_\varepsilon(x_j)$ cover N. Then the distance from N to the set $\{x_1, \ldots, x_p\}$ is at most ε.*

PROOF. The proof is obvious.

COROLLARY 2.2. *The space $[M]$ is compact.*

PROOF. Let $x_1, \ldots, x_k$ be an ε-net in M. Then the subsets of the set $x_1, \ldots, x_k$ form an ε-net in $[M]$.

Now let us give a constructive description of the convergence in $[M]$.

LEMMA 2.3. *Let $N_j, N \in [M]$. Then $N_j \to N$ provided the following two conditions hold:*

$1°$ *for any $n \in N$ and any $\varepsilon > 0$, the set $B_\varepsilon(n) \cap N_j$ is not empty, starting from some index j;*

$2°$ *for any $m \notin N$ there exists $\varepsilon > 0$ such that the intersection $B_\varepsilon(m) \cap N_j$ is empty, starting from some index j.*

Furthermore, let $N_j \in [M]$ be a sequence. Let Σ be the set of all its limit points. (The elements of the set Σ are closed subsets of M.) Let

$$Y = \bigcap_{K \in \Sigma} K, \qquad Z = \bigcup_{K \in \Sigma} K.$$

LEMMA 2.4. a) *We have* $y \in Y$ *if for any* $\varepsilon > 0$, *starting from some index* j, *the set* $B_\varepsilon(y) \cap N_j$ *is not empty.*

b) *We have* $z \in Z$ *if for any* $\varepsilon > 0$ *there exist arbitrarily large indices* j *such that the set* $B_\varepsilon(z) \cap N_j$ *is not empty.*

The proof is obvious.

Lemma 2.4 can also be rewritten in the following form.

LEMMA 2.5. a) $Y = \bigcap_{\Delta \subset \mathbb{N}} \overline{\bigcup_{j \in \Delta} N_j}$, *where* Δ *ranges over all infinite subsets of* $\mathbb{N}$,

b) $Z = \bigcap_{i \in \mathbb{N}} \overline{\bigcup_{j > i} N_j}$.

2.4. Construction of a Hausdorff quotient space. As above, let M be a compact metric space and $M = \bigcup M_\alpha$ ($\alpha \in A$) a partition. Denote by S the set of all subsets of M that are the unions of elements of the partition. Let the partition satisfy the following property: if $N \in S$, then its closure $\overline{N}$ also satisfies $\overline{N} \in S$. This property certainly holds if we consider the partition of the space M into the orbits of some group G.

Furthermore, let $\mathcal{M} \subset M$ be an open dense subset and $\mathcal{M} \in S$. Consider the set $\mathcal{A} \subset A$ of all $\alpha \in A$ such that $M_\alpha \subset \mathcal{M}$. Consider the subset $\widetilde{\mathcal{A}} \subset [M]$ that consists of all sets of the form $\overline{M_\alpha}$, $\alpha \in \mathcal{A}$. Denote by $\overline{\mathcal{A}}$ the closure of $\widetilde{\mathcal{A}}$ in $[M]$. By construction, $\overline{\mathcal{A}}$ is a compact metric space that contains $\widetilde{\mathcal{A}}$ as a dense subset. We call the space $\overline{\mathcal{A}}$ the *Hausdorff quotient space* of the space M.

Certainly, the set $\overline{\mathcal{A}}$ depends not only on the metric space M, but also on the subset $\mathcal{M}$.

EXAMPLE 2.6. Let the multiplicative group $\mathbb{R}^*$ of positive numbers act on $\overline{\mathbb{C}} = \mathbb{C} \cup \infty$ by multiplication of vectors by numbers. The orbits of $\mathbb{R}^*$ are the open rays starting from the origin 0 and the points 0 and ∞. If $\mathcal{M} = \overline{\mathbb{C}} \setminus \{0, \infty\}$, then the corresponding space $\overline{\mathcal{A}}$ is homeomorphic to the circle. The same assertion holds if $\mathcal{M}$ is obtained from $\overline{\mathbb{C}} \setminus \{0, \infty\}$ by deleting a finite number of rays. However, if $\mathcal{M} = \overline{\mathbb{C}}$, then the Hausdorff quotient is the disjoint union of two points and a circle.

Informally, we are interested in the situation where $\mathcal{M}$ is the union of the "generic" sets M_α.

2.5. Description of the Hausdorff quotient $\overline{\mathcal{A}}$ in terms of the non-Hausdorff quotient $\mathcal{A}$. Let $\alpha_j \in \mathcal{A}$ (we emphasize that α_j belongs not just to A, but to a certain subset of A). We say that the sequence α_j is *rigidly convergent* if all its limit points are limits. (Recall that a limit point of a sequence is a limit of a subsequence.)

We say that a subset $T \subset A$ is a *limit set* if there exists a rigidly convergent sequence $\alpha_j \in \mathcal{A}$ such that T is the set of limits of the sequence α_j.

EXAMPLE 2.7 (see subsection 2.2). Let Gr_4^2 be the set of all two-dimensional subspaces of $\mathbb{C}^2 \oplus \mathbb{C}^2$. Let $\mathrm{Gr}_4^2/\mathbb{C}^*$ be the space of orbits of the group $\mathbb{C}^*$ (that acts by multiplication of a linear relation by numbers).

Consider the sequence V_j in Gr_4^2 whose elements are the subspaces that consist of points of the form $(x, y; x, jy) \in \mathbb{C}^2 \oplus \mathbb{C}^2$. In other words, V_j is the graph of the

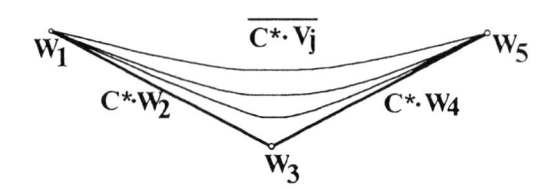

FIGURE 2.1

operator $\begin{pmatrix} 1 & 0 \\ 0 & j \end{pmatrix}$. By abuse of language, we can say that the graphs V_j are points of the space $\mathrm{Gr}_4^2/\mathbb{C}^*$. Consider the following five sequences in Gr_4^2:

$$S_j^{(1)} = (1/j^2)V_j, \quad S_j^{(2)} = (1/j)V_j, \quad S_j^{(3)} = (1/\sqrt{j})V_j, \quad S_j^{(4)} = V_j, \quad S_j^{(5)} = jV_j.$$

Note that all $S_j^{(k)}$ (where $k \in \{1, 2, 3, 4, 5\}$) are representatives of the $\mathbb{C}^*$-orbit of the element V_j.

The limits of the sequences $S_j^{(1)}, \ldots, S_j^{(5)}$ in the Grassmannian Gr_4^2 are the subspaces $W_1, \ldots, W_5$, respectively, consisting of vectors of the form

$$W_1 : (x, y; 0, 0), \ W_2 : (x, y; 0, y), \ W_3 : (x, 0; 0, y), \ W_4 : (x, 0; x, y), \ W_5 : (0, 0; x, y).$$

Thus, the points $W_1, \ldots, W_5$, regarded as elements of $\mathrm{Gr}_4^2/\mathbb{C}^*$, are limits of the sequence $V_j \in \mathrm{Gr}_4^2/\mathbb{C}^*$.

In $\mathrm{Gr}_4^2/\mathbb{C}^*$, there are no other limit points of the sequence V_j. Therefore, $\{W_1, \ldots, W_5\}$ is a limit set.

PROPOSITION 2.8. *Let $N \subset M$ be a closed set. The following conditions are equivalent*:
  a) *N is an element of the Hausdorff quotient $\overline{\mathcal{A}}$;*
  b) *there is a limit set $T \subset A$ such that $N = \bigcup_{\beta \in T} M_\beta$.*

PROOF. Let $\overline{M_{\alpha_j}}$ converge to N in the Hausdorff metric. Then by Lemma 2.5 we have $N \in S$, i.e., N is composed of sets M_β. We can readily see that here the index β ranges over a limit set.

EXAMPLE 2.9. Let us return to the previous example. The sets $\overline{\mathbb{C}^* \cdot V_j}$ are complex curves isomorphic to the Riemann sphere $\overline{\mathbb{C}} = \mathbb{C}P^1$. This sequence of curves converges, in the Hausdorff metric, to the union of two complex curves with a common point. In Figure 2.1 we show the arrangement of $\mathbb{C}^*$-orbits that correspond to the points V_j and W_1, W_2, W_3, W_4, and W_5.

2.6. **Example.** Let $\overline{\mathbb{C}} = \mathbb{C} \cup \infty$ be the Riemann sphere. Let $M = (\overline{\mathbb{C}})^n$. Let the group $\mathbb{C}^*$ act on M by the formula $(x_1, \ldots, x_n) \mapsto (\lambda x_1, \ldots, \lambda x_n)$. The quotient $\overline{\mathbb{C}}^n/\mathbb{C}^*$ is not a Hausdorff space. We shall construct the Hausdorff quotient.

For $\mathcal{M} \subset M$ we take the set $(\mathbb{C}^*)^n$. Let $\mathcal{A}$ be the set of orbits of $\mathbb{C}^*$ in $(\mathbb{C}^*)^n$.

EXAMPLE 2.10. Consider the following sequence of orbits in $(\mathbb{C}^*)^4/\mathbb{C}^*$ (we indicate representatives of the orbits in $(\mathbb{C}^*)^4$):

$$z_j = (1, j, 2j, j^2) \in (\mathbb{C}^*)^4.$$

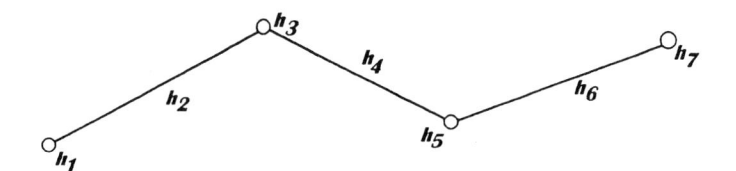

FIGURE 2.2

Then z_j is rigidly convergent, and the set of its limits (a limit set) is

$$h_1 = (0,0,0,0), \quad h_2 = (0,0,0,1), \quad h_3 = (0,0,0,\infty), \quad h_4 = (0,1,2,\infty),$$
$$h_5 = (0,\infty,\infty,\infty), \quad h_6 = (1,\infty,\infty,\infty), \quad h_7 = (\infty,\infty,\infty,\infty).$$

The points h_1, h_3, h_5, and h_7 are stable under the action of $\mathbb{C}^*$. The orbits of the points h_2, h_4, and h_6 are isomorphic to $\mathbb{C}^*$. Figure 2.2 illustrates the contiguity of the orbits.

THEOREM 2.11. a) *Every sequence*

$$z^{(j)} = (z_1^{(j)}, \ldots, z_n^{(j)}) \in (\mathbb{C}^*)^n$$

(we write down representatives of $\mathbb{C}^*$-orbits) that is rigidly convergent in $(\mathbb{C}^*)^n/\mathbb{C}^*$ must have the following form. There exists a family of subsets

$$\emptyset = L_0 \subset L_1 \subset \cdots \subset L_k = \{1, \ldots, n\}$$

of the n-tuple $\{1, \ldots, n\}$ such that $L_{j+1} \neq L_j$ and
  1^+ *for each p and any $\alpha \in L_p$, $\beta \notin L_p$ we have $\lim_{j \to \infty} z_\alpha^{(j)}/z_\beta^{(j)} = \infty$.*
  2^+ *for any $\alpha, \beta \in L_p \setminus L_{p+1}$ there exists a limit $\lim_{j \to \infty} z_\alpha^{(j)}/z_\beta^{(j)} \in \mathbb{C}^*$.*
  b) *Under the conditions above, the limit set for the sequence $z^{(j)}$ consists of elements of two types, $u^0, \ldots, u^k$ and $v^1, \ldots, v^k$:*
  $1°$ $u^p = (u_1^p, \ldots, u_n^p)$, *where*

$$u_\alpha^p = \begin{cases} 0, & \alpha \notin L_p, \\ \infty, & \alpha \in L_p. \end{cases}$$

  $2°$ *Choose $\gamma \in L_p \setminus L_{p-1}$. Then $v^p = (v_1^p, \ldots, v_n^p)$, where*

$$v_\alpha^p = \begin{cases} \infty, & \alpha \in L_{p-1}, \\ \lim_{j \to \infty} z_\alpha^{(j)}/z_\gamma^{(j)}, & \alpha \in L_p \setminus L_{p-1}, \\ 0, & \alpha \notin L_p. \end{cases}$$

REMARK 2.12. For the case considered in Example 2.10 we have a chain of subsets

$$\emptyset \subset \{4\} \subset \{2,3,4\} \subset \{1,2,3,4\},$$

and the corresponding limit set is

$$u^0 = h_1, \quad u^1 = h_3, \quad u^2 = h_5, \quad u^3 = h_7, \quad v^1 = h_2, \quad v^2 = h_4, \quad v^3 = h_6.$$

PROOF OF THEOREM 2.11. Since the Hausdorff quotient space is compact, any sequence $z^{(j)} \in (\mathbb{C}^*)^n/\mathbb{C}^*$ contains a rigidly convergent subsequence. Now we shall perform the promised construction, and the rigidly convergent sequence will have the form described in assertion a) of the theorem.

If our sequence is initially not of the form described in assertion a), then our procedure will make it possible to choose two different rigidly convergent subsequences with different limit sets.

Let us proceed with the description of the extracting procedure.

A point $z^{(j)} = (z_1^{(j)}, \ldots, z_n^{(j)})$ can be regarded as a point of the projective space $\mathbb{C}P^{n-1}$. From the sequence $z^{(j)}$ we extract a subsequence $u^{(j)} = (u_1^{(j)}, \ldots, u_n^{(j)})$ that is convergent in $\mathbb{C}P^{n-1}$, and let $r = (r_1, \ldots, r_n)$ be its limit. Let $L_1 \subset \{1, \ldots, n\}$ be the set of indices α such that $r_\alpha \neq 0$. For convenience of notation, assume that $L_1 = \{1, \ldots, \beta\}$. Furthermore, consider the sequence of vectors $v^{(j)} = (u_{\beta+1}^{(j)}, \ldots, u_n^{(j)})$, which we regard as a sequence in $\mathbb{C}P^{n-\beta-1}$ with homogeneous coordinates $u_{\beta+1} : u_{\beta+2} : \cdots : u_n$.

Choose a convergent subsequence $w^{(j)} = (w_{\beta+1}^{(j)}, \ldots, w_n^{(j)})$ of $v^{(j)}$. Let $q = (q_{\beta+1}, \ldots, q_n)$ be its limit. Let K be the set of indices α such that $q_\alpha \neq 0$ and let $L_2 = L_1 \cup K$.

Now we can continue the arguments with the remaining coordinates, and so on. This completes the proof. □

Now we consider an arbitrary filtration

$$\mathfrak{A}: \varnothing \subset A_1 \subset \cdots \subset A_k = \{1, \ldots, n\}.$$

To any such filtration we assign the set $R_\mathfrak{A}$ whose elements are the k-tuples $\{P^{(1)}, \ldots, P^{(k)}\}$ satisfying the following condition: each $P^{(\mu)} \in (\overline{\mathbb{C}})^n/\mathbb{C}^*$ has the form $\mathbb{C}^* \cdot (x_1^{(\mu)}, \ldots, x_n^{(\mu)})$, where

$$x_\alpha^{(\mu)} = 0 \quad \text{for } \alpha \notin A_k,$$
$$x_\alpha^{(\mu)} = \infty \quad \text{for } \alpha \in A_{k-1},$$
$$x_\alpha^{(\mu)} \neq 0, \infty \quad \text{for } \alpha \in A_k \setminus A_{k-1}.$$

The set $R_\mathfrak{A}$ is a complex variety isomorphic to $(\mathbb{C}^*)^{n-k}$.

Theorem 2.11 implies the following assertion.

THEOREM 2.13. *The Hausdorff quotient space coincides with the union of the sets $R_\mathfrak{A}$ over all filtrations $\mathfrak{A}$ of the set $\{1, \ldots, n\}$.*

Denote by $\overline{\mathbb{T}^{n-1}}$ the space thus obtained. By construction, $\overline{\mathbb{T}^{n-1}}$ is a compact metric space. It turns out that, in fact, $\overline{\mathbb{T}^{n-1}}$ is a smooth complex algebraic variety.

Now we shall construct a complex analytic atlas on $\overline{\mathbb{T}^{n-1}}$. The charts of this atlas are indexed by the maximal filtrations (or the linear orderings, equivalently) of the set $\{1, \ldots, n\}$:

$$\varnothing \subset A_1 \subset A_2 \subset \cdots \subset A_n = \{1, \ldots, n\},$$

where A_l consists of l elements. Let us describe the chart corresponding to the filtration

(2.1) $$\mathfrak{h}: \varnothing \subset \{1\} \subset \{2\} \subset \cdots \subset \{1, \ldots, n\}$$

(the other charts can be obtained from this one by using the action of the symmetric group). Define the mapping $\Delta\colon \mathbb{C}^{n-1} \to \overline{\mathbb{T}^{n-1}}$ as follows. Let $z = (z_1, \ldots, z_{n-1}) \in \mathbb{C}^{n-1}$ have the form

$$z = (z_1, \ldots, z_{\alpha_1-1}, 0, z_{\alpha_1+1}, \ldots, z_{\alpha_2-1}, 0, z_{\alpha_2+1}, \ldots),$$

where all z_γ are nonzero, except for $z_{\alpha_1}, z_{\alpha_2}, \ldots$. Then $\Delta(z)$ belongs to the set $R_{\mathfrak{A}}$ that corresponds to the filtration

$$\mathcal{W}\colon \varnothing \subset \{1, \ldots, \alpha_1\} \subset \{1, \ldots, \alpha_2\} \subset \cdots,$$

and we have $\Delta(z) = (P_1, P_2, \ldots)$, where $P_s \in (\mathbb{C})^n$ has the form

$$z = \Big(\underbrace{\infty, \ldots, \infty}_{\alpha_{s-1} \text{ times}}, 1, z_{\alpha_{s-1}+1}, z_{\alpha_{s-1}+1} z_{\alpha_{s-1}+2}, \ldots, \prod_{\alpha_{s-1} < m < \alpha_s} z_{\alpha_m}, 0, 0, \ldots \Big).$$

Now we consider the chart that corresponds to a maximal filtration

(2.2) $$\mathfrak{b}\colon \varnothing \subset B_1 \subset B_2 \subset \cdots \subset B_n = \{1, \ldots, n\}.$$

Denote by $i \prec j$ the ordering on the set $\{1, \ldots, n\}$ given by the filtration (2.2), namely, we set $i \prec j$ if there exists B_α such that $i \in B_\alpha$ and $j \notin B_\alpha$.

Let

(2.3) $$\varnothing \subset \{1, \ldots, q_1\} \subset \{1, \ldots, q_2\} \subset \cdots,$$

be the intersection of the filtrations (2.1) and (2.2).

As above, let $z_1, \ldots, z_{n-1}$ be the coordinates in the chart corresponding to the filtration (2.1) and $u_1, \ldots, u_{n-1}$ the coordinates in the chart corresponding to the filtration (2.2). The overlap functions

$$z_1 = z_1(u_1, \ldots, u_{n-1}), \quad \ldots, \quad z_{n-1} = z_{n-1}(u_1, \ldots, u_{n-1})$$

are defined on the entire space $\mathbb{C}^{n-1}$, except for the hyperplanes $u_j = 0$, where j ranges over the entire set $\{1, \ldots, n-1\}$ except for the points $q_1, q_2, \ldots$ (see (2.3)). These functions can readily be calculated:

(2.4) $$z_j = \prod_{\alpha \prec j} u_\alpha \Big/ \prod_{\alpha \prec j-1} u_\alpha.$$

Note that the variables u_{q_i} never occur in the denominator of formulas (2.4) (note that $z_{q_i} = u_{q_i}$); therefore, the overlap functions are indeed holomorphic on the domain under consideration.

2.7 Remark: Universalization of separated quotient. In subsection 2.4 for each $\mathcal{M} \in S$ we constructed separated quotient $\overline{\mathcal{A}} = \overline{\mathcal{A}}(\mathcal{M})$ of the space M. It is possible to construct separated quotient which does not depend on choice of subset $\mathcal{M}$. For this aim consider the space

$$\overline{\mathcal{A}}_{\text{univ}} = \cap_{\mathcal{M} \in S} \overline{\mathcal{A}}(\mathcal{M}),$$

where intersection is given by all open dense sets $\mathcal{M} \in S$. All separated quotients which we consider in this paper coincide with $\overline{\mathcal{A}}_{\text{univ}}$.

§3. Hinges

3.1. Definition. A *hinge* in $\mathbb{C}^n$ is a sequence

$$\mathcal{P} = (P_1, \ldots, P_k)$$

of n-dimensional linear relations $\mathbb{C}^n \rightrightarrows \mathbb{C}^n$ (*links of the hinge*) defined up to multiplication by scalar factors (which are different for different P_j) and satisfying the following conditions.

(3.1) $\quad\quad 1^+ \quad \operatorname{Ker} P_j = \operatorname{Dom} P_{j+1},$

(3.2) $\quad\quad \quad \operatorname{Im} P_j = \operatorname{Indef} P_{j+1},$

(3.3) $\quad\quad 2^+ \quad P_j \neq \operatorname{Ker} P_j \oplus \operatorname{Indef} P_j,$

(3.4) $\quad\quad 3^+ \quad \operatorname{Indef} P_1 = 0,$

(3.5) $\quad\quad \quad \operatorname{Ker} P_k = 0.$

Denote by Hinge_n the set of all hinges in $\mathbb{C}^n$.

REMARK 3.1. Certainly, here the main condition is stated in 1^+. Condition 3^+ is condition 1^+ interpreted for $j = 0$ and $j = k$. Condition 2^+ is not very essential and can be replaced by other conditions which are not worse by any means. For the sake of being definite, we chose one of the possibilities (which are formally nonequivalent but are essentially the same) in the statement; see Theorem 3.6 and the discussion in subsection 5.2. We would also like to note the following.

REMARK 3.2. By 2^+ we have the strict inclusions

$$\operatorname{Ker} P_j \supset \operatorname{Ker} P_{j+1}, \quad\quad \operatorname{Dom} P_j \supset \operatorname{Dom} P_{j+1},$$
$$\operatorname{Im} P_j \subset \operatorname{Im} P_{j+1}, \quad\quad \operatorname{Indef} P_j \subset \operatorname{Indef} P_{j+1}.$$

This means that the number k of links of the hinge is at most n.

EXAMPLE 3.3. The graph of an invertible operator is a hinge. The graph of a noninvertible operator does not satisfy condition 3^+.

EXAMPLE 3.4. Consider a two-link hinge $\mathcal{P} = (P_1, P_2)$. By condition 3^+, P_1 is the graph of an operator $A \colon \mathbb{C}^n \to \mathbb{C}^n$, and P_2 is the graph of an operator from the second copy of $\mathbb{C}^n$ into the first one. By condition 1^+, we have $AB = 0$ and $BA = 0$.

EXAMPLE 3.5. The subspaces W_2 and W_4 from Example 2.7 form a hinge.

3.2. Space of hinges as a Hausdorff quotient of the Grassmannian. Consider the action of the group $\mathbb{C}^*$ on Gr_{2n}^n described in subsection 2.2. For the set $\mathcal{M} \subset \operatorname{Gr}_{2n}^n$, we take the group $GL_n(\mathbb{C})$; to be more exact, consider the set of graphs of the invertible operators $\mathbb{C}^n \to \mathbb{C}^n$.

THEOREM 3.6. *For the subset $\mathcal{M} = GL_n(\mathbb{C})$, the limit sets in $\operatorname{Gr}_{2n}^n / \mathbb{C}^*$, are precisely the sets of the following form:*

(3.6) $\quad\quad\quad\quad Q_0, P_1, Q_1, P_2, \ldots, P_k, Q_k,$

where $\mathcal{P} = (P_1, \ldots, P_k)$ is a hinge and

$$Q_j = \operatorname{Ker} P_j \oplus \operatorname{Im} P_j = \operatorname{Dom} P_{j+1} \oplus \operatorname{Indef} P_{j+1} \subset \mathbb{C}^n \oplus \mathbb{C}^n.$$

COROLLARY 3.7. *The metric space* Hinge_n *is compact. The group* $PGL_n(\mathbb{C})$ *is a dense open subset of* Hinge_n.

Below we shall return to the proof of the theorem and the corollary; now we continue discussing the definition of hinge.

3.3. Another interpretation of hinges.
Let $\mathcal{P} = (P_1, \ldots, P_n)$ be a hinge in $\mathbb{C}^n$. For the summands in the sum $\mathbb{C}^n \oplus \mathbb{C}^n$, we introduce the following notation:

$$V := \mathbb{C}^n \oplus 0, \qquad W := 0 \oplus \mathbb{C}^n.$$

Furthermore, define subspaces Y_s and Z_s for $s \in \{0, \ldots, k\}$ as follows:

$$Y_j = \mathrm{Ker}\, P_j = \mathrm{Dom}\, P_{j+1}, \qquad j = 1, \ldots, k-1,$$
$$Y_0 = V, \qquad Y_k = 0,$$
$$Z_j = \mathrm{Im}\, P_j = \mathrm{Indef}\, P_{j+1}, \qquad j = 1, \ldots, k-1,$$
$$Z_0 = 0, \qquad Z_k = W.$$

We obtained two flags:

(3.7) $$V = Y_0 \supset Y_1 \supset \cdots \supset Y_k = 0,$$
(3.8) $$0 = Z_0 \subset Z_1 \subset \cdots \subset Z_k = V.$$

The linear relation P_j defines an invertible linear operator

$$A_j \colon \mathrm{Dom}\, P_j / \mathrm{Ker}\, P_j \to \mathrm{Im}\, P_j / \mathrm{Indef}\, P_j.$$

In particular,

(3.9) $$\dim Y_{j-1}/Y_j = \dim Z_j/Z_{j-1}.$$

Thus, every hinge defines the two flags (3.7) and (3.8) satisfying condition (3.9), and the collection of invertible operators

(3.10) $$A_j \colon Y_{j-1}/Y_j \to Z_j/Z_{j-1}$$

defined up to a factor.

Conversely, let the flags (3.7) and (3.8) and a collection of invertible operators (3.10) defined up to a factor be given. Let us choose subspaces $R_j \subset Y_{j-1}$ and $Q_j \subset Z_j$ such that

$$Y_{j-1} = Y_j \oplus R_j, \qquad Z_j = Z_{j-1} \oplus Q_j.$$

Then the operator A_j defines an operator $A'_j \colon R_j \to Q_j$. Furthermore, consider a linear relation $P_j \colon V \rightrightarrows W$ defined as the sum of the three subspaces:

$$Y_j \subset V, \qquad Z_{j-1} \subset W, \qquad \mathrm{graph}(A'_j) \subset R_j \oplus Q_j \subset V \oplus W.$$

We can readily see that the collection $(P_1, \ldots, P_k)$ is a hinge.

Thus, we have obtained a canonical one-to-one correspondence between the space of hinges and the space of collections (3.7), (3.8), and (3.10). These collections are said to be *framed pairs of flags*.

3.4. **Orbits of the group** $PGL_n(\mathbb{C}) \times PGL_n(\mathbb{C})$. As usual, denote by $PGL_n(\mathbb{C})$ the quotient group of the group $GL_n(\mathbb{C})$ by its center $\mathbb{C}^*$.

The group $G_n = PGL_n(\mathbb{C}) \times PGL_n(\mathbb{C})$ acts on Hinge_n in an obvious way:

$$(g_1, g_2) \colon \mathcal{P} \mapsto g_1^{-1} \mathcal{P} g_2 = (g_1^{-1} P_1 g_2, \ldots, g_1^{-1} P_k g_2).$$

PROPOSITION 3.8. *The hinges $\mathcal{P} = (P_1, \ldots, P_k)$ and $\mathcal{R} = (R_1, \ldots, R_k)$ belong to the same orbit of the group $PGL_n(\mathbb{C}) \times PGL_n(\mathbb{C})$ if and only if the sets of numbers*

(3.11) $$\dim \operatorname{Ker} P_1, \ \ldots, \ \dim \operatorname{Ker} P_{k-1}$$

and

$$\dim \operatorname{Ker} R_1, \ \ldots, \ \dim \operatorname{Ker} R_{k-1}$$

coincide. The dimension of the orbit that contains the hinge $\mathcal{P} = (P_1, \ldots, P_k)$ is equal to $n^2 - k$.

Let us give an equivalent formulation of this statement in terms given in subsection 3.3.

PROPOSITION 3.9. *The only $PGL_n(\mathbb{C}) \times PGL_n(\mathbb{C})$-invariant of a framed pair of flags is the set of numbers*

(3.12) $$\dim Y_1, \ \ldots, \ \dim Y_{k-1}.$$

The dimension of the corresponding orbit is equal to $n^2 - k$.

PROOF. Clearly, the set of invariants given in (3.11) (or in (3.12), which is the same) completely determines the orbit. Let us choose a canonical representative on any orbit. Namely, for $0 = j_0 < j_1 < \cdots < j_k = n$ we define a *canonical hinge* $\mathcal{R}(j_1, \ldots, j_{k-1}) = (R_1, \ldots, R_k)$, where $R_s \subset \mathbb{C}^n \oplus \mathbb{C}^n$ consists of all vectors of the form $(x_1, \ldots, x_n; y_1, \ldots, y_n)$ such that

(3.13) $\quad x_\alpha = 0, \quad \alpha > j_{s+1}; \qquad y_\beta = 0, \quad \beta \leqslant j_s; \qquad x_\gamma = y_\gamma, \quad j_s < \gamma \leqslant j_{s+1}.$

The stabilizer of this hinge consists of the pairs of block matrices of order

$$(\lambda_1 + \cdots + \lambda_k) \times (\lambda_1 + \cdots + \lambda_k),$$

where $\lambda_s = j_s - j_{s-1}$, of the form

$$\begin{pmatrix} A_1 & * & * & \cdots \\ 0 & A_2 & * & \cdots \\ 0 & 0 & A_3 & \cdots \\ \vdots & \vdots & \vdots & \ddots \end{pmatrix}, \quad \begin{pmatrix} t_1 A_1^{-1} & 0 & 0 & \cdots \\ * & t_2 A_2^{-1} & 0 & \cdots \\ * & * & t_3 A_3^{-1} & \cdots \\ \vdots & \vdots & \vdots & \ddots \end{pmatrix},$$

where $t_j \in \mathbb{C}^*$. The dimension of the stabilizer is equal to $n^2 + k$, and thus the assertion on the dimension of the orbit is proved.

3.5. **Proof of Theorem 3.6.** First we show that any set of the form (3.6) is a limit set in the sense of subsection 2.5. To this end, it suffices to restrict ourselves to the case of hinges of the form $\mathcal{R}(j_1, \ldots, j_{k-1})$ (see the proof of Proposition 3.9).

Let $a_1 > \cdots > a_k$. Consider the sequence

$$S_t = \begin{pmatrix} e^{a_1 t} E_{\lambda_1} & & \\ & e^{a_2 t} E_{\lambda_2} & \\ & & \ddots \end{pmatrix},$$

where $t = 1, 2, 3, \ldots$ and $\lambda_s = j_s - j_{s-1}$ (we denote by E_λ the identity matrix of size $\lambda \times \lambda$). Our further arguments proceed as in Example 2.7. By considering the sequence of matrices $e^{-a_q t} S_t$ we obtain, as the limit, the linear relation (3.13). Furthermore, choose b_q so that $a_{q-1} < b_q < a_q$. By considering the sequence of matrices $e^{-b_q t} S_t$ we obtain, as the limit, a linear relation R_q that consists of vectors of the form $(x_1, \ldots, x_n; y_1, \ldots, y_n)$ such that

$$\begin{cases} x_\alpha = 0, & \alpha > \lambda_q, \\ y_\alpha = 0, & \alpha \leqslant \lambda_q. \end{cases}$$

The set of linear relations $(R_0, Q_1, R_1, \ldots, R_{k-1}, Q_k, R_k)$ has the desired form.

Conversely, consider a sequence $A_j \in GL_n(\mathbb{C})$ regarded as a sequence in the (non-Hausdorff) quotient space $\mathrm{Gr}_{2n}^n / \mathbb{C}^*$. It suffices to show that A_j contains a rigidly convergent subsequence such that the set of its limits has the form (3.6).

Represent A_j in the form $A_j = B_j \Delta_j C_j$, where B_j and C_j are unitary matrices and Δ_j is a diagonal matrix:

$$\Delta_j = \begin{pmatrix} \delta_1^{(j)} & & \\ & \delta_2^{(j)} & \\ & & \ddots \end{pmatrix}, \qquad \delta_1^{(j)} \geqslant \delta_2^{(j)} \geqslant \cdots > 0.$$

Without loss of generality we may assume that the sequences B_j and C_j converge in the unitary group (otherwise we can pass to a subsequence).

Now the convergence is completely determined by the middle factor, and we shall watch this factor only. To the vector $\delta^{(j)} = \{\delta_1^{(j)} \geqslant \cdots \geqslant \delta_n^{(j)} > 0\}$ (composed of the eigenvalues of the matrix Δ_j) we can apply the procedure described in the proof of Theorem 2.11.

Namely, from the sequence $\delta^{(j)} = (\delta_1^{(j)} : \cdots : \delta_n^{(j)}) \in \mathbb{R}P^{n-1}$ we extract a subsequence $\tau^{(j)} = (\tau_1^{(j)}, \ldots, \tau_n^{(j)})$ that converges in $\mathbb{R}P^{n-1}$. Let $(p_1, \ldots, p_\alpha, 0, \ldots, 0)$ be its limit. Furthermore, from the sequence $(\tau_{\alpha+1}^{(j)} : \cdots : \tau_n^{(j)}) \in \mathbb{R}P^{n-\alpha-1}$ we extract a subsequence that is convergent in $\mathbb{R}P^{n-\alpha-1}$, and so on.

Finally, from the sequence Δ_j we extract a subsequence $\Xi^{(\mu)}$ of the form

$$\Xi^{(\mu)} = \begin{pmatrix} a_1^{(\mu)} D_1^{(\mu)} & & \\ & a_2^{(\mu)} D_2^{(\mu)} & \\ & & \ddots \end{pmatrix},$$

where:

1. the matrices $D_m^{(\mu)}$ are diagonal; for chosen m and $\mu \to \infty$, the sequence $D_m^{(\mu)}$ has a limit, and this limit is an invertible matrix;

2. we have $a_m^{(\mu)} > 0$, and $\lim_{\mu \to \infty} a_m^{(\mu)} / a_{m+1}^{(\mu)} = \infty$ for any m.

The set of limits of the sequence $\Xi^{(\mu)} = \Delta_{j_r}$ clearly has the form (3.6) (see Example 2.7), and this completes the proof of the theorem.

3.6. Contiguity of the orbits. Thus, the metric space Hinge_n is compact. As was shown in subsection 3.4, this space is the union of 2^{n-1} orbits of the group $PGL_n(\mathbb{C}) \times PGL_n(\mathbb{C})$. Recall that the orbits are indexed by the collections of numbers
$$\Sigma(\mathcal{P}) = (\dim \operatorname{Ker} P_1, \ldots, \dim \operatorname{Ker} P_{k-1}).$$

THEOREM 3.10. *The orbit of a hinge $\mathcal{R}$ is contained in the closure of the orbit of a hinge $\mathcal{R}'$ if and only if $\Sigma(\mathcal{R}') \subset \Sigma(\mathcal{R})$.*

COROLLARY 3.11. *The group $PGL_n(\mathbb{C})$ is an open dense set in Hinge_n.*

PROOF. The proof repeats the arguments of subsection 3.5. We only need to perform the selection procedure not for matrices but for points of a given orbit. By means of unitary matrices, the flags (3.7) and (3.8) can be transformed into canonical position, and then the problem is essentially reduced to the extraction of operators (3.10).

§4. Projective embedding

4.1. Semigroup GL_n^*. Let $V = \mathbb{C}^n$. Consider the semigroup GL_n^* formed by all linear relations in $\mathbb{C}^n$ of dimension n together with *null*.

Note that GL_n^* is a semigroup indeed (see relation (1.3)). We also note that the product of linear relations of dimension n itself can be not of dimension n. However, if the product has a "wrong" dimension, then this product in the category GA is equal to *null*.

Furthermore, we note that for $P \in GL_n^*$, the operator $\lambda(P)$ (see §1) preserves the subspaces $\Lambda^j \mathbb{C}^n \subset \Lambda \mathbb{C}^n$. Denote by $\lambda_j(P)$ the restriction of the operator $\lambda(P)$ to $\Lambda^j \mathbb{C}^n$ (certainly, if P is the graph of an operator A, then $\lambda_j(P) = \Lambda^j(A)$).

LEMMA 4.1. *Let $P \in GL_n^*$ be a linear relation.*

a) *The operator $\lambda_j(P)$ is nonzero if and only if j satisfies the condition*

(4.1) $$\dim \operatorname{Indef} P \leqslant j \leqslant \dim \operatorname{Im} P.$$

b) *If j satisfies the condition $\dim \operatorname{Indef} P < j < \dim \operatorname{Im} P$, then the operator $\lambda_j(P)$ uniquely determines a linear relation P up to a factor.*

The assertion is quite clear from the explicit construction of the operators $\lambda_j(P)$.

We are mainly interested in the case of $j = \dim \operatorname{Indef} P$ and $j = \dim \operatorname{Im} P$.

Case A): $j = \dim \operatorname{Indef} P$.

Let $e_1, \ldots, e_j$ be a basis in $\operatorname{Indef} P$ and let $f_1, \ldots, f_j$ be a basis in the space of linear functionals that annihilate $\operatorname{Dom} P$ (note that the condition $\dim P = n$ implies $\dim \operatorname{Dom} P = n - \dim \operatorname{Indef} P$). Then, as can be readily verified, the operator $\lambda_j(P)$ coincides with the restriction of the operator

(4.2) $$a(e_1) \cdots a(e_j) a^+(f_1) \cdots a^+(f_j)$$

to the subspace $\Lambda^j V \subset \Lambda V$.

We stress that in our case the operator $\lambda_j(P)$ is completely determined by the subspaces $\operatorname{Indef} P$ and $\operatorname{Dom} P$. We also stress that the rank of the operator (4.2)

FIGURE 4.1

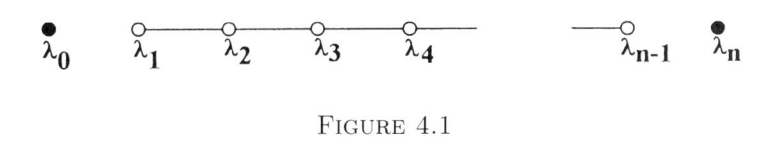

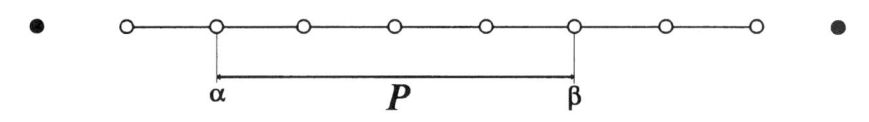

FIGURE 4.2

is equal to one. The image of this operator is the line spanned by the multivector $e_1 \wedge \cdots \wedge e_j \in \Lambda^j V$, and the kernel (of codimension one) coincides with the kernel of the linear functional $f_1 \wedge \cdots \wedge f_j \in \Lambda^j V^\circ = (\Lambda^j V)^\circ$.

Case B): $j = \dim \operatorname{Im} P$.

Let $e_1, \ldots, e_j$ be a basis in $\operatorname{Im} P$ and let $f_1, \ldots, f_j$ be a basis in the space of linear functionals that annihilate $\operatorname{Ker} P$. Then the operator $\lambda_j(P)$ coincides with the restriction of the operator

$$(4.3) \qquad a(e_1) \cdots a(e_j) a^+(f_1) \cdots a^+(f_j)$$

to the subspace $\Lambda^j V$.

We stress that the operators (4.2) and (4.3) coincide.

Case C): Q=Ker $Q \oplus$ Indef Q.

Consider a linear relation Q of the form

$$Q = \operatorname{Ker} Q \oplus \operatorname{Indef} Q \subset \mathbb{C}^n \oplus \mathbb{C}^n.$$

Let $\dim \operatorname{Indef} Q = j$, let $e_1, \ldots, e_j$ be a basis in $\operatorname{Indef} Q$ and let $f_1, \ldots, f_j$ be a basis in the space of linear functionals annihilating $\operatorname{Ker} Q$. Then we have

$$(4.4) \qquad \lambda_j(Q) = a(e_1) \cdots a(e_j) a^+(f_1) \cdots a^+(f_j), \qquad \lambda_\alpha(Q) = 0, \quad \alpha \neq j.$$

Consider the Dynkin diagram of the group $A_{n-1} = GL_n(\mathbb{C})$ shown in Figure 4.1. Here the circles mark the fundamental representations λ_j of the group $GL_n(\mathbb{C})$, i.e., the representations in $\Lambda^j V$. It is convenient to add two black circles, from the left and from the right, that correspond to λ_0 and λ_n.

Let $P \in GL_n^*$ be a linear relation. By the *domain of action* of P we mean the set of all j that satisfy (4.1). We shall depict the domain of action in Figure 4.2, where $\alpha = \dim \operatorname{Indef} P$ and $\beta = \dim \operatorname{Im} P$. Outside of the domain of action of P, the operators $\lambda_j(P)$ are equal to 0 and on the boundary (i.e., for $j = \alpha$ and for $j = \beta$), the operators $\lambda_j(P)$ have rank 1.

4.2. The operators $\lambda_j(\mathcal{P})$. Let $\mathcal{P} = (P_1, \ldots, P_k)$ be a hinge. Choose some $j \in \{1, \ldots, n-1\}$. Consider the sequence of operators

$$(4.5) \qquad \lambda_j(P_1), \ldots, \lambda_j(P_k).$$

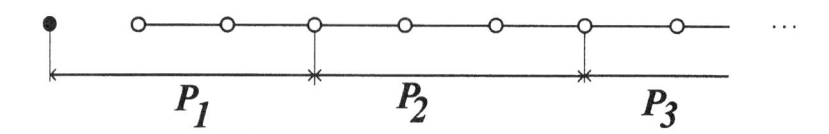

FIGURE 4.3

THEOREM 4.2. *There are exactly two possibilities concerning* (4.5).
1^+ *Exactly one term of the sequence* (4.5) *differs from* 0.
2^+ *There is a unique α such that $\lambda_j(P_\alpha) \neq 0$ and $\lambda_j(P_{\alpha+1}) \neq 0$. In this case*

$$\dim \mathrm{Dom}\,(P_\alpha) = j = \dim \mathrm{Indef}\,(P_{\alpha+1}),$$

the operators $\lambda_j(P_\alpha)$ and $\lambda_j(P_{\alpha+1})$ have rank 1 and coincide, up to a scalar factor.

PROOF. Indeed, by the definition of a hinge ($\mathrm{Im}\, P_\alpha = \mathrm{Dom}\, P_{\alpha+1}$), the domains of action of the linear relations $P_1, \ldots, P_k$ (see (4.1)) border on each other as shown in Figure 4.3.

Consider the linear relations P_μ and $P_{\mu+1}$. By the definition of hinge, we have $\mathrm{Im}\, P_\mu = \mathrm{Indef}\, P_{\mu+1}$ and $\mathrm{Ker}\, P_\mu = \mathrm{Dom}\, P_{\mu+1}$. Let $\alpha = \dim P_\mu$. By the remarks in subsection 4.1 (see (4.2) and (4.3)) we have $\lambda_\alpha(P_\mu) = s \cdot \lambda_\alpha(P_{\mu+1})$, $s \in \mathbb{C}^*$, and this proves the theorem.

We define the operator $\lambda_j(\mathcal{P}) \colon \Lambda^j \mathbb{C}^n \to \Lambda^j \mathbb{C}^n$ (which can be determined up to a scalar factor) as the nonzero element of the set

$$\lambda_j(P_1), \ldots, \lambda_j(P_k).$$

REMARK 4.3. Let us supplement the hinge $\mathcal{P} = (P_1, \ldots, P_k)$ to obtain the set $(Q_0, P_1, Q_1, \ldots, P_k, Q_k)$; see Theorem 3.6. Choose some μ. Denote the number

$$\dim \mathrm{Im}\, P_\mu = \dim \mathrm{Im}\, Q_\mu = \dim \mathrm{Indef}\, Q_\mu = \dim \mathrm{Indef}\, P_{\mu+1}$$

by α. Then, by the remarks in subsection 4.1, we have

$$\lambda_\alpha(P_\mu) = s \cdot \lambda_\alpha(Q_\mu) = t \cdot \lambda_\alpha(P_{\mu+1}), \qquad s, t \in \mathbb{C}^*.$$

The domains of action of the linear relations $Q_0, P_1, Q_1, \ldots$ are arranged as shown in Figure 4.4.

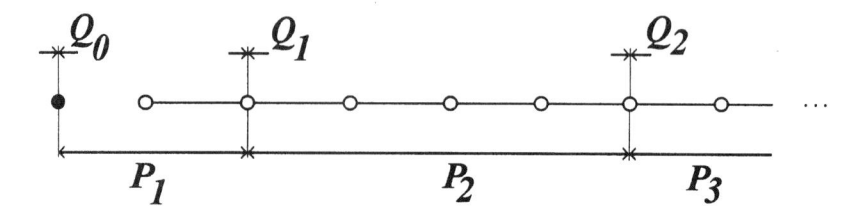

FIGURE 4.4

4.3. Projective embedding.
Let W be a linear subspace and let $Op(W)$ be the set of linear operators on W. Denote by $\mathbb{P}(Op(W))$ the set of nonzero operators defined up to a factor. Consider the mapping

$$\text{Hinge}_n \to \prod_{j=1}^{n-1} \mathbb{P}(Op(\Lambda^j \mathbb{C}^n))$$

given by the formula

$$\mathcal{P} = (P_1, \ldots, P_k) \mapsto \sigma(\mathcal{P}) = (\lambda_1(\mathcal{P}), \ldots, \lambda_{n-1}(\mathcal{P})).$$

THEOREM 4.4. *The mapping σ is continuous.*

PROOF. Let us prove that σ maps convergent sequences into convergent sequences. Let $\mathcal{P}_\alpha \in \text{Hinge}_n$ converge to $\mathcal{P}$. Without loss of generality we may assume that $\mathcal{P}$ is a canonical hinge of the form $\mathcal{R}(j_1, \ldots, j_k)$ (see subsection 3.4). Furthermore, without loss of generality we may assume that all points $\mathcal{P}_\alpha$ belong to the same orbit of the group $GL_n \times GL_n$ (otherwise we can pass to a subsequence). Let $\mathcal{R}(h_1, \ldots, h_s)$ be the canonical representative of this orbit. We stress that the set $h_1, \ldots, h_s$ is a subset of the collection $j_1, \ldots, j_k$. Furthermore, let the collection h_ν be empty (the general case differs from this one by complication of notation only), i.e., let the sequence $\mathcal{P}$ consist of invertible operators, which will be denoted by A_α.

The sequence A_α is representable in the form $A_\alpha = B_\alpha \Delta_\alpha C_\alpha$, where B_α and C_α are sequences of unitary matrices that tend to the identity matrix and Δ_α has the form

$$\Delta_\alpha = \begin{pmatrix} a_1^{(\alpha)} D_1^{(\alpha)} & & \\ & a_2^{(\alpha)} D_2^{(\alpha)} & \\ & & \ddots \end{pmatrix},$$

where

1°. $D_m^{(\alpha)}$ is a diagonal matrix of the size $(j_m - j_{m-1})$, and for any m, the sequence $D_m^{(\alpha)}$ tends to the identity matrix as $\alpha \to \infty$.

2°. $a_m^{(\alpha)} > 0$, and for any m we have $\lim_{\alpha \to \infty} a_m^{(\alpha)}/a_{m+1}^{(\alpha)} = \infty$.

The convergence $\Lambda^j(\Delta_\alpha) \to \lambda_j(\mathcal{R}(j_1, j_2, \ldots))$ is more or less obvious.

We illustrate the assertion by the following example.

EXAMPLE 4.5. Consider the sequence of operators

$$A_j = \begin{pmatrix} j^2 & & & \\ & j & & \\ & & j & \\ & & & 1 \end{pmatrix} : \mathbb{C}^4 \to \mathbb{C}^4.$$

In Hinge_n, this sequence is convergent to a (three-link) hinge $\mathcal{R}(1,3)$. The sequence $j^{-2} A_j$ is convergent to the operator in $\mathbb{C}^4$ with the matrix

$$R = \begin{pmatrix} 1 & & & \\ & 0 & & \\ & & 0 & \\ & & & 0 \end{pmatrix}.$$

The sequence $j^{-3}\Lambda^2 A_j$ is convergent to the operator S in $\Lambda^2\mathbb{C}^4$ that has the form $S(e_1 \wedge e_2) = e_1 \wedge e_2$, $S(e_1 \wedge e_3) = e_1 \wedge e_3$, and $S(e_\alpha \wedge e_\beta) = 0$ for all other pairs e_α, e_β. The sequence $j^{-4}\Lambda^3 A_j$ is convergent to the operator T given by the formula $T(e_1 \wedge e_2 \wedge e_3) = e_1 \wedge e_2 \wedge e_3$, and $T(e_\alpha \wedge e_\beta \wedge e_\gamma) = 0$ for all other triples e_α, e_β, e_γ.

We can readily see that the collection of operators R, S, and T really corresponds to the hinge $\mathcal{R}(1,3)$.

4.4. Smoothness. It follows from our constructions that Hinge_n is a projective algebraic variety.

THEOREM 4.6. *The variety* Hinge_n *is smooth.*

Before passing to the proof of the theorem, we consider the following example.

EXAMPLE 4.7 (*diagonal hinges*). We return to the situation described in subsection 2.6. Construct the natural embedding $(\overline{\mathbb{C}})^n \to \text{Gr}_{2n}^n$. Let us express a point of the jth copy of $\overline{\mathbb{C}}$ as the ratio a_j/b_j, where $a_j, b_j \in \mathbb{C}$ and at least one of these numbers is nonzero. Then to the point $a_1/b_1, \ldots, a_n/b_n$ we assign the subspace of $\mathbb{C}^n \oplus \mathbb{C}^n$ that consists of the vectors of the form $(b_1x_1, b_2x_2, \ldots, b_nx_n; a_1x_1, a_2x_2, \ldots, a_nx_n)$. The constructed embedding commutes with the action of the group $\mathbb{C}^*$, and therefore it induces a mapping of the Hausdorff quotients

$$\overline{\mathbb{T}^{n-1}} \to \text{Hinge}_n.$$

Furthermore, we note that the (smooth) variety $\overline{\mathbb{T}^{n-1}}$ is exactly the closure of the group of diagonal matrices in the space Hinge_n.

PROOF OF THEOREM 4.6. It suffices to introduce smooth coordinates in a neighborhood of an arbitrary point $\mathcal{R}(j_1, \ldots, j_s)$. Let $\varkappa_\alpha = j_\alpha - j_{\alpha-1}$. Consider a subgroup G of $GL_n \times GL_n$, where G consists of pairs of block matrices that have the size $(\varkappa_1 + \varkappa_2 + \cdots) \times (\varkappa_1 + \varkappa_2 + \cdots)$ and are of the form

$$\begin{pmatrix} E & & & \cdots \\ * & E & & \cdots \\ * & * & E & \cdots \\ \vdots & \vdots & \vdots & \ddots \end{pmatrix}; \quad \begin{pmatrix} B_1 & * & * & \cdots \\ & B_2 & * & \cdots \\ & & B_3 & \cdots \\ \vdots & \vdots & \vdots & \ddots \end{pmatrix}.$$

This subgroup is supplementary to the stabilizer of the point $\mathcal{R}(j_1, \ldots, j_s)$. Let us present a smooth transverse section to the orbits of this subgroup. To this end we consider the set D of operators of the form

$$\begin{pmatrix} x_1 E_{\varkappa_1} & & \\ & x_2 E_{\varkappa_2} & \\ & & \ddots \end{pmatrix}.$$

We can readily see that the closure $\overline{D}$ of the set D in the space of hinges is isomorphic to the variety $\overline{\mathbb{T}^s}$ from subsection 2.6. This is the desired section.

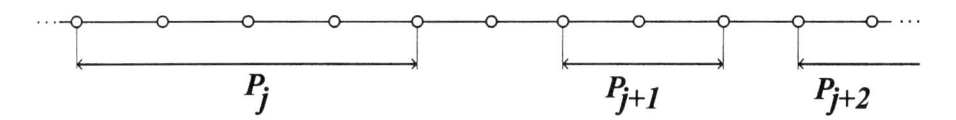

FIGURE 5.1

§5. Semigroup of hinges

5.1. Weak hinges. By a *weak hinge* in $\mathbb{C}^n$ we mean a family $\mathcal{P} = (P_1, \ldots, P_k)$ of linear relations of dimension $\leqslant n$ that are determined up to a factor and satisfy the relations
$$\operatorname{Ker} P_j \supset \operatorname{Dom} P_{j+1} \quad \text{and} \quad \operatorname{Im} P_j \subset \operatorname{Indef} P_{j+1}.$$

EXAMPLE 5.1. A HINGE IS A WEAK HINGE. Sometimes we shall speak of an *exact hinge* instead of the term "hinge".

EXAMPLE 5.2. The empty set is a weak hinge.

EXAMPLE 5.3. Let $\mathcal{P} = (P_1, \ldots, P_k)$ be a hinge. Then for any set of indices $0 < i_1 < \cdots < i_s \leqslant k$, the set $(P_{i_1}, \ldots, P_{i_s})$ is a weak hinge.

We have the following chains of inclusions:
$$\operatorname{Dom} P_1 \supset \operatorname{Ker} P_1 \supset \operatorname{Dom} P_2 \supset \operatorname{Ker} P_2 \supset \cdots,$$
$$\operatorname{Indef} P_j \subset \operatorname{Im} P_1 \subset \operatorname{Indef} P_2 \subset \operatorname{Im} P_2 \subset \cdots.$$

Therefore, the domains of action of linear relations are roughly arranged as shown in Figure 5.1.

We stress that our definition (in contrast to those in subsection 3.1) does not forbid the case $P_\alpha = \operatorname{Ker} P_\alpha \oplus \operatorname{Indef} P_\alpha$.

5.2. Equivalence of weak hinges. Now we give a (technical and not very important) definition of the equivalence of weak hinges. Two weak hinges are equivalent if one of them can be obtained from the other by applying (possibly many times) the two operations, of addition and deletion of a linear relation, that are described below.

First operation. If the domains of action are arranged as shown in Figure 5.2 (for P_{j-1}, the domain of action is a singleton), then the linear relation P_{j-1} can be deleted (i.e., $(\ldots, P_{j-2}, P_{j-1}, P_j, \ldots) \sim (\ldots, P_{j-2}, P_j, \ldots)$). Note that in this case we have
$$P_{j-1} = \operatorname{Ker} P_{j-1} \oplus \operatorname{Indef} P_{j-1}, \qquad \operatorname{Ker} P_{j-1} = \operatorname{Dom} P_j,$$
$$\operatorname{Im} P_{j-1} = \operatorname{Indef} P_{j-1} = \operatorname{Indef} P_j.$$

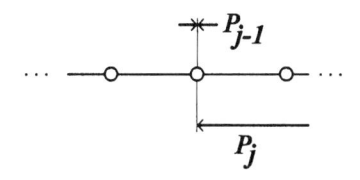

FIGURE 5.2

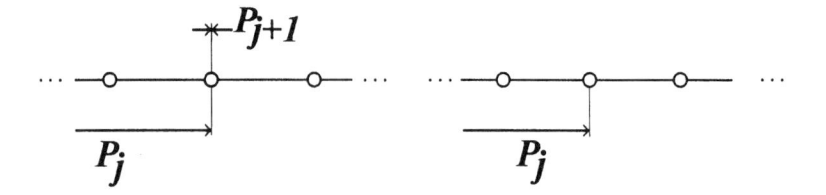

FIGURE 5.3

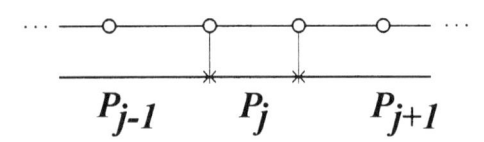

FIGURE 5.4

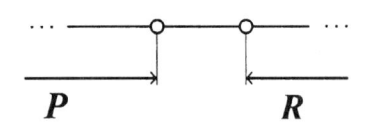

FIGURE 5.5

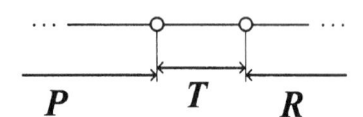

FIGURE 5.6

Conversely, if Q is a term of a weak hinge, then we can place the term $R = \operatorname{Dom} Q \oplus \operatorname{Indef} Q$ before Q. Similarly, the pictures shown in Figure 5.3 are equivalent.

Second operation. Let the domain of action of P_j consist of two points and the domains of action of P_{j-1} and P_{j+1} tightly border on the domain of action of P_j (see Figure 5.4). Then the link P_j can be deleted from the weak hinge. Conversely, let P and R be neighboring terms of a weak hinge and suppose that $\dim \operatorname{Ker} P - \dim \operatorname{Dom} R = 1$, i.e., the domains of the action are arranged as shown in Figure 5.5. Then $\dim \operatorname{Dom} R + \dim \operatorname{Im} P = n - 1$ and the linear relation T, of dimension n, such that $\operatorname{Indef} T = \operatorname{Im} P$ and $\operatorname{Ker} T = \operatorname{Dom} R$ is defined uniquely up to a factor. Now we can add T to the weak hinge (see Figure 5.6).

5.3. Multiplication of hinges. Denote by $\widetilde{\operatorname{Hinge}}_n$ the set of all weak hinges in $\mathbb{C}^n$ defined up to equivalence. Let

$$\mathcal{P} = (P_1, \ldots, P_k), \mathcal{R} = (R_1, \ldots, R_l) \in \widetilde{\operatorname{Hinge}}_n.$$

Consider the set of all possible products $A_{ij} = P_i R_j$ different from *null*.

THEOREM 5.4. *The set A_{ij} (being appropriately ordered) is a weak hinge.*

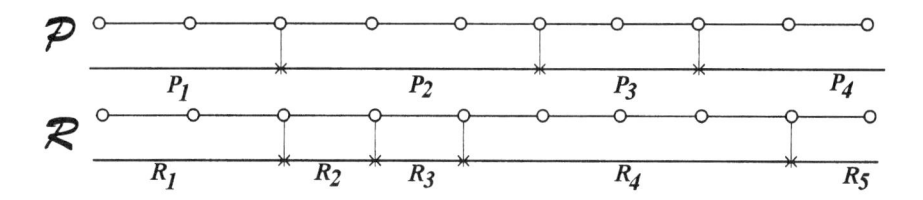

FIGURE 5.7

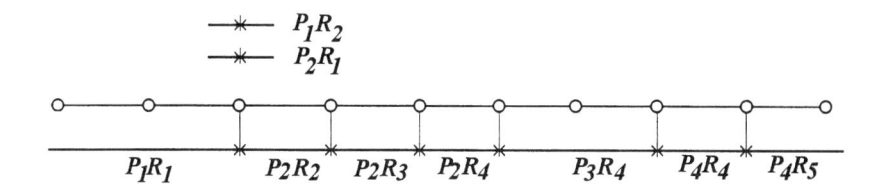

FIGURE 5.8

FIGURE 5.9

EXAMPLE 5.5. In $\mathbb{C}^{11}$ consider two hinges $\mathcal{P} = (P_1, \ldots, P_4)$ and $\mathcal{R} = (R_1, \ldots, R_5)$ with the domains of action of the links shown in Figure 5.7. If these hinges are in general position, then the product has the form shown in Figure 5.8. If these hinges are not in general position, then any of the links can be skipped.

In Figure 5.9 the crosses denote the pairs $\{i, j\}$ for which (in the case of general position) we have $P_i R_j \neq null$.

Let us pass to the proof of the theorem.

LEMMA 5.6. *If $P_i R_j \neq null$, then the domains of action of P_i and R_j have a nonzero intersection.*

PROOF. Let the domains of action be disjoint, and, for the sake of being definite, let the domains of action be arranged as shown in Figure 5.10.

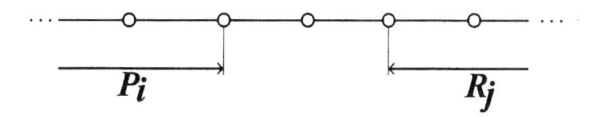

FIGURE 5.10

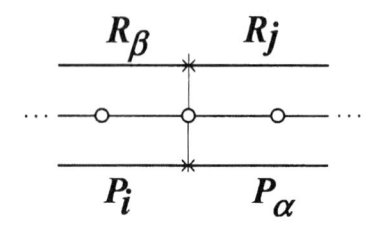

FIGURE 5.11

Then we have
$$\operatorname{Ker} P_i \supset \operatorname{Dom} R_j, \qquad \operatorname{Im} P_i \subset \operatorname{Indef} R_j,$$
both these inclusions being strict, and
$$\dim \operatorname{Ker} P_i + \dim \operatorname{Indef} R_j > \dim \operatorname{Ker} P_i + \dim \operatorname{Im} P_i = n.$$

Therefore, $\operatorname{Ker} P_i$ and $\operatorname{Indef} R_j$ have nonzero intersection, i.e., $P_i R_j = \text{null}$.

Lemma 5.6 implies the following assertion.

LEMMA 5.7. *Let $P_i R_j \ne \text{null}$ and $P_\alpha R_\beta \ne \text{null}$. Then at least one of the following three possibilities holds.*
  a) $i \leqslant \alpha$, $j \leqslant \beta$.
  b) $i \geqslant \alpha$, $j \geqslant \beta$.
  c) *The domains of action of P_i, P_α, R_j, and R_β are arranged as shown in Figure 5.11.*

Note that the last case is of no interest because in this case $S = P_i R_j = P_\alpha R_\beta$ is $S = \operatorname{Ker} S \oplus \operatorname{Indef} S$. Moreover, since we have $P_i R_j \ne \text{null}$ and $P_\alpha R_\beta \ne \text{null}$, the linear relation S can be deleted by the first rule of subsection 5.2.

Thus, let $i \leqslant \alpha$ and $j \leqslant \beta$. Assume that $T = P_i R_j \ne \text{null}$ and $S = P_\alpha R_\beta \ne \text{null}$. We must show that either $T = S$ or
$$\operatorname{Ker} T \supset \operatorname{Dom} S, \qquad \operatorname{Im} T \subset \operatorname{Indef} S.$$

It suffices to consider three cases:
  a) $i < \alpha$, $j < \beta$.
  b) $i = \alpha$, $j < \beta$.
  c) $i < \alpha$, $j = \beta$.

In the first case we have

(5.1)
$$\operatorname{Ker} T \supset \operatorname{Ker} R_j \supset \operatorname{Dom} R_\beta \supset \operatorname{Dom} S,$$
$$\operatorname{Im} T \subset \operatorname{Im} P_i \subset \operatorname{Indef} P_\alpha \subset \operatorname{Indef} S.$$

In the second case we have the same chain (5.1) together with
$$\operatorname{Im} T = P_i(\operatorname{Im} R_j) \subset P_i(\operatorname{Indef} R_\alpha) \subset \operatorname{Indef} S.$$

The third case is similar to the second one. This completes the proof of the theorem.

Thus, we see that the set $\widetilde{\operatorname{Hinge}}_n$ is a semigroup.

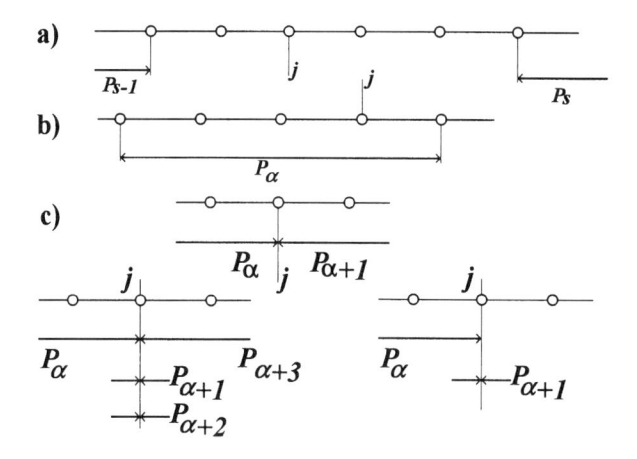

FIGURE 5.12

5.4. Fundamental representations of the semigroup $\widetilde{\text{Hinge}}_n$.
Let
$$j \in \{1, \ldots, n-1\}.$$

Define the representation $\lambda_j(\,\cdot\,)$ of the semigroup $\widetilde{\text{Hinge}}_n$ in the space $\Lambda^j \mathbb{C}^n$.

Let $\mathcal{P} = (P_1, \ldots, P_k) \in \widetilde{\text{Hinge}}_n$. Consider the set of operators

(5.2) $$\lambda_j(P_1), \ldots, \lambda_j(P_k).$$

The following assertion is obvious.

PROPOSITION 5.8. *For any $\mathcal{P}$, one of the following three properties holds.*
a) *All operators (5.2) are equal to 0.*
b) *There exists an α such that $\lambda_j(P_\alpha) \ne 0$, and this α is unique.*
c) *There exist indices $\alpha, \alpha+1, \ldots, \beta$ such that $\lambda_j(P_\sigma) = 0$ for $\sigma < \alpha$ and for $\sigma > \beta$, and the operators $\lambda_j(P_\alpha), \ldots, \lambda_j(P_\beta)$ are of rank one and coincide up to a factor.*

We illustrate cases a), b), and c) in Figure 5.12.

The last pictures show why many (> 2) links P_s such that $\lambda_j(P_s) \ne 0$ can occur. The reason for this occurrence is obvious: for any linear relation P entering the hinge, we can always add an arbitrary number of identical terms $Q = \operatorname{Ker} P \oplus \operatorname{Im} P$.

Now we assume that $\lambda_j(\mathcal{P})$ is an operator defined up to a factor and satisfying the following conditions.
1) If case a) of the theorem holds, then $\lambda_j(\mathcal{P}) = 0$.
2) If we have case b) or c), then $\lambda_j(\mathcal{P})$ is a nonzero term of the sequence $\lambda_j(P_\alpha)$.

We can readily see that $\mathcal{P} \mapsto \lambda_j(\mathcal{P})$ is a projective representation of the semigroup $\widetilde{\text{Hinge}}_n$.

5.5. Topology on $\widetilde{\text{Hinge}}_n$.
Let V be a linear space. Denote by $\mathbb{P}^\circ V = \mathbb{P} V \cup 0$ the set of vectors $v \in V$ defined up to a factor, i.e., $\mathbb{P}^\circ V$ is the projective space $\mathbb{P} V$ to which we add 0.

The topology on $\mathbb{P}^\circ V$ is defined by the following condition: a set $S \subset \mathbb{P}^\circ V$ is closed if and only if $0 \in S$ and $S \cap \mathbb{P} V$ is closed in $\mathbb{P} V$.

REMARK 5.9. This is not a Hausdorff topology. The closure of any point $h \in \mathbb{P}V$ is the two-point set that consists of h and 0.

We have a mapping $\widetilde{\mathrm{Hinge}}_n \to \prod_{j=1}^{n-1} \mathbb{P}^\circ(\Lambda^j \mathbb{C}^n)$ given by the formula

$$\mathcal{P} \mapsto (\lambda_1(\mathcal{P}), \ldots, \lambda_{n-1}(\mathcal{P})). \tag{5.3}$$

PROPOSITION 5.10. *The mapping (5.3) is an embedding.*

PROOF. In subsection 5.2, the equivalence of weak hinges was defined in such a way that (5.3) is an embedding.

Thus, $\widetilde{\mathrm{Hinge}}_n$ is embedded into $\prod_{j=1}^{n-1} \mathbb{P}^\circ(\Lambda^j \mathbb{C}^n)$, and this embedding induces a topology on $\widetilde{\mathrm{Hinge}}_n$. We can readily verify that it is not a Hausdorff topology.

EXAMPLE 5.11. Let us describe the closure of a point

$$\mathcal{P} = (P_1, \ldots, P_k) \in \widetilde{\mathrm{Hinge}}_n. \tag{5.4}$$

Let us supplement the collection $\mathcal{P}$ to an equivalent hinge

$$\widehat{\mathcal{P}} = (S_1, P_1, T_1, S_2, P_2, T_2, \ldots), \tag{5.5}$$

where $S_j = \mathrm{Dom}\, P_j \oplus \mathrm{Indef}\, P_j$ and $T_j = \mathrm{Ker}\, P_j \oplus \mathrm{Im}\, P_j$. Then the closure of the point $\mathcal{P}$ consists of all possible weak hinges obtained from $\widehat{\mathcal{P}}$ by deleting some set of links. In particular, the closure of any point contains an empty hinge.

5.6. Generators.

PROPOSITION 5.12. *The semigroup $\widetilde{\mathrm{Hinge}}_n$ is generated by the group PGL_n and by the canonical hinges $\mathcal{R}(1), \ldots, \mathcal{R}(n-1)$.*

We omit the (more or less clear) proof. We only note that for $i_1 < \cdots < i_k$ we have $\mathcal{R}(i_1) \cdots \mathcal{R}(i_k) = \mathcal{R}(i_1, \ldots, i_k)$.

5.7. Central extension of the semigroup $\widetilde{\mathrm{Hinge}}_n$.

An object with a non-Hausdorff topology can create the impression of something pathological. In fact, the semigroup $\widetilde{\mathrm{Hinge}}_n$ has something like a Hausdorff central extension, which we will now describe.

Let $Op(H)$ be the semigroup of operators in a linear space H. We introduce the semigroup $\widehat{\mathrm{Hinge}}_n$ as the subsemigroup of $\prod_{j=1}^{n-1} Op(\Lambda^j \mathbb{C}^n)$ that consists of the following collections of operators:

$$(s_1 \cdot \lambda_1(\mathcal{P}), \ldots, s_{n-1} \cdot \lambda_{n-1}(\mathcal{P})), \tag{5.6}$$

where $\mathcal{P} \in \widetilde{\mathrm{Hinge}}_n$ and $s_j \in \mathbb{C}^*$.

The projection

$$\widehat{\mathrm{Hinge}}_n \to \widetilde{\mathrm{Hinge}}_n \tag{5.7}$$

is defined in an obvious way, namely, to the collection (5.6) we assign the weak hinge $\mathcal{P}$. The preimages of different points $\mathcal{P} \in \widetilde{\mathrm{Hinge}}_n$ in $\widehat{\mathrm{Hinge}}_n$ have different dimensions (namely, the fiber over $\mathcal{P}$ is $(\mathbb{C}^*)^m$, where m is the number of nonzero operators $\lambda_1(\mathcal{P}), \ldots, \lambda_{n-1}(\mathcal{P})$), and this is the reason why the quotient space $\widetilde{\mathrm{Hinge}}_n = \widehat{\mathrm{Hinge}}_n/(\mathbb{C}^*)^{n-1}$ is a non-Hausdorff space.

It would be of interest to obtain explicit formulas for the product of collections of the form (5.6). This question can clearly be reduced to the problem of explicitly describing the category GA^* (see subsection 1.8).

§6. Representations

6.1. Representations of the group SL_n. Consider an irreducible representation of the group $SL_n(\mathbb{C})$ with numerical labels $a_1, \ldots, a_{n-1}$ (see Figure 6.1).

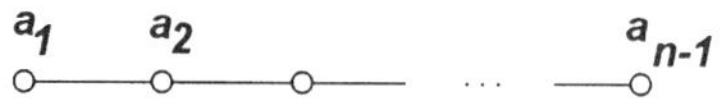

FIGURE 6.1

We recall the construction of this representation. Consider the space

$$(6.1) \qquad W = \bigotimes_{j=1}^{n-1} (\Lambda^j(\mathbb{C}^n))^{\otimes a_j}$$

The group SL_n acts on this space by the formula $g \mapsto (\Lambda^j g)^{\otimes a_j}$. Let $h_j = e_1 \wedge \cdots \wedge e_j \in \Lambda^j(\mathbb{C}^n)$, and let

$$v = v_a = \bigotimes_{j=1}^{n-1} (h_j)^{\otimes a_j} \in W.$$

Consider the cyclic span $V(a) = V(a_1, \ldots, a_{n-1})$ of the vector v (i.e., the minimal SL_n-invariant subspace containing v). Then the irreducible representation $\rho_a = \rho_{a_1,\ldots,a_{n-1}}$ with numerical labels $a = (a_1, \ldots, a_{n-1})$ is the restriction of the representation (6.1) to the subspace $V(a_1, \ldots, a_{n-1})$. The vector v_a is called the *highest weight vector*.

6.2. Projective embedding of Hinge_n. Now we construct a mapping

$$\pi_a = \pi_{a_1,\ldots,a_{n-1}} \colon \mathrm{Hinge}_n \to \mathbb{P}(Op(V(a_1, \ldots, a_{n-1}))).$$

Let $\mathcal{P} = (P_1, \ldots, P_l) \in \mathrm{Hinge}_n$. Consider the operator

$$(6.2) \qquad \prod_{j=1}^{n-1} \lambda_j(\mathcal{P})^{\otimes a_j}$$

in W.

LEMMA 6.1. *The subspace $V(a)$ is invariant with respect to the operator* (6.2).

PROOF. $V(a)$ is invariant under SL_n, and SL_n is dense in Hinge_n.

Denote by $\pi_a(\mathcal{P})$ the restriction of the operator (6.2) to $V(a)$.

LEMMA 6.2. $\pi_a(\mathcal{P}) \neq 0$ *for all* $\mathcal{P} \in \mathrm{Hinge}_n$.

PROOF. It suffices to verify that $\pi_a(\mathcal{R}(i_1, \ldots, i_k)) \neq 0$ for all canonical hinges $\mathcal{R}(i_1, \ldots, i_k)$. However, we can readily see that $\mathcal{R}(i_1, \ldots, i_k) v_a = s \cdot v_a$, $s \in \mathbb{C}^*$, which completes the proof.

EXAMPLE 6.3. The operator $\pi_a(\mathcal{R}(1,\ldots,n-1))$ is the projection onto the highest weight vector v_a.

Denote by $\rho_a(SL_n)$ the set of all operators of the form $\rho_a(g)$, where $g \in SL_n$.

THEOREM 6.4. *The closure of the set $\rho_a(SL_n)$ in $\mathbb{P}Op(V(a))$ coincides with the image of the mapping*

(6.3) $$\pi_a\colon \mathrm{Hinge}_n \to \mathbb{P}Op(V(a)).$$

PROOF. By Lemma 6.2, we have $\pi_a(\mathrm{Hinge}_n) \subset \mathbb{P}Op(V(a))$. The mapping π_a is clearly continuous and the set Hinge_n is compact.

PROPOSITION 6.5. *If all $a_1,\ldots,a_{n-1}$ are nonzero, then the mapping (6.3) is an embedding.*

We shall prove this assertion later.

REMARK 6.6. If there are zeros among the numbers $a_1,\ldots,a_{n-1}$, then the mapping π_a is not an embedding. Namely, in this case the mapping $\pi_a = \sigma \circ \tau$, i.e.,

$$\mathrm{Hinge}_n \xrightarrow{\tau} \prod_{j=1}^{n-1} \mathbb{P}(Op(\Lambda^j \mathbb{C}^n)) \xrightarrow{\sigma} \mathbb{P}(Op(V(a)))$$

can be factored through

(6.4) $$\prod_{j:a_j \neq 0} \mathbb{P}(Op(\Lambda^j \mathbb{C}^n)).$$

In fact, in this case, the image of the mapping π_a is homeomorphic (and also equivalent as a variety) to the image of Hinge_n in (6.4).

6.3. Representation $\widetilde{\pi_a}$ of the semigroup $\widetilde{\mathrm{Hinge}_n}$. Let $\mathcal{P} \in \widetilde{\mathrm{Hinge}_n}$. Consider (see (6.1)) the operator $\bigotimes_{j=1}^{n-1} \lambda_j(\mathcal{P})^{\otimes a_j}$ in W. Denote by $\widetilde{\pi_a}(\mathcal{P})$ the restriction of this operator to the subspace $V(a)$. Clearly, $\mathcal{P} \mapsto \widetilde{\pi_a}(\mathcal{P})$ is a projective representation of the semigroup Hinge_n.

Let $\mathbb{P}^\circ H$ be the space of vectors from H defined up to a scalar factor.

THEOREM 6.7. *The closure of the image $\rho_a(SL_n)$ in $\mathbb{P}^\circ Op(V(a))$ coincides with $\widetilde{\pi_a}(\widetilde{\mathrm{Hinge}_n})$.*

This assertion repeats Theorem 6.4.

REMARK 6.8. Let all $a_j \neq 0$. Let $\mathcal{P}$ be a weak hinge that is not equivalent to any exact hinge. Then $\widetilde{\pi_a}(\mathcal{P}) = 0$. In other words, the closure of $\rho_a(SL_n)$ in $\mathbb{P}^\circ Op(V(a))$ is as follows. This is the semigroup that consists of exact hinges and 0. If $\mathcal{P}$ and $\mathcal{R}$ are exact hinges, then $\pi_a(\mathcal{R})\pi_a(\mathcal{P})$ is equal to 0 if and only if $\mathcal{RP}$ is a weak hinge that is not equivalent to any exact hinge.

REMARK 6.9. Let there be zeros among the numbers $a_1,\ldots,a_{n-1}$. Assume that $\mathcal{P} = (P_1,\ldots,P_n)$ is a hinge. Then $\widetilde{\pi_a}(\mathcal{P}) \neq 0$ if and only if the union of the domains of action of the relations P_j contains all α such that $a_\alpha \neq 0$.

6.4. **Proof of Proposition 6.5.** Let Q be the orbit of the canonical hinge $\mathcal{R}(1,\ldots,n-1)$ under the action of the group $PGL_n \times PGL_n$. Clearly, for any $\mathcal{P} \in \text{Hinge}_n$ and any $S \in Q$ either we have $\mathcal{P}S \in Q$ or $\mathcal{P}S$ is not an exact hinge (and for the generic points we have the first case).

Furthermore, if $\mathcal{P}, \mathcal{P}' \in \text{Hinge}_n$ and for any $S \in Q$ we have $\mathcal{P}S = \mathcal{P}'S$, then $\mathcal{P} = \mathcal{P}'$. In other words, a hinge $\mathcal{P}$ is completely defined by its products with the elements of the orbit Q (and it also suffices to consider the generic elements of Q).

Let us show that the mapping $\pi_a \colon \text{Hinge}_n \to \mathbb{P}Op(V(a))$ defines an isomorphic embedding of the orbit Q in $\mathbb{P}Op(V(a))$. The group $PGL_n \times PGL_n$ acts on $Op(V(a))$ by multiplication by the operators $\rho_a(g)$ from the left and from the right. The operator $\pi_a(\mathcal{R}(1,\ldots,n-1))$ is the projection onto the highest weight vector v_a. Therefore, the stabilizer of the operator $\pi_a(\mathcal{R}(1,\ldots,n-1))$ in $PGL_n \times PGL_n$ is the product of two Borel subgroups $B \times B' \subset PGL_n \times PGL_n$, where B is the upper triangular subgroup and B' is the lower triangular subgroup. This stabilizer coincides with that of $\mathcal{R}(1,\ldots,n-1)$ in $PGL_n \times PGL_n$ (see subsection 3.4). Thus, we have proved that the mapping π_a on Q is injective.

Now let $\pi_a(\mathcal{P}) = \pi_a(\mathcal{P}')$. Then $\pi_a(\mathcal{P}S) = \pi_a(\mathcal{P}'S)$ for all hinges S and, in particular, for $S \in Q$. However, this implies $\mathcal{P} = \mathcal{P}'$.

6.5. **Representations $\widehat{\pi_a}$ of the semigroup $\widehat{\text{Hinge}}_n$.** Let the family of operators

$$\mathfrak{b} = (B_1, \ldots, B_{n-1}) \in \prod_{j=1}^{n-1} Op(\Lambda^j \mathbb{C}^n)$$

belong to $\widehat{\text{Hinge}}_n$. Define the operator $\bigotimes_{j=1}^{n-1} B_j^{\otimes a_j}$ in the space W. Denote by $\widehat{\pi_a}(\mathfrak{b})$ the restriction of this operator to $V(a)$. Clearly, $\mathfrak{b} \mapsto \widehat{\pi_a}(\mathfrak{b})$ is a *linear* representation of the semigroup $\widehat{\text{Hinge}}_n$.

Now let ν be an arbitrary (in general, reducible) representation of the group SL_n in a linear space H. Let us decompose ν into irreducible representations. In any irreducible component, we have the action of the semigroup $\widehat{\text{Hinge}}_n$, and hence $\widehat{\text{Hinge}}_n$ acts on H.

REMARK 6.10. For the semigroup $\widetilde{\text{Hinge}}_n$, this construction is impossible, because in the category of projective representations we have no direct sum operation.

Chapter II. Supplementary remarks

§7. Action of the semigroup $\widetilde{\text{Hinge}}_n$ on the flag space

7.1. **Space $\mathcal{F}_n$.** Denote by $\mathcal{F}_n$ the space of all flags

(7.1) $$0 \subset V_1 \subset V_2 \subset \cdots \subset V_k \subset \mathbb{C}^n$$

(it is assumed that all inclusions are strict). Let $0 < i_1 < \cdots < i_k < n$. Denote by $\mathcal{F}_n(i_1, \ldots, i_k)$ the space of flags (7.1) such that $\dim V_1 = i_1$, $\dim V_2 = i_2$, ... The spaces $\mathcal{F}_n(i_1, \ldots, i_k)$ are smooth varieties.

Let a number set $i_1, i_2, \ldots$ contain a number set $j_1, j_2, \ldots$. Then a projection

$$\pi_{j_1,j_2,\ldots}^{i_1,i_2,\ldots} \colon \mathcal{F}_n(i_1, \ldots, i_k) \to \mathcal{F}_n(j_1, \ldots, j_k)$$

is well defined; namely, from the flag $\mathcal{L} \in \mathcal{F}_n(i_1, \ldots, i_k)$ we delete all subspaces except for the subspaces of the dimensions $j_1, j_2, \ldots$.

Let us define a non-Hausdorff topology on the space

$$\mathcal{F}_n = \bigcup_{k,\, 0 < i_1 < \cdots < i_k < n} \mathcal{F}_n(i_1, \ldots, i_k).$$

Let $\mathcal{L} \in \mathcal{F}(\alpha_1, \ldots, \alpha_k)$ and $\mathcal{L}_j \in \mathcal{F}(\beta_1^j, \ldots, \beta_{m_j}^j)$. Then the relation $\mathcal{L} = \lim_{j \to \infty} \mathcal{L}_j$ means that the following two conditions are satisfied.

1. Starting from some number j, the collections $(\beta_1^j, \ldots, \beta_{m_j}^j)$ contain the collection $(\alpha_1, \ldots, \alpha_k)$.
2. The projections of $\mathcal{L}_j$ to $\mathcal{F}(\alpha_1, \ldots, \alpha_k)$ converge to $\mathcal{L}$ in the topology of $\mathcal{F}(\alpha_1, \ldots, \alpha_k)$.

EXAMPLE 7.1. Consider a flag $\mathcal{L}\colon 0 \subset V_1 \subset \cdots \subset V_k \subset \mathbb{C}^n$. Then the closure of the point $\mathcal{L}$ consists of all points (flags) $V_{i_1} \subset \cdots \subset V_{i_\alpha}$ obtained from $\mathcal{L}$ by deleting some subspaces.

7.2. Action of the semigroup $\widetilde{\mathrm{Hinge}}_n$ on the space $\mathcal{F}_n$. Let

$$\mathcal{P} = (P_1, \ldots, P_k) \in \widetilde{\mathrm{Hinge}}_n,$$

and let $\mathcal{L}\colon V_1 \subset V_2 \subset \cdots \subset V_s$ be an element of $\mathcal{F}$. Consider the set Σ of all pairs (i, j) such that

(7.2) $$\operatorname{Ker} P_i \cap V_j = 0,$$

(7.3) $$\operatorname{Dom} P_i + V_j = \mathbb{C}^n.$$

THEOREM 7.2. *The collection of subspaces $P_i V_j$, where the pair (i, j) ranges over Σ, form a flag.*

We omit the simple proof of this theorem, which is similar to the proof of Theorem 5.4.

Denote this flag by $\mathcal{PL}$. We can readily see that $\mathcal{P}\colon \mathcal{L} \mapsto \mathcal{PL}$ is an action of the semigroup $\widetilde{\mathrm{Hinge}}_n$ on $\mathcal{F}_n$.

7.3. Hausdorff bundle over $\mathcal{F}_n$. Denote by $\widehat{\mathcal{F}}_n$ the set of collections

(7.4) $$(h_1, \ldots, h_{n-1}) \in \prod_{j=1}^{n-1} \Lambda^j \mathbb{C}^n$$

such that h_j have the form

$$h_1 = t_1 \cdot v_1, \quad h_2 = t_2 \cdot v_1 \wedge v_2, \quad \ldots, \quad h_{n-1} = t_{n-1} \cdot v_1 \wedge \cdots \wedge v_{n-1},$$

where $t_j \in \mathbb{C}$ and $v_1, \ldots, v_{n-1}$ are linearly independent vectors in $\mathbb{C}^n$. We allow the possibility $t_j = 0$, i.e., some vectors h_j can be zero.

Let us define a projection $\widehat{\mathcal{F}}_n \to \mathcal{F}_n$.

Let $h_{i_1}, \ldots, h_{i_k}$ be all nonzero vectors among $\{h_j\}$. Let V_{i_k} be a subspace spanned by the vectors $v_1, \ldots, v_{i_k}$. Then to the element (7.4) there corresponds the flag

(7.5) $$0 \subset V_{i_1} \subset \cdots \subset V_{i_k} \subset \mathbb{C}^n.$$

Note that the fiber over the point is of dimension k. As in subsection 5.7, the fibers over different points have different dimensions, and therefore the quotient $\mathcal{F}_n = \widehat{\mathcal{F}_n}/(\mathbb{C}^*)^{n-1}$ is a non-Hausdorff space.

7.4. Action of the semigroup $\widehat{\text{Hinge}}_n$ on $\widehat{\mathcal{F}_n}$. Let $\mathfrak{b} = (B_1, \ldots, B_{n-1}) \in \prod_{j=1}^{n-1} Op(\Lambda^j \mathbb{C}^n)$ be an element of $\widehat{\text{Hinge}}_n$ and let $h = (h_1, \ldots, h_{n-1}) \in \prod_{j=1}^{n-1} \Lambda^j V$ be an element of $\widehat{\mathcal{F}_n}$. Then the element $\mathfrak{b}h$ is defined by the formula $\mathfrak{b}h = (Bh_1, \ldots, Bh_n)$.

§8. Complete symmetric varieties

Let G be a semisimple Lie group. Let σ be an involution on g (that is, σ is an automorphism and $\sigma^2 = \text{id}$). Let H be the set of fixed points of σ. The homogeneous spaces G/H, where G and H are groups of the type described above, are called *symmetric spaces*.

In this section we are interested in the case of complex groups G and H. The completions of G/H constructed below are called *complex symmetric varieties*.

8.1. Quasi-inverse hinge. Recall that we denote by $P^\square$ the linear relation quasi-inverse to P (see subsection 1.1). Let $\mathcal{P} = (P_1, \ldots, P_k)$ be a weak hinge. The *quasi-inverse hinge* $\mathcal{P}^\square$ is defined by

$$\mathcal{P}^\square = (P_k^\square, \ldots, P_1^\square).$$

We can readily see that for any $\mathcal{P}, \mathcal{R} \in \widetilde{\text{Hinge}}_n$ we have

(8.1) $$(\mathcal{P}\mathcal{R})^\square = \mathcal{R}^\square \mathcal{P}^\square$$

8.2. Transposed hinge. Let $V \simeq \mathbb{C}^n$ be a linear space endowed with nondegenerate symmetric bilinear form

(8.2) $$\langle x, y \rangle = \sum x_j y_j.$$

Introduce a skew-symmetric bilinear form on $V \oplus V$:

(8.3) $$\{(x, u), (y, v)\} = \sum x_j v_j - \sum u_j y_j.$$

Let $P\colon V \rightrightarrows V$ be a linear relation. Denote by P^t the orthogonal complement to P.

EXAMPLE 8.1. Let P be the graph of a linear operator A. Then P^t is the graph of the transposed linear operator A^t. Indeed, A and A^t are related by the identity $\langle x, A^t y \rangle - \langle Ax, y \rangle = 0$, which exactly means that the vectors $(x, Ax) \in P$ and $(y, A^t y) \in P^t$ are orthogonal with respect to the form (8.3).

REMARK 8.2. For any linear relations P and Q, we have $(PQ)^t = Q^t P^t$.

Furthermore, let $\mathcal{P} = (P_1, \ldots, P_k) \in \widetilde{\text{Hinge}}_n$ be a weak hinge. Then the *transposed* hinge $\mathcal{P}^t$ is defined by the formula $\mathcal{P}^t = (P_1^t, \ldots, P_k^t)$. We can readily see that

(8.4) $$(\mathcal{P}\mathcal{R})^t = \mathcal{R}^t \mathcal{P}^t$$

for all $\mathcal{P}, \mathcal{R} \in \widetilde{\text{Hinge}}_n$.

8.3. Completion of the group $O_n(\mathbb{C})$.

Denote by $\overline{O_n}$ the space of exact hinges $\mathcal{P} \in \text{Hinge}_n$ satisfying the condition $\mathcal{P}^\square = \mathcal{P}^t$. Denote by $\widetilde{O_n}$ the space of weak hinges satisfying the same condition.

It follows from (8.1) and (8.4) that the set $\widetilde{O_n}$ is closed with respect to the multiplication of hinges, that is, $\widetilde{O_n}$ is a semigroup.

The group $PO_n(\mathbb{C})$ (that is, the quotient group of the group $O_n(\mathbb{C})$ by its center) can be embedded in $\widetilde{O_n}$ in an obvious way. Namely, $O_n(\mathbb{C})$ consists of one-link hinges (Q), where Q ranges over the graphs of orthogonal operators. It is obvious that the group $O_n(\mathbb{C})$ is dense in $\overline{O_n}$ and in $\widetilde{O_n}$.

PROPOSITION 8.3. *The space $\overline{O_n}$ is a smooth algebraic variety.*

PROOF. Consider the mapping $\sigma\colon \text{Hinge}_n \to \text{Hinge}_n$ given by the formula $\sigma(\mathcal{P}) = (\mathcal{P}^\square)^t$. Then $\sigma^2 = \text{id}$, and $\overline{O_n}$ is the set of fixed points of σ. Therefore, $\overline{O_n}$ is a smooth variety.

REMARK 8.4. Let us discuss in detail the conditions that must be satisfied by a hinge

(8.5) $$\mathcal{P} = (P_1, \ldots, P_k) \in \widetilde{O_n}.$$

For all j we have

(8.6) $$P_j^t = P_{k-j}^\square.$$

Thus, the hinge $\mathcal{P}$ is completely determined by its terms:

(8.7) $$(P_1, \ldots, P_s),$$

where $s = k/2$ for k even and $s = (k+1)/2$ for k odd. Let us discuss the conditions satisfied by the linear relations $(P_1, \ldots, P_s)$. There are two cases.

    a) Let k be odd and $s = (k+1)/2$ (i.e., P_s is the middle term of the hinge (8.5)). In this case, relation (8.6) means that $P_s^t = P_s^\square$. The last relation is equivalent to the condition that P_s is a morphism of the category GD (see subsection 1.7).

    b) Consider any other P_j with $j \leqslant s$. Then P_j and $(P_j^\square)^t$ occur in the same hinge, and hence

$$\text{Ker } P_j \supset \text{Dom}((P_j^\square)^t) \quad \text{and} \quad \text{Im } P_j \subset \text{Indef}((P_j^\square)^t).$$

These conditions are equivalent to the following three conditions:

1*. $\text{Ker } P_j$ is co-isotropic, i.e., $\text{Ker } P_j \supset (\text{Ker } P_j)^\perp$, where the symbol $\perp$ stands for the orthogonal complement with respect to (8.2);

2*. $\text{Im } P_j$ is isotropic, i.e., $\text{Im } P_j \subset (\text{Im } P_j)^\perp$;

3*. $\text{Ker } P_j \supset \text{Im } P_j$.

Conversely, if we have a weak hinge (8.7) satisfying conditions a) and b), then it is an element (8.5) of the semigroup $\widetilde{O_n}$.

8.4. Complete quadrics.

Now we construct a completion of the space

$$PGL_n(\mathbb{C})/PO_n(\mathbb{C}).$$

Denote by $\widetilde{PGL_n/O_n}$ ($\overline{PGL_n/O_n}$, respectively) the space of weak hinges (exact hinges, respectively) $\mathcal{P} = (P_1, \ldots, P_k)$ such that $\mathcal{P} = \mathcal{P}^t$ (where the transposition is the same as in subsection 8.2). This condition can be written in the form $P_j^t = P_j$ for all j.

REMARK 8.5. The condition $P = P^t$ is equivalent to the assumption that P is a maximal isomorphic subspace (*Lagrangian subspace*) of the space $\mathbb{C}^n \oplus \mathbb{C}^n$ endowed with skew-symmetric form (8.3).

The semigroup $\widetilde{O_n}$ acts on $\widetilde{PGL_n/O_n}$ by the formula $\mathcal{P} \mapsto \mathcal{RPR}^t$, where $\mathcal{P} \in \widetilde{PGL_n/O_n}$ and $\mathcal{R} \in \widetilde{O_n}$.

The *Study–Semple space of complete quadrics* is $\overline{PGL_n/O_n}$.

8.5. Complete quadrics as a Hausdorff quotient.
Consider the Lagrangian Grassmannian $\mathcal{L}$ in the space $\mathbb{C}^n \oplus \mathbb{C}^n$. The group $\mathbb{C}^*$ acts on $\mathcal{L}$ by multiplication of a linear relation by a number (under this operation, a Lagrangian linear relation maps into a Lagrangian one).

The graphs of invertible operators $A: \mathbb{C}^n \to \mathbb{C}^n$ form an open dense set $\mathcal{L}^\circ$ in $\mathcal{L}$; here the condition $\mathrm{graph}(A) \in \mathcal{L}$ is equivalent to the condition $A = A^t$.

Note that the quadrics in $\mathbb{C}P^{n-1}$ are in a one-to-one correspondence with symmetric matrices A defined up to a scalar factor; i.e., $\mathcal{L}^\circ/\mathbb{C}^*$ can be regarded as the space of all quadrics.

Applying the construction of Hausdorff quotient space to $\mathcal{L}$ and its open set $\mathcal{L}^\circ$, we obtain exactly the space $\overline{PGL_n/O_n}$.

8.6. Another description of complete quadrics.
Let $\mathcal{P} = (P_1, \ldots, P_k)$, $P_j = P_j^t$, be an exact hinge. Then the following flag is defined:

$$\mathbb{C}^n \supset \mathrm{Ker}\, P_1 \supset \cdots \supset \mathrm{Ker}\, P_k \supset 0.$$

Furthermore, note that the form (8.3) defines a nondegenerate pairing between the two summands in $\mathbb{C}^n \oplus \mathbb{C}^n$. We can readily see that $\mathrm{Im}\, P_j \subset \{0\} \oplus \mathbb{C}^n$ is the annihilator of $\mathrm{Ker}\, P_j \subset \mathbb{C}^n \oplus \{0\}$, and therefore the spaces $\mathrm{Ker}\, P_j/\mathrm{Ker}\, P_{j+1}$ and $\mathrm{Im}\, P_{j+1}/\mathrm{Im}\, P_j$ are dual to each other. A linear relation P_{j+1} induces a nondegenerate linear operator $\mathrm{Ker}\, P_j/\mathrm{Ker}\, P_{j+1} = \mathrm{Dom}\, P_{j+1}/\mathrm{Ker}\, P_{j+1} \to \mathrm{Im}\, P_{j+1}/\mathrm{Im}\, P_j = \mathrm{Im}\, P_{j+1}/\mathrm{Indef}\, P_{j+1}$, and therefore we obtain a nondegenerate quadratic form on the space $\mathrm{Ker}\, P_j/\mathrm{Ker}\, P_{j+1}$.

Thus, the following collection of data can be regarded as a point of the space of complete quadrics:
1*. a flag $0 = V_0 \subset V_1 \subset \cdots \subset V_k \subset V_{k+1} = \mathbb{C}^n$;
2*. for any $j \in \{0, 1, \ldots, k\}$, a nondegenerate quadratic form Q_j on the quotient space V_{j+1}/V_j defined up to a scalar factor.

8.7. Completion of the group $Sp_{2n}(\mathbb{C})$.
Consider the space $V = \mathbb{C}^{2n}$ with nondegenerate skew-symmetric bilinear form

$$\{x, y\} = \sum_{i=1}^{n} x_i y_{i+n} - \sum_{j=n+1}^{2n} x_j y_{j-n}.$$

In the space $\mathbb{C}^{2n} \oplus \mathbb{C}^{2n}$ we introduce a symmetric bilinear form $M((x,y),(u,v)) = \{x, v\} - \{y, u\}$.

Let $P: \mathbb{C}^{2n} \rightrightarrows \mathbb{C}^{2n}$ be a linear relation. Denote by P^s the orthogonal complement to P with respect to the form M.

Now let $\mathcal{P} = (P_1, \ldots, P_k) \in \widehat{\text{Hinge}}_n$. Then the hinge $\mathcal{P}^s$ is defined by $\mathcal{P}^s = (P_1^s, \ldots, P_k^s)$.

Furthermore, the semigroup $\widetilde{Sp_{2n}}$ (the variety $\overline{Sp_{2n}}$) is defined as the set of weak hinges (exact hinges, respectively) that satisfies the condition $\mathcal{P}^s = \mathcal{P}^\square$.

Remark 8.4 (about the collections $\mathcal{P} \in \widetilde{O_n}$) remains valid for $\mathcal{P} \in \widetilde{Sp}$. The only difference is that, in our case, the question is in the category C and not in the categories B and GD.

8.8. Space $\overline{PGL_n/Sp_{2n}}$. We can naturally regard $PGL_n(\mathbb{C})/Sp_{2n}(\mathbb{C})$ as the space of nondegenerate skew-symmetric forms on $\mathbb{C}^{2n}$ defined up to a scalar factor. The space $\widetilde{PGL_n/Sp_{2n}}$ ($\overline{PGL_n/Sp_{2n}}$, respectively) is defined as the space of weak hinges (exact hinges, respectively) such that $\mathcal{P} = \mathcal{P}^s$.

All remarks of 8.4–8.6 on complete quadrics hold for $\overline{PGL_n/Sp_{2n}}$ as well.

8.9. Completion of $GL_{p+q}(\mathbb{C})/GL_p(\mathbb{C}) \times GL_q(\mathbb{C})$. We can naturally regard the space $X_{p,q} = GL_{p+q}(\mathbb{C})/GL_p(\mathbb{C}) \times GL_q(\mathbb{C})$ as the space of operators A on $\mathbb{C}^{p+q}$ such that $A^2 = E$ and q eigenvalues of A are equal to 1 and the other p are equal to (-1).

Let V_+ and V_- be the eigen subspaces of the operator A that correspond to the eigenvalues ± 1. Clearly, A is completely determined by the subspaces V_+ and V_-, and thus $X_{p,q}$ can be regarded as the set of pairs of subspaces $L, M \subset \mathbb{C}^{p+q}$ such that $\dim L = q$, $\dim M = p$, and $L \cap M = 0$.

Denote by $\overline{X_{p,q}} = \overline{GL_{p+q}/GL_p \times GL_q}$ the closure of $X_{p,q} \subset GL_{p+q}$ in Hinge. Assume for simplicity that $q \neq p$ and, for the sake of being definite, that $q > p$. Then $\overline{X_{p,q}} \setminus X_{p,q}$ consists of the hinges

(8.8) $$\mathcal{P} = (P_1, \ldots, P_k, \ldots, P_{2k+1})$$

such that
1. $P_j^\square = P_{2k+1-j}$;
2. the natural operator A: $\text{Dom } P_k / \text{Ker } P_k \to \text{Im } P_k / \text{Indef } P_k$ satisfies the condition $A^2 = E$; moreover, $\dim \text{Ker}(A - E) - \dim \text{Ker}(A + E) = q - p$.

8.10. The completions of the symmetric spaces $Sp_{2(n+k)}(\mathbb{C})/Sp_{2n}(\mathbb{C}) \times Sp_{2k}(\mathbb{C})$ and $O_{n+k}(\mathbb{C})/O_n(\mathbb{C}) \times O_k(\mathbb{C})$ can be constructed in just the same way as in the previous subsection; however, the hinge $\mathcal{P}$ must belong to $\overline{Sp_{2(n+k)}}$ and $\overline{O_{n+k}}$, respectively.

8.11. Space $\overline{PO_{2n}/GL_n}$. Let us describe the symmetric space

(8.9) $$O_{2n}(\mathbb{C})/GL_n(\mathbb{C}).$$

Consider a $2n$-dimensional complex space V endowed with a nondegenerate bilinear form. It is natural to regard all possible decompositions of the space

(8.10) $$V = V_+ \oplus V_-$$

into the direct sum of maximal isotropic subspaces as points of the space in (8.9). Indeed, the stabilizer of the pair $\{V_+, V_-\}$ is the group of matrices of the form $\begin{pmatrix} g & \\ & (g^t)^{-1} \end{pmatrix}$, $g \in GL_n(\mathbb{C})$, and the group $O(2n, \mathbb{C})$ transitively acts on the set of decompositions (8.10).

Furthermore, consider an operator $A \in O(2n, \mathbb{C})$ that is equal to iE on V_+ and to $(-iE)$ on V_-. Then we have $A^2 = -E$, and the set of these operators is in a one-to-one correspondence with the set of decompositions (8.10).

Let us take the closure of this set in $\overline{O_{2n}}$. We obtain the set of hinges $\mathcal{P} = (P_1, \ldots, P_k)$ such that $\mathcal{P}^t = \mathcal{P}^\square$ (this is a condition on $\mathcal{P} \in \overline{O_{2n}}$) and $\mathcal{P}^\square = -\mathcal{P}$. In other words, $\mathcal{P}$ satisfies the conditions $P_j^t = -P_j$ and $P_j^\square = -P_{k-j}$.

8.12. The completion of the space $Sp_{2n}(\mathbb{C})/GL_n(\mathbb{C})$ is constructed in the same way.

8.13. **Smoothness.** The smoothness of all the above varieties is proved by the same arguments as in the proof of Proposition 8.3.

§9. Real forms of complete symmetric varieties. The Satake–Furstenberg boundary

9.1. **Real hinges.** Until now we discussed complex hinges only. However, we can also consider hinges over reals or over quaternions. For example, arguments just like those used in the complex case show that the set of all hinges in $\mathbb{R}^n$ is a smooth real analytic variety. Now we shall present a more involved example of a real form of a complex symmetric variety; namely, we consider a real form of complex quadrics.

9.2. **Boundaries of the spaces $SL(n, \mathbb{R})/SO(p, n-p)$.** In $\mathbb{R}^n$ we take the bilinear form $\langle x, y \rangle = \sum x_j y_j$. In the space $\mathbb{R}^{2n}$ we consider the skew-symmetric bilinear form $\{(x,y),(u,v)\} = \sum x_j v_j - \sum y_j u_j$.

Let $\mathcal{L}$ be a Lagrangian Grassmannian in $\mathbb{R}^n \oplus \mathbb{R}^n$. Let the multiplication group $\mathbb{R}^*$ of reals act on $\mathcal{L}$ by multiplication by scalars.

Consider an open dense subset $\mathcal{L}^\circ$ in $\mathcal{L}$ that consists of the graphs of the invertible operators $A \colon \mathbb{R}^n \to \mathbb{R}^n$; note that these operators satisfy the relation $A = A^t$.

Furthermore, apply the construction of Hausdorff quotient space to the $\mathcal{L}$ and to the open subset $\mathcal{L}^\circ$. We obtain the set C_n of exact real hinges $\mathcal{P} = (P_1, \ldots, P_k)$ that satisfy the condition $P_j = P_j^t$ (where P_j^t was defined in subsection 8.2).

The set $\mathcal{L}^\circ$ consists of the one-link hinges.

The group $GL_n(\mathbb{R})$ acts on $\mathcal{L}^\circ$ by the transformations

(9.1) $$A \mapsto g^t A g,$$

and on the space C_n by the transformations $\mathcal{P} \mapsto g^t \mathcal{P} g$.

An arbitrary real symmetric matrix A can be reduced, by transformations (9.1), to the form

$$J_s = \left.\left(\begin{array}{cccccc} 1 & & & \vdots & & \\ & \ddots & & \vdots & & \\ & & 1 & \vdots & & \\ \cdots & \cdots & \cdots & \vdots & \cdots & \cdots \\ & & & \vdots & -1 & \\ & & & \vdots & & \ddots \\ & & & \vdots & & & -1 \end{array}\right)\right\}\begin{array}{c} j \\ \\ \\ n-s \end{array}$$

and the stabilizer of J_s is the group $O(s, n-s)$. The matrix A, in our case, is defined up to a factor, and thus the canonical forms J_s and J_{n-s} are equivalent.

Thus, the space $\mathcal{L}^\circ$ is the disjoint union of open orbits (symmetric spaces)

(9.2) $$PGL(n, \mathbb{R})/SO(s, n-s),$$

where $s \in \{0, 1, \ldots, [(n+1)/2]\}$.

The closure of the orbit

(9.3) $$PGL(n, \mathbb{R})/PO(n)$$

is called the *Furstenberg–Satake compactification* of the Riemannian symmetric space (9.3), and the closures of the orbits (9.2) are compactifications of pseudo-Riemannian symmetric spaces $PGL(n, \mathbb{R})/SO(s, n-s)$.

9.3. The Satake–Furstenberg compactification.
Now we shall give a more detailed description of the boundary of the Riemannian symmetric space $G/K = SL(n, \mathbb{R})/SO(n)$.

Consider a linear relation $\mathbb{R}^n \rightrightarrows \mathbb{R}^n$ such that $P = P^t$. Then $\operatorname{Im}(P)$ is the orthogonal complement of $\operatorname{Ker}(P)$, and $\operatorname{Indef}(P)$ is the orthogonal complement to $\operatorname{Dom}(P)$ (with respect to the standard inner product in $\mathbb{R}^n$). Hence, a linear relation P defines a nondegenerate pairing

(9.4) $$\operatorname{Dom}(P)/\operatorname{Ker}(P) \times \operatorname{Im}(P)/\operatorname{Indef}(P) \to \mathbb{R}.$$

Moreover, a linear relation P defines an operator

(9.5) $$\operatorname{Dom}(P)/\operatorname{Ker}(P) \to \operatorname{Im}(P)/\operatorname{Indef}(P).$$

Hence, a symmetric linear relation P defines a nondegenerate symmetric bilinear form $\mathfrak{q}_P$ on the space $\operatorname{Dom}(P)/\operatorname{Ker}(P)$.

We say that a symmetric linear relation P is *nonnegative definite* if the form

$$[(v, w); (v', w')] := \langle v, w' \rangle + \langle v', w \rangle$$

is nonnegative definite on the subspace P. This condition holds if and only if the quadratic form defined by the bilinear form $\mathfrak{q}_P$ is positive.

REMARK 9.1. Let a linear relation P be the graph of an operator A. Then P is nonnegative definite if and only if A is nonnegative definite.

By our definition, a point of the *Satake–Furstenberg compactification* of the space $SL(n, \mathbb{R})/SO(n)$ is determined by the following data:

1* an integer $s \in \{1, \ldots, n-1\}$;
2* a hinge $\mathcal{P} = (P_1, \ldots, P_s)$ such that all linear relations P_j are nonnegative definite and satisfy the condition $P_j = P_j^t$ for all j.

Consider a point of the Satake–Furstenberg compactification (i.e., let the data of the form 1*–2* be given). Introduce the subspaces $V_j = \operatorname{Ker}(P_j) = \operatorname{Dom}(P_{j+1})$. Then the form related to $[\,\cdot\,,\,\cdot\,]$ defines a positive definite form on the quotient space $\operatorname{Dom}(P_j)/\operatorname{Ker}(P_j)$. We see that a point of the Satake–Furstenberg boundary can be defined by the following data:

1* an integer $s \in \{1, \ldots, n-1\}$;
2* a flag $0 \subset V_1 \subset \cdots \subset V_s \subset \mathbb{R}^n$, where all subspaces $0, V_1, \ldots, V_s, \mathbb{R}^n$, are distinct, and
3* for any $j \in \{1, \ldots, s\}$, a positive definite quadratic form Q_j on the quotient space $\operatorname{Dom}(P_j)/\operatorname{Ker}(P_j)$.

§10. Category of hinges

10.1. Spaces $\widetilde{\mathrm{Hinge}}_n(p,q)$. By a *weak n-dimensional hinge* $\mathcal{P}\colon \mathbb{C}^p \rightrightarrows \mathbb{C}^q$ we mean the collection $\mathcal{P} = (P_1, \ldots, P_s)$ of n-dimensional linear relations $P_j\colon \mathbb{C}^p \rightrightarrows \mathbb{C}^q$ such that $\operatorname{Ker} P_j \supset \operatorname{Dom} P_{j+1}$ and $\operatorname{Im} P_j \subset \operatorname{Indef} P_{j+1}$. Denote by $\widetilde{\mathrm{Hinge}}_n(p,q)$ the space of all n-dimensional weak hinges $\mathbb{C}^p \rightrightarrows \mathbb{C}^q$ and denote by $\widetilde{\mathrm{Hinge}}(p,q)$ the space

$$\widetilde{\mathrm{Hinge}}(p,q) = \bigcup_{n=0}^{p+q} \widetilde{\mathrm{Hinge}}_n(p,q).$$

10.2. Multiplication. The multiplication

$$\widetilde{\mathrm{Hinge}}(p,q) \times \widetilde{\mathrm{Hinge}}(q,r) \to \widetilde{\mathrm{Hinge}}(p,r)$$

is defined in just the same way as the multiplication of weak hinges. Let $\mathcal{P} = (P_1, \ldots, P_k) \in \widetilde{\mathrm{Hinge}}(p,q)$ and $\mathcal{R} = (R_1, \ldots, R_l) \in \widetilde{\mathrm{Hinge}}(q,r)$. Consider the set of all products $Q_{ij} = R_i P_j$ that differ from *null*. Then, after an appropriate ordering, the set Q_{ij} becomes a weak hinge $\mathbb{C}^p \rightrightarrows \mathbb{C}^r$.

Let us define the category *HINGE*. Its objects are linear spaces $\mathbb{C}^p$, and the set of morphisms of $\mathbb{C}^p$ into $\mathbb{C}^q$ is $\widetilde{\mathrm{Hinge}}(p,q)$.

10.3. Action on the exterior powers. Let $\mathcal{P} = (P_1, \ldots, P_k) \in \widetilde{\mathrm{Hinge}}_n(p,q)$. Let us define the operator $\lambda_j(\mathcal{P})\colon \Lambda^j \mathbb{C}^p \to \Lambda^{j+n-p}\mathbb{C}^q$. To this end we consider the sequence of operators

$$(10.1) \qquad \lambda_j(P_1), \ldots, \lambda_j(P_k)\colon \Lambda^j\mathbb{C}^p \to \Lambda^{j+n-p}\mathbb{C}^q,$$

where $\lambda_j(P_s)$ is the restriction of the operator $\lambda(P_s)$ (see §1) to $\Lambda^j \mathbb{C}^p$. By definition, we take for $\lambda_j(\mathcal{P})$ the nonzero term of the sequence (10.1) if this nonzero term exists, and set $\lambda_j(\mathcal{P}) = 0$ otherwise.

If $\mathcal{P} \in \widetilde{\mathrm{Hinge}}_n(p,q)$ and $\mathcal{R} \in \widetilde{\mathrm{Hinge}}_m(q,r)$, then $\lambda_j(\mathcal{RP}) = t \cdot \lambda_{j+n-p}(\mathcal{R}) \lambda_j(\mathcal{P})$, where $t \in \mathbb{C}^*$.

10.4. Action on flags. Let $0 \subset V_1 \subset \cdots \subset V_k \subset \mathbb{C}^p$ be a flag in the space $\mathbb{C}^p$, i.e., an element of the space $\mathcal{F}_p$, in the notation of §7. Let $\mathcal{P} = (P_1, \ldots, P_k) \in \widetilde{\mathrm{Hinge}}_n(p,q)$. Furthermore, consider the set Σ of all pairs (i,j) such that $\operatorname{Dom} P_i + V_j = \mathbb{C}^p$ and $\operatorname{Ker} P_i \cap V_j = 0$. Then the set of subspaces $P_i V_j$, where (i,j) ranges over Σ, forms a flag in $\mathbb{C}^q$; that is, a hinge $\mathcal{P}$ defines a mapping $\mathcal{F}_p \to \mathcal{F}_q$.

Chapter III. Other boundaries

§11. Velocity compactifications of symmetric spaces

In this section we consider the symmetric spaces $SL(n,\mathbb{R})/SO(n)$ only.

We recall that a point of the space $SL(n,\mathbb{R})/SO(n)$ can be identified with a positive definite real matrix that is defined up to a scalar factor.

11.1. The simplest velocity compactification. Consider a positive definite matrix $A \in Q = SL(n,\mathbb{R})/SO(n)$. Let $a_1 \geqslant \cdots \geqslant a_n$ be the eigenvalues of A. Let $\lambda_j = \ln a_j$. Denote by $\Lambda(A)$ the set

$$(11.1) \qquad \Lambda(A) = (\lambda_1, \ldots, \lambda_n), \qquad \lambda_1 \geqslant \lambda_2 \geqslant \cdots.$$

The matrix A is defined up to a factor, and hence $\Lambda(A)$ is defined up to an additive constant:

$$(11.2) \qquad (\lambda_1, \ldots \lambda_n) \sim (\lambda_1 + \sigma, \ldots, \lambda_n + \sigma).$$

We denote by Σ_n the space of all sets $\Lambda(A)$ (see (11.2)). We can readily see that $\Lambda(A)$ is an $(n-1)$-dimensional simplicial cone. We can assume that $\lambda_n = 0$, and hence the cone Σ_n can be regarded as the space of sets $\{\lambda_1 \geqslant \cdots \geqslant \lambda_{n-1} \geqslant 0\}$. We denote by $\Delta_n = \partial \Sigma_n$ the $(n-2)$-dimensional simplex $1 \geqslant \mu_2 \geqslant \mu_3 \geqslant \cdots \geqslant \mu_{n-1} \geqslant 0$ and set $\mu_1 = 1$ and $\mu_n = 0$. Then Δ_n is called the *velocity simplex*. Consider the natural projection $\pi \colon (\Sigma_n \setminus 0) \to \Delta_n$ defined by the rule

$$\pi(\lambda_1, \ldots \lambda_{n-1}, 0) = (\lambda_2/\lambda_1, \lambda_3/\lambda_1, \ldots, \lambda_{n-1}/\lambda_1).$$

Now we define the compactification $\overline{\Sigma}_n = \Sigma_n \cup \Delta_n$ of Σ_n as follows. We say that a sequence $L_j = (\lambda_1^{(j)}, \ldots, \lambda_n^{(j)}) \in \Sigma_n$ converges to an element $M \in \Delta_n$ if the following two conditions hold:

1) $\lambda_1^{(j)} - \lambda_n^{(j)} \to \infty$ as $j \to \infty$, and
2) the sequence $\pi(L_j) \in \Delta_n$ converges to M.

Moreover, we introduce the *velocity compactification* of the symmetric space $SL(n, \mathbb{R})/SO(n)$ by the relation $\overline{Q}^{\mathrm{vel}} = (SL(n, \mathbb{R})/SO(n)) \cup \Delta_n$, where a sequence A_j in Q is convergent to $M \in \Delta_n$ whenever $\Lambda(A_j)$ converges to M with respect to the topology of $\overline{\Sigma}_n$.

11.2. The polyhedron of Karpelevich velocities. Now we describe a more delicate compactification of the simplicial cone Σ_n (namely, the compactification by Karpelevich velocities). Consider a sequence $\lambda^{(j)} = \{\lambda_1^{(j)} \geqslant \cdots \geqslant \lambda_n^{(j)}\} \in \Sigma_n$. Let this sequence have the limit $1 \geqslant \mu_2 \geqslant \ldots \geqslant \mu_{n-1} \geqslant 0$ in Δ_n. It may happen that some of the numbers μ_i are equal, i.e., $\mu_k = \mu_{k+1} = \cdots = \mu_l$. In this case we can separate velocities that correspond to the subset $\{\lambda_k^{(j)} \geqslant \cdots \geqslant \lambda_l^{(j)}\} \in \Sigma_{l-k+1}$ by the same rule as above.

Definition of the polyhedron. Denote by $I_{\alpha,\beta}$ the set $\{\alpha, \alpha+1, \ldots, \beta\} \subset \mathbb{N}$.

Consider an interval $I_{\alpha,\beta} = \{\alpha, \alpha+1, \ldots, \beta\}$ and denote by $\Sigma(I_{\alpha,\beta})$ the simplicial cone $\lambda_\alpha \geqslant \lambda_{\alpha+1} \geqslant \cdots \geqslant \lambda_\beta$, where the elements of the cone $\Sigma(I_{\alpha,\beta})$ are defined up to an additive constant (see (11.2)). Moreover, introduce the simplex $\Delta(I_{\alpha,\beta})$ defined by the inequalities $1 = \mu_\alpha \geqslant \mu_{\alpha+1} \geqslant \cdots \geqslant \mu_{\beta-1} \geqslant \mu_\beta = 0$ and consider the compactification $\overline{\Sigma}(I_{\alpha,\beta}) = \Sigma(I_{\alpha,\beta}) \cup \Delta(I_{\alpha,\beta})$.

REMARK 11.1. For the case $\alpha = \beta$, the set $\Sigma(I_{\alpha,\alpha}) = \overline{\Sigma}(I_{\alpha,\alpha})$ is a singleton (a real defined up to an additive constant).

For $k \leqslant \alpha \leqslant \beta \leqslant l$ we define the mapping $\pi_{\alpha,\beta}^{k,l} \colon \Sigma(I_{k,l}) \to \Sigma(I_{\alpha,\beta})$ by the formula $\pi_{\alpha,\beta}^{k,l}(\lambda_k, \ldots, \lambda_l) = (\lambda_\alpha, \ldots, \lambda_\beta)$. We introduce two polyhedra:

$$\Xi(k,l) := \prod_{\alpha,\beta \,:\, k \leqslant \alpha \leqslant \beta \leqslant l} \Sigma(I_{\alpha,\beta}), \qquad \overline{\Xi}(k,l) := \prod_{\alpha,\beta \,:\, k \leqslant \alpha \leqslant \beta \leqslant l} \overline{\Sigma}(I_{\alpha,\beta}).$$

We clearly have $\Xi(k,l) \subset \overline{\Xi}(k,l)$. Consider the (diagonal) embedding $i \colon \Sigma(I_{k,l}) \to \Xi(k,l)$ (which is the product of the mappings $\pi_{\alpha,\beta}^{k,l}$).

We define the *polyhedron of Karpelevich velocities* $\mathcal{K}(k,l)$ as the closure of the subset $i(\Sigma(I_{k,l}))$ in $\overline{\Xi}(k,l)$.

A CRITERION FOR THE CONVERGENCE OF A SEQUENCE OF INTERIOR POINTS TO A BOUNDARY POINT. Consider a sequence $\Lambda^{(j)} = \{\lambda_k^{(j)}, \lambda_{k+1}^{(j)}, \ldots, \lambda_l^{(j)}\}$. Then the sequence $\Lambda^{(j)}$ is convergent in $\mathcal{K}(k,l)$ if and only if all sequences of the form $\pi_{\alpha,\beta}^{k,l}(\Lambda^{(j)}) = (\lambda_\alpha^{(j)}, \ldots, \lambda_\beta^{(j)})$ are convergent in $\overline{\Sigma}(I_{\alpha,\beta})$.

The polyhedron of Karpelevich velocities is thus defined. Now we give an explicit (but cumbersome) description of its combinatorial structure.

11.3. Combinatorial description of the polyhedron of Karpelevich velocities.

TREE-PARTITIONS. Consider the set $I_{k,l} := \{k, k+1, \ldots, l\}$. By a *partition* of $I_{k,l}$ we mean a representation of $I_{k,l}$ in the form $I_{k,m_1} \cup I_{m_1+1,m_2} \cup \cdots \cup I_{m_{s-1}+1,l}$, where $s > 1$. A system $\mathfrak{a}$ formed by subsets of $I_{k,l}$ is called a *tree partition* if the following conditions hold:
  a) $I_{k,l} \in \mathfrak{a}$;
  b) any element $J \in \mathfrak{a}$ has the form $I_{\alpha,\beta} = \{\alpha, \alpha+1, \ldots, \beta\}$;
  c) If $J_1, J_2 \in \mathfrak{a}$, then we have either $J_1 \cap J_2 = \varnothing$ or one of the conditions $J_1 \supset J_2$ and $J_2 \subset J_1$;
  d) for any $J = I_{\alpha,\beta} \in \mathfrak{a}$ we have one of the following two possibilities:
  (1) 1^* there is no $K \in \mathfrak{a}$ such that $K \subset J$ (in this case, $I_{\alpha,\beta}$ is said to be *irreducible*);
  (2) 2^* $J = I_{\alpha,\beta}$ is the

(11.3) $$I_{\alpha,\beta} = I_{\alpha,\gamma_1} \cup I_{\gamma_1+1,\gamma_2} \cup I_{\gamma_2+1,\gamma_3} \cup \cdots \cup I_{\gamma_{s-1}+1,\beta},$$

where $I_{\alpha,\gamma_1}, I_{\gamma_1+1,\gamma_2}, \ldots, I_{\gamma_{s-1}+1,\beta} \in \mathfrak{a}$ and there is no $K \in \mathfrak{a}$ such that $I_{\alpha,\beta} \supset K \supset I_{\gamma_{i-1}+1,\gamma_i}$, $K \neq I_{\alpha,\beta}, I_{\gamma_{i-1},\gamma_i}$. In this case, J is said to be *reducible* and (11.3) is called the *canonical decomposition* of J.

REMARK 11.2. Let $I_{\alpha,\beta} \in \mathfrak{a}$. Let $\mathfrak{b}$ be the set of all $J \subset I_{\alpha,\beta}$ such that $J \in \mathfrak{a}$. Then $\mathfrak{b}$ is a tree partition of $I_{\alpha,\beta}$.

REMARK 11.3. In other words, a tree partition can be defined by the following data: a partition of the segment $I_{k,l} \subset \mathbb{N}$ into subsegments, partitions of some subsegments, etc.

We denote by $TP(k,l)$ the set of all tree partitions of $I_{k,l}$. Introduce the canonical partial ordering on $TP(k,l)$. Let $\mathfrak{a}, \mathfrak{b} \in TP(k,l)$. We say that $\mathfrak{a} > \mathfrak{b}$ whenever $J \in \mathfrak{a}$ implies $J \in \mathfrak{b}$ (i.e., $\mathfrak{b} \supset \mathfrak{a}$).

The partially ordered set $TP(k,l)$ contains a unique maximal element $\mathfrak{a}_0$. This is the tree partition that consists of a single element $I_{k,l}$.

An element $\mathfrak{b} \in TP(k,l)$ is minimal if the following two conditions hold:
  a) any irreducible element of $\mathfrak{b}$ is a singleton;
  b) if $J \in \mathfrak{b}$ is reducible, then the canonical decomposition of J contains exactly two elements ($s = 2$ in (11.3)).

Consider a partition $\mathfrak{t}$ of $I_{\alpha,\beta}$:

(11.4) $$I_{\alpha,\beta} = I_{\alpha,\gamma_1} \cup I_{\gamma_1+1,\gamma_2} \cup I_{\gamma_2+1,\gamma_3} \cup \cdots \cup I_{\gamma_{s-1}+1,\beta}.$$

Denote by $\widetilde{\Delta}(I_{\alpha,\beta}|\mathfrak{t})$ the open simplex

(11.5) $$1 = \mu_\alpha = \cdots = \mu_{\gamma_1} > \mu_{\gamma_1+1}$$
$$= \mu_{\gamma_1+2} = \cdots = \mu_{\gamma_2} > \cdots > \mu_{\gamma_{s-1}+1} = \cdots = \mu_\beta = 0$$

and by $\Delta(I_{\alpha,\beta}|\mathfrak{t})$ the compact simplex

(11.6) $$1 = \mu_\alpha = \cdots = \mu_{\gamma_1} \geqslant \mu_{\gamma_1+1}$$
$$= \mu_{\gamma_1+2} = \cdots = \mu_{\gamma_2} \geqslant \ldots \geqslant \mu_{\gamma_{s-1}+1} = \cdots = \mu_\beta = 0.$$

In $\Delta(I_{\alpha,\beta}|\mathfrak{t})$ and $\widetilde{\Delta}(I_{\alpha,\beta}|\mathfrak{t})$ we introduce the natural coordinates

$$\tau_2 := \mu_{\gamma_1+1} = \cdots = \mu_{\gamma_2}, \quad \ldots, \quad \tau_{s-1} := \mu_{\gamma_{s-2}+1} = \cdots = \mu_{\gamma_{s-1}}.$$

REMARK 11.4. If s=2, then $\Delta(J|\mathfrak{t}) = \widetilde{\Delta}(J|\mathfrak{t})$ is a singleton $\{1 > 0\}$.

REMARK 11.5. We have $\Delta(I_{\alpha,\beta}) = \bigcup_\mathfrak{t} \widetilde{\Delta}(I_{\alpha,\beta}|\mathfrak{t})$, where the union is taken over all partitions of $I_{\alpha,\beta}$.

Let us choose a tree partition $\mathfrak{a} \in TP(k,l)$. For any element $J \in \mathfrak{a}$ we consider its canonical decomposition $\mathfrak{t}$ and denote the simplex $\widetilde{\Delta}(J|\mathfrak{t})$ by $\widetilde{\Delta}(\mathfrak{a}, J)$. For any $\mathfrak{a} \in TP(k,l)$ we define the *face* $F(\mathfrak{a})$ as follows:

(11.7) $$F(\mathfrak{a}) = \Big(\prod_{J=I_{\alpha,\beta} \in \mathfrak{a} \text{ is irreducible}} \Sigma(I_{\alpha,\beta}) \Big) \prod_{J \in \mathfrak{a} \text{ is redicible}} \widetilde{\Delta}(\mathfrak{a}, J).$$

REMARK 11.6. For the trivial tree partition $\mathfrak{a}_0$, we have $F(\mathfrak{a}_0) = \Sigma(I_{k,l})$. If $\mathfrak{b}$ is a minimal tree partition, then $F(\mathfrak{b})$ is a singleton.

Now we can represent the *polyhedron of Karpelevich velocities* $\mathcal{K}(k,l)$ as the union $\mathcal{K}(k,l) = \bigcup_{\mathfrak{a} \in TP(k,l)} F(\mathfrak{a})$. Let us define a certain topology of a compact metric space on $\mathcal{K}(k,l)$. The face $F(\mathfrak{a}_0) = \Sigma(I_{k,l})$ will be an open dense subset of $\mathcal{K}(k,l)$.

REMARK 11.7. Let $l = k$. Then $\mathcal{K}(k,k)$ is a singleton. Let $l = k+1$. Then there are two tree partitions of the set $\{k, k+1\}$: the trivial tree partition $\mathfrak{a}_0$ and the minimal tree partition $\mathfrak{a}_1$ with the elements $(k, k+1)$, (k), and $(k+1)$. The face $F(\mathfrak{a}_0)$ is the closed semi-axis $\lambda_1 > 0$. The face $F(\mathfrak{a}_1)$ is a singleton. Hence, $\mathcal{K}(k, k+1)$ is the segment $[0, \infty]$.

THE CONVERGENCE OF INTERIOR POINTS TO A BOUNDARY POINT. For the definition of convergence we proceed by induction. We assume that the convergence is defined for all Karpelevich polyhedra $\mathcal{K}(\alpha, \beta)$ such that $\beta - \alpha < l - k$. We define the convergence of a sequence

$$x^{(j)} = \{x_k^{(j)} \geqslant \cdots \geqslant x_l^{(j)}\} \in \Sigma(I_{k,l}) = F(\mathfrak{a}_0)$$

in two stages.

Step 1. The convergence of $x^{(j)}$ in $\overline{\Sigma}(I_{k,l})$ is necessary for the convergence of this sequence in $\mathcal{K}(k,l)$. Let y be the limit of $x^{(j)}$ in $\overline{\Sigma}(I_{k,l})$.

If $y \in \Sigma(k,l)$, then the sequence is said to be convergent in $\mathcal{K}(k,l)$, and y is called the limit of $x^{(j)}$ in $\mathcal{K}(k,l)$.

Step 2. Let $y \notin \Sigma(I_{k,l})$. Then y belongs to an open simplex $\widetilde{\Delta}(I_{k,l}|\mathfrak{t})$, that is, y is of the form

$$\{1 = y_k = \cdots = y_{\gamma_1} > y_{\gamma_1+1} = \cdots = y_{\gamma_2} > \cdots > y_{\gamma_{s-1}+1} = \cdots = \gamma_l = 0\}.$$

In this case, we say that the sequence $x^{(j)}$ is convergent in $\mathcal{K}(k,l)$ if and only if all sequences $x_{[\psi]}^{(j)} := (x_{\gamma_\psi+1}^{(j)}, \ldots, x_{\gamma_{\psi+1}}^{(j)}) \in \Sigma(I_{\gamma_\psi+1, \gamma_{\psi+1}})$ converge in the corresponding polyhedra of Karpelevich velocities $\mathcal{K}(\gamma_\psi+1, \gamma_{\psi+1})$ (their convergence being defined by the induction assumption).

This concludes the definition.

EXAMPLE 11.8. Let $k = 1$ and $l = 8$. Consider a sequence $x^j = (x_1^j, \ldots, x_8^j)$, where

$$x_1^{(j)} = 2j^3, \quad x_2^{(j)} = j^3, \quad x_3^{(j)} = j^2 + j + 2, \quad x_4^{(j)} = j^2 + j + 1,$$
$$x_5^{(j)} = j^2 + j, \quad x_6^{(j)} = 2j, \quad x_7^{(j)} = j, \quad x_8^{(j)} = 0.$$

Then the associated tree partition has the form

$$(1\ 2\ 3\ 4\ 5\ 6\ 7\ 8)$$
$$(1)\ (2)\ (3\ 4\ 5\ 6\ 7\ 8)$$
$$(3\ 4\ 5\ )\ (6\ 7\ 8)$$
$$(6)\ (7)\ (8)$$

and the limit of $x^{(j)}$ in $\overline{\Sigma}(I_{1,8})$ is the set

(11.8) $\qquad \{1 > 1/2 > 0 = 0 = 0 = 0 = 0 = 0\} \in \Delta(I_{1,8}).$

The sequence $x^{(j)}$ defines the sequence $y^{(j)} = (x_3^{(j)}, \ldots, x_8^{(j)}) \in \Sigma(I_{3,8})$. The limit of $y^{(j)}$ in $\overline{\Sigma}(I_{3,8})$ is the set

(11.9) $\qquad \{1 \geqslant 1 \geqslant 1 \geqslant 0 \geqslant 0 \geqslant 0\} \in \Delta(I_{3,8}).$

Moreover, we have the sequences $z^{(j)} = (x_3^{(j)}, x_4^{(j)}, x_5^{(j)}) \in \Sigma(I_{3,5})$ and $u^{(j)} = (x_6^{(j)}, x_7^{(j)}, x_8^{(j)}) \in \Sigma(I_{6,8})$. We have $z^{(j)} = (j^2+j+2, j^2+j+1, j^2+j) = (2,1,0)$ (recall that the collection $z^{(j)}$ is defined up to an additive constant), and $\lim z^{(j)}$ is the set

(11.10) $\qquad \{2 > 1 > 0\} \in \Sigma(I_{3,5}).$

Finally, $u^{(j)} = (2j, j, 0)$, and the limit of $u^{(j)}$ in $\overline{\Sigma}(I_{6,8})$ is the point

(11.11) $\qquad \{1 > 1/2 > 0\} \in \widetilde{\Delta}(I_{6,8}).$

The limit of the sequence $x^{(j)}$ is the family of sets (11.8)–(11.11).

THE TOPOLOGY ON THE BOUNDARY OF $I_{k,l}$. This topology must satisfy the following property: the closure of $F(\mathfrak{a})$ consists of all faces $F(\mathfrak{b})$ such that $\mathfrak{b} < \mathfrak{a}$.

We assume that the topology is defined for all polyhedra $\mathcal{K}(\alpha, \beta)$ such that $\beta - \alpha < l - k$.

Let us define the convergence of a sequence $Z^{(j)} \in F(\mathfrak{a})$ in two stages.

Step 1. Let $h^{(j)} = \{1 = h_k^{(j)} \geqslant h_{k+1}^{(j)} \geqslant \cdots \geqslant h_l^{(j)} = 0\}$ be the component of $Z^{(j)}$ related to the factor $\widetilde{\Delta}(\mathfrak{a}, I_{k,l})$ in the product (11.7). Then the convergence of $h^{(j)}$ in $\Delta(\mathfrak{a}, I_{k,l})$ is necessary for the convergence of this sequence in $\mathcal{K}(k,l)$. Let u be the limit of $h^{(j)}$ in $\Delta(\mathfrak{a}, I_{k,l})$.

Step 2. Consider the partition of $I_{k,l}$ related to $\mathfrak{a}$, that is, $I_{k,l} = I_{k,\gamma_1} \cup I_{\gamma_1+1,\gamma_2} \cup \cdots \cup I_{\gamma_{s-1}+1,l}$. Then the set u has the form

$$u = \{1 = u_k = \cdots = u_{\gamma_1} \geqslant u_{\gamma_1+1} = \cdots = u_{\gamma_2} \geqslant \cdots\}.$$

Introduce the indices $\tau_1, \tau_2, \ldots$ such that

$$\{1 = u_k = \cdots = u_{\tau_1} > u_{\tau_1+1} = \cdots = u_{\tau_2} > \cdots\}.$$

Then $\{\tau_1, \tau_2, \ldots\}$ is a subset of $\{\gamma_1, \gamma_2, \ldots\}$), and hence any segment of the form $I_{\tau_\alpha+1,\tau_{\alpha+1}}$ is the union of segments $I_{\gamma_m+1,\gamma_{m+1}}$.

On any set of the form $\{\tau_\alpha+1, \tau_\alpha+2, \ldots, \tau_{\alpha+1}\}$, we introduce the tree partition $\mathfrak{b}_\alpha$ induced by the tree partition $\mathfrak{a}$. The sequence $Z^{(j)}$ induces a sequence $Z^{(j)}_{[\alpha]}$ on any face $F(\mathfrak{b}_\alpha) \subset \mathcal{K}(\tau_\alpha+1, \tau_{\alpha+1})$.

The sequence $Z^{(j)}$ is said to be convergent if and only if any sequence $Z^{(j)}_{[\alpha]}$ in the Karpelevich polyhedron $\mathcal{K}(\tau_\alpha+1, \tau_{\alpha+1})$ is convergent.

11.4. The compactification of the symmetric space by the Karpelevich velocities. Consider the boundary $\partial \mathcal{K}(1,n) := \mathcal{K}(1,n) \setminus \Sigma(I_{1,n})$ of the polyhedron $\mathcal{K}(1,n)$ and define the compactification $(SL(n,\mathbb{R})/SO(n)) \cup (\partial \mathcal{K}(1,n))$ of the symmetric space $SL(n,\mathbb{R})/SO(n)$. Let $x^{(j)}$ be a sequence in $SL(n,\mathbb{R})/SO(n)$ and let $y \in \partial \mathcal{K}(I_{1,n})$. Then we have $x^{(j)} \to y$ if and only if the following conditions hold:
1) the distance $d(x^{(j)},0)$ satisfies the relation $d(x^{(j)},0) \to \infty$, and
2) we have $\Lambda(x^{(j)}) \to y$ in the topology of $\mathcal{K}(I_{1,n})$ (where $\Lambda(\cdot)$ is defined by formula (11.1)).

§12. Tits building on the matrix sky

In this section we consider the symmetric spaces $SL(n,\mathbb{R})/SO(n)$ only. Recall that the points of the space $SL(n,\mathbb{R})/SO(n)$ can be identified with real positive definite matrices defined up to a scalar factor.

We recall that any geodesic curve in the space $SL(n,\mathbb{R})/SO(n)$ has the form

$$(12.1) \qquad \gamma(s) = A \begin{pmatrix} \exp(\lambda_1 s) & & & \\ & \exp(\lambda_2 s) & & \\ & & \ddots & \\ & & & \exp(\lambda_n s) \end{pmatrix} A^t,$$

where

$$(12.2) \qquad A \in SL(n,\mathbb{R}), \qquad \lambda_1 \geqslant \cdots \geqslant \lambda_n.$$

In this section, by a *geodesic* we mean an oriented geodesic, without specifying any parametrization.

12.1. **The matrix sky (the visibility boundary).** Consider a noncompact Riemannian symmetric space G/K. Choose a point $x_0 \in G/K$ (for the case under consideration, $G/K = SL(n,\mathbb{R})/SO(n)$, it is natural to take $x_0 = E$). Let T_{x_0} be the tangent space at the point x_0 (in our case, the tangent space is the space of symmetric matrices defined up to a scalar summand, that is, $Q \simeq Q + \lambda E$). Let S be the space of rays in T_{x_0} that start from the origin (that is, $S = (T_{x_0} \setminus 0)/\mathbb{R}_+^*$, where $\mathbb{R}_+^*$ is the multiplicative group of positive reals). Let $v \in S$ and let $\widetilde{v} \in T_{x_0}$ be a tangent vector on the ray v. Let $\gamma_v = \gamma_v(t)$ be the geodesic such that $\gamma_v(0) = x_0$ and $\gamma'_v(0) = \tilde{v}$. We are not interested in the parametrization of the geodesic γ, but its direction is essential for us.

Let Sk be another copy of the sphere S. The points of the sphere Sk are regarded as points at infinity of G/K. The sphere Sk is called the *matrix sky* or the *visibility boundary*. Let us describe the topology on the space

$$\overline{(G/K)}^{\mathrm{vis}} := G/K \cup Sk.$$

We equip the spaces G/K and Sk with the standard topology. Let y_j be a sequence in G/K. Let $v \in Sk$. Let $\gamma^{(j)}$ be a geodesic joining the points x_0 and y_j. Consider the vectors $v_j \in S$ such that $\gamma^{(j)} = \gamma_{v_j}$. The sequence $y_j \in G/K$ is said to be convergent to a point $v \in Sk$ whenever the following conditions hold:
 1) $\rho(x_0, y_j) \to \infty$, and
 2) $v_j \to v$ in the natural topology of the sphere Sk.

12.2. The projection of the matrix sky to the velocity simplex. Let $G/K = SL(n, \mathbb{R})/SO(n)$. Consider a geodesic γ that starts from $x_0 = E$. Then γ has the form

$$(12.3) \qquad \gamma(s) = A \begin{pmatrix} \exp(\lambda_1 s) & & & \\ & \exp(\lambda_2 s) & & \\ & & \ddots & \\ & & & \exp(\lambda_n s) \end{pmatrix} A^t, \qquad s \in \mathbb{R},$$

where $A \in SO(n)$ and $\lambda_1 \geqslant \cdots \geqslant \lambda_n = 0$. Let $\Delta = \Delta_n$ be the simplex $1 \geqslant \mu_2 \geqslant \cdots \geqslant \mu_{n-1} \geqslant 0$ (see subsection 11.1). To any geodesic γ we assign the point

$$D(\gamma): 1 \geqslant \frac{\lambda_2}{\lambda_1} \geqslant \frac{\lambda_3}{\lambda_1} \geqslant \cdots \geqslant \frac{\lambda_{n-1}}{\lambda_1} \geqslant 0$$

of the simplex Δ. We obtain a map D from Sk to velocity simplex Δ.

It is clear that $D(\gamma)$ is the limit of the geodesic γ in the simplest velocity compactification of $SL(n,\mathbb{R})/SO(n)$. We say that $D(\gamma) \in \Delta$ is the *velocity of the geodesic* γ.

12.3. The projection of the matrix sky to the space of flags. Let $\mathcal{F}$ be the space of all flags

$$0 = V_0 \subset V_1 \subset \cdots \subset V_s = \mathbb{R}^n$$

in $\mathbb{R}^n$ ($s = 0, 1, \ldots, n$), see §7. Denote by $\mathcal{F}_{\mathrm{complete}}$ the space of complete flags in $\mathbb{R}^n$ (i.e., $s = n$).

Let a geodesic γ be given by the expression (12.3). Assume that the collection $\lambda_1, \ldots, \lambda_n$ has the form

$$(12.4) \qquad \lambda_1 = \lambda_2 = \cdots = \lambda_{s_1} > \lambda_{s_1+1} = \lambda_{s_1+2} = \cdots = \lambda_{s_2} > \cdots.$$

Let T_α be the subspace in $\mathbb{R}^n$ formed by the vectors $(x_1, \ldots, x_{s_\alpha}, 0, 0, \ldots)$. Let $V_\alpha = AT_\alpha$ (see (12.3)). By $F(\gamma)$ denote the flag

$$(12.5) \qquad V_1 \subset V_2 \subset V_3 \subset \cdots.$$

We obtain the mapping $F \colon Sk \to \mathcal{F}$. We can readily see that the geodesic γ is defined by the pair

$$(D(\gamma); F(\gamma)) \in \Delta \times \mathcal{F}.$$

A pair (velocity (12.4), flag (12.5)) is not arbitrary and must satisfy the condition $\dim V_j = s_j$.

12.4. Limits of geodesics on the matrix sky. Consider an arbitrary geodesic γ given by (12.1)–(12.2) (generally speaking, $E \notin \gamma$). Introduce a geodesic κ_s joining the points $x_0 = E$ and $\gamma(s)$, where s is a parameter on γ. Let us calculate the limit $\lim_{s \to \infty} \kappa(s)$.

To this end we represent the matrix $A \in GL(n,\mathbb{R})$ in the form $A = UB$, where $U \in O(n)$ and B is an upper triangular matrix. It is easy to prove that the limit of the family of geodesics γ_s is the geodesic $\sigma(t)$ given by the formula

$$\sigma(t) = U \begin{pmatrix} \exp(\lambda_1 t) & & \\ & \ddots & \\ & & \exp(\lambda_n t) \end{pmatrix} U^{-1}.$$

This remark has several simple consequences.

A. *The construction of the matrix sky does not depend on the point x_0.*

Indeed, consider two points x_0 and x_1 and denote the corresponding matrix skies by $Sk(x_0)$ and $Sk(x_1)$. Consider a geodesic γ starting at x_1. Then γ has a limit on $Sk(x_0)$. This defines the canonical mapping $\psi_{10}\colon Sk(x_1) \to Sk(x_0)$. We also have the canonical mapping $\psi_{01}\colon Sk(x_0) \to Sk(x_1)$. We can easily show that $\psi_{01} \circ \psi_{10} = \mathrm{id}$ and $\psi_{10} \circ \psi_{01} = \mathrm{id}$, and we obtain the canonical bijection $Sk(x_0) \leftrightarrow Sk(x_1)$.

B. *In particular, for any point $x \in G/K$ and any point $y \in Sk$, there exists a unique geodesic joining x and y.*

C. *The group G/K naturally acts on the space $(G/K)^{\mathrm{vis}}$.*

Indeed, the group G acts on the space of geodesics. $\square$

Note that for any $g \in G$ and any $\gamma \in Sk$ we have

$$D(g \cdot \gamma) = D(\gamma), \qquad F(g \cdot \gamma) = g \cdot F(\gamma).$$

12.5. A simplicial structure on the matrix sky. Consider a complete flag $L \in \mathcal{F}_{\mathrm{complete}}$ of the form $L: 0 \subset W_1 \subset \cdots \subset W_{n-1} \subset \mathbb{R}^n$, where $\dim W_j = j$.

Now for each $L \in \mathcal{F}_{complete}$ we will construct a canonical embedding $\sigma_L\colon \Delta \to Sk$. Consider an orthonormal basis $e_1, \ldots, e_n \in \mathbb{R}^n$ such that $e_j \in W_j$ and e_j is orthogonal to W_{j-1}. Let $M = \{1 = \mu_1 \geqslant \mu_2 \geqslant \cdots \geqslant \mu_{n-1} \geqslant \mu_n = 0\}$ be a point of Δ. Consider a family of operators $R(s)$ defined by

$$R(s)e_k = \exp(\mu_k s)e_k$$

Then $R(s)$ is a geodesic and we define $\sigma_L(M)$ as the limit of this geodesic.

The map $\sigma_L\colon \Delta \to Sk$ satisfies the conditions:
1) $D \circ \sigma_L$ is the identity mapping $\Delta \to \Delta$; and
2) the image of the mapping $F \circ \sigma_L\colon \Delta \to \mathcal{F}$ consists of the subflags of the flag L.

We obtain the tiling of the sphere Sk by the simplexes $\sigma_L(\Delta)$. These simplexes are indexed by the points L of the space of *complete* flags. We can readily show that this tiling satisfies the following conditions.

a) Let $g \in SL(n,\mathbb{R})$. Then $\sigma_{gL}(\Delta) = g \cdot \sigma_L(\Delta)$.
b) If $L \neq L'$, then the interiors of the simplexes $\sigma_L(\Delta)$ and $\sigma_{L'}(\Delta)$ are disjoint.

c) Let $L\colon V_1 \subset V_2 \subset \cdots \subset V_{n-1}$ and $L'\colon V'_1 \subset V'_2 \subset \cdots \subset V'_{n-1}$ be complete flags. If $V_j \ne V'_j$ for all j, then $\sigma_L(\Delta) \cap \sigma_{L'}(\Delta) = \varnothing$. Otherwise the intersection $\Lambda = \sigma_L(\Delta) \cap \sigma_{L'}(\Delta)$ is a common face of the simplexes $\sigma_L(\Delta)$ and $\sigma_{L'}(\Delta)$. Now let us describe the face Λ. Let $\alpha_1,\ldots,\alpha_s$ be the indices j such that $V_j = V'_j$ (i.e., $V_{\alpha_i} = V'_{\alpha_i}$ and $V_j \ne V'_j$ for all $j \ne \alpha_i$). Consider the face
$$1 = \lambda_1 = \cdots = \lambda_{\alpha_1} \geqslant \lambda_{\alpha_1+1} = \lambda_{\alpha_1+2} = \cdots = \lambda_{\alpha_2} \geqslant \cdots$$
of the simplex Δ. Then $\Lambda = \Sigma_L(N) = \Sigma_{L'}(N)$.

Now the sphere Sk is endowed with the structure of a Tits building (see [**Tit**]).

12.6. The Tits metric on the matrix sky. Let $y_1, y_2 \in \sigma_L(\Delta)$. We introduce the distance $d(y_1, y_2)$ between y_1 and y_2 as the angle between the geodesics $x_0 y_1$ and $x_0 y_2$. Let $z, u \in Sk$. Consider a chain $z = z_1, z_2, \ldots, z_\beta = u$ ($z_j \in Sk$) such that for any j, the points z_j, z_{j+1} belong to the same element of our tiling.

Let us define the *Tits metric* $D(\,\cdot\,,\,\cdot\,)$ on Sk by the formula
$$D(z,u) = \inf\left(\sum_j d(z_j, z_{j+1})\right),$$
where the infimum is taken over all chains $z_1, \ldots, z_\beta$.

REMARK 12.1. The topology on the sphere Sk introduced by the Tits metric is not equivalent to the standard topology of the sphere.

EXAMPLE 12.2. Let $n = 3$ and $G/K = SL(3,\mathbb{R})/SO(3)$. Then Sk is a 4-dimensional sphere S^4 and $\dim \Delta = 1$, and thus the simplexes $\sigma_L(\Delta)$ are segments. Let us describe the simplicial structure on $Sk = S^4$. Let P be the space of all 1-dimensional linear subspaces of $\mathbb{R}^3$ and Q the space of 2-dimensional subspaces of $\mathbb{R}^3$ (certainly, $P \simeq Q$ are projective planes). We intend to construct a graph, say, Γ. The set of vertices of Γ is $P \cup Q$. Assume that $p \in P$, $q \in Q$, and $p \subset q$. Then p and q are joined by an edge, and all edges have this form. Assume that the length of any edge is $\pi/3$. Then the graph Γ is isometric to the sphere $Sk = S^4$ endowed with the Tits metric.

12.7. Abel subspaces. Let A be an orthogonal matrix. Consider the subvariety $R[A] \subset SL(n,\mathbb{R})/SO(n)$ that consists of all matrices of the form
$$\psi_A(s_1,\ldots,s_{n-1}) = A \begin{pmatrix} \exp(s_1) & & & & \\ & \exp(s_2) & & & \\ & & \ddots & & \\ & & & \exp(s_{n-1}) & \\ & & & & 1 \end{pmatrix} A^{-1},$$
where $s_1,\ldots,s_{n-1} \in \mathbb{R}$.

The mapping $(s_1,\ldots,s_{n-1}) \mapsto \psi(s_1,\ldots,s_{n-1})$ is an isometric embedding
$$\mathbb{R}^{n-1} \to SL(n,\mathbb{R})/SO(n)$$
(with respect to the standard metrics in $\mathbb{R}^{n-1}$ and in $SL(n,\mathbb{R})/SO(n)$).

Consider the trace $S[A]$ of the space $R[A]$ on the surface Sk. It is obvious that $S[A]$ is the union of $(n-1)!$ simplexes $\sigma_L(\Delta)$. These simplexes are separated by the hyperplanes $s_i = s_j$.

§13. Hybridization: The Dynkin–Olshanetsky and Karpelevich boundaries

13.1. Hybridization. Let $i_1 \colon G/K \to X$ and $i_2 \colon G/K \to Y$ be embeddings of a symmetric space G/K into compact metric spaces X and Y. Let the images of these embeddings in X and Y be dense.

Consider the embedding $i_1 \times i_2 \colon G/K \to X \times Y$ defined by the formula $h \mapsto (i_1(h), i_2(h))$, where $h \in G/K$. Let Z be the closure of the image of G/K in $X \times Y$. Then Z is a new compactification of G/K. We say that Z is the *hybrid* of X and Y.

Now we apply this construction for the case in which X is a velocity compactification and Y is the Satake–Furstenberg compactification.

13.2. The Dynkin–Olshanetsky boundary. Consider the hybrid Z of the simplest velocity compactification (see subsection 11.1) and the Satake–Furstenberg compactification for some noncompact Riemannian symmetric space. We again restrict ourselves to the case $G/K = SL(n,\mathbb{R})/O(n)$.

A point of the space Z is given by the following data:

0^* $s \in \{1,\ldots,n-1\}$;

1^* a hinge $\mathcal{P} = (P_1,\ldots,P_s)$ such that $P_j = P_j^t$ and P_j are nonnegative definite for all $j \in \{1,\ldots,s\}$ (see §9); and

2^* a point of the simplex Δ_s of the form $1 \geqslant \mu_2 \geqslant \cdots \geqslant \mu_{s-1} \geqslant 0$.

Let $x^{(j)} \in SL(n,\mathbb{R})/O(n)$ be an unbounded sequence and $a_1^{(j)} \geqslant \cdots \geqslant a_n^{(j)}$ the set of eigenvalues of $x^{(j)}$. Let $\lambda_\alpha^{(j)} = \ln a_\alpha^{(j)}$. Then the point $\Lambda(x^{(j)}) := (\lambda_1^{(j)}, \lambda_2^{(j)}, \ldots)$ is a point of the simplicial cone Σ_n (see subsection 12.1), and the sequence $x^{(j)} \in SL(n,\mathbb{R})/O(n)$ converges in Z if and only if $x^{(j)}$ converges in the Furstenberg–Satake compactification and $\Lambda(x^{(j)})$ converges in the velocity simplex $\overline{\Sigma}_n = \Sigma_n \cup \Delta$.

Now let us calculate the data 0^*–2^* corresponding to the limit $\lim x^{(j)}$. Let $\mathcal{P} = (P_1,\ldots,P_s)$ be the limit of $x^{(j)}$ in the Satake–Furstenberg compactification. Let $\gamma_j = \dim \mathrm{Im}(P_j)$. Let $(\tau_2,\ldots,\tau_{s-1})$ be the limit of $\Lambda(x^{(j)})$ in the simplex Δ. Then the set $\tau_2 \geqslant \tau_3 \geqslant \ldots$ has the form

(13.1) $\qquad 1 = \tau_1 = \cdots = \tau_{\gamma_1} \geqslant \tau_{\gamma_1+1} = \cdots = \tau_{\gamma_2} \geqslant 4.1\cdots$.

We assume that

(13.2) $\qquad \mu_j := \tau_{\gamma_{j-1}+1} = \cdots = \tau_{\gamma_j}$

and obtain data of the form 0^*–2^*.

13.3. The projection of the Dynkin–Olshanetsky boundary to the matrix sky. Let the data 0^*–2^* be given. Consider the new data:

1^+ the flag $\mathrm{Ker}(P_1) \supset \mathrm{Ker}(P_2) \supset \mathrm{Ker}(P_3) \supset \cdots$; and

2^+ the set of numbers $\tau_2,\ldots,\tau_{s-1}$ defined by formula (13.2).

These data define a point of the matrix sky (see subsections 12.2–12.3).

13.4. Limits of geodesics. Let us consider a geodesic given by

$$\gamma(s) = A \begin{pmatrix} \exp(\lambda_1 s) & & & \\ & \exp(\lambda_2 s) & & \\ & & \ddots & \\ & & & 1 \end{pmatrix} A^t,$$

where $A \in SL(n,\mathbb{R})$ and $\lambda_1 \geqslant \cdots \geqslant \lambda_n = 0$. Let $\mathcal{P} = (P_1, \ldots, P_s)$ be the limit of γ in the space of hinges, and let the limit of γ in the velocity simplex Δ_n be $\tau_2, \ldots, \tau_{n-1}$, where $\tau_j = \lambda_j/\lambda_1$.

Let $\gamma_\alpha = \dim \mathrm{Im}(P_\alpha)$. We introduce the numbers $\mu_\alpha := \tau_{\gamma_{\alpha-1}+1} = \cdots = \tau_{\gamma_\alpha}$. Now we obtain data of the form 13.2 0^*–2^*.

REMARK 13.1. Some points of the Dynkin–Olshanetsky boundary are not limits of geodesics. For example, the point defined by the data 13.2 0^*–2^* is the limit of a geodesic if and only if $1 > \mu_2 > \cdots > \mu_{n-1} > 0$.

13.5. The Karpelevich compactification. The Karpelevich compactification is the hybrid of the compactification by the Karpelevich velocities and of the Satake–Furstenberg compactification. Namely, a boundary point of the Karpelevich compactification is given by following data:

0^* $s \in \{2, \ldots, n-1\}$;
1^* a hinge $\mathcal{P} = (P_1, \ldots, P_s)$ such that $P_j = P_j^t$ are positive definite for all $j \in \{1, \ldots, s\}$ (see §10); and
2^* a point of the boundary of the polyhedron $\mathcal{K}(1,s)$ of Karpelevich velocities (see subsection 11.2).

The topology on the Karpelevich compactification can be defined in an obvious way. The natural projection $\partial \mathcal{K}(1,s) \to \Delta(I_{1,s})$ defines a projection of the Karpelevich boundary to the Dynkin–Olshanetsky boundary.

§14. The space of geodesics and sea urchins

14.1. The space of geodesics. Consider the noncompact Riemannian symmetric space $G/K = SL(n,\mathbb{R})/SO(n)$. Denote by $\mathfrak{G}$ the space of all oriented geodesics in G/K.

Here the definition of the topology on the space $\mathfrak{G}$ is rather delicate. We shall describe the topology that seems to be the most natural. Consider a collection of integers $A = (\alpha_0, \ldots, \alpha_\sigma)$ such that $1 = \alpha_0 \leqslant \alpha_1 \leqslant \cdots \leqslant \alpha_\sigma = n$. Denote by $\Delta(A)$ the open simplex

$$1 = \lambda_1 = \cdots = \lambda_{\alpha_1} > \lambda_{\alpha_1+1} = \cdots = \lambda_{\alpha_2} > \cdots > \lambda_{\alpha_{\sigma-1}+1} = \cdots = \lambda_n = 0.$$

For different A, the simplexes $\Delta(A)$ are disjoint, and the union $\cup_A \Delta(A)$ coincides with the simplex Δ_n. Consider a geodesic $\gamma \in \mathfrak{G}$. The velocity of this geodesic belongs to some simplex $\Delta(A)$. The space of all geodesics with given velocity $\Lambda \in \Delta(A)$ is an $SL(n,\mathbb{R})$-homogeneous space. The stabilizer $G(A)$ of the geodesic γ (considered up to conjugacy) depends only on the set A (and depends neither on Λ nor on the geodesic itself): $G(A) = \mathbb{R}_+^* \times \prod O(\alpha_{j+1} - \alpha_j)$. Denote by $\mathfrak{G}(A)$ the space of all geodesics whose velocities are elements of $\Delta(A)$. Then

$$\mathfrak{G}(A) \simeq \Delta(A) \times SL(n,\mathbb{R})/G(A).$$

We endow this space with the standard direct product topology and the space

$$\mathfrak{G} = \bigcup_A \mathfrak{G}(A) \simeq \bigcup_A \Delta(A) \times SL(n,\mathbb{R})/G(A)$$

with the topology of disjoint union.

REMARK 14.1. Thus, the space of geodesics is disconnected. This is natural. Indeed, let $A_0 = \{0, 1, \ldots, n\}$. Consider the set of limits of the geodesic $\gamma \in \mathfrak{G}(A_0)$ on the matrix sky. Then this set is open and dense. The set of limits of $\gamma \in \mathfrak{G}(A)_0$ on the Satake–Furstenberg boundary is compact. Namely, it is the minimal compact $SL(n, \mathbb{R}^n)$-orbit on the boundary.

14.2. The space of geodesics as a boundary of the symmetric space.
Let us define a natural topology on the space $\mathfrak{R} = G/K \cup \mathfrak{G}$. We equip the space G/K with the standard topology. The space $\mathfrak{G}$ is equipped with the above topology, and thus the space $\mathfrak{G}$ is closed in $\mathfrak{R}$. Choose a point $b_0 \in G/K$. Let $x_j \in G/K$ be an unbounded sequence. The sequence x_j converges in $\mathfrak{R}$ if and only if it satisfies the following conditions:
1) the sequence of geodesics $b_0 x_j$ converges, and we denote by y the limit of this sequence on the matrix sky; and
2) there exists a limit z of the sequence of geodesics $x_j y$.

By the limit of the sequence x_j we mean the geodesic z.

REMARK 14.2. In our case, the dimension $\dim \mathfrak{G}$ of the boundary is given by the formula $\dim \mathfrak{G} = 2 \dim G/K - 2$, which is greater than $\dim G/K$ (even in the case $G/K = SL(2, \mathbb{R})/SO(2)$, i.e., the Lobachevsky plane).

REMARK 14.3. The space $\mathfrak{R}$ is not compact (since $\mathfrak{G}$ is not compact)

14.3. Sea urchins.
Recall that for any geodesic $\gamma \in \mathfrak{G}$ we can define a velocity $\{\mu_2, \mu_3, \ldots\}$, which is a point of the simplex Δ (see subsection 11.1). We denote by $\mathfrak{G}^{\mathrm{rat}}$ the space of geodesics with rational velocities (i.e., the numbers μ_j are rational). Consider the so-called *sea urchin*, i.e., the set $\mathfrak{R}^{\mathrm{rat}} := G/K \cup \mathfrak{G}^{\mathrm{rat}} \subset \mathfrak{R}$. We are not interested in the topology on the sea urchin (it seems natural to endow the set of velocities with the discrete topology, consider the ordinary topology on the space of geodesics with a given velocity, and introduce the natural (see subsection 14.2) convergence of sequences in G/K to geodesics).

14.4. Projective universality.
Let ρ_j be a finite family of linear irreducible representations of the group G in the spaces V_j. We assume that for any j there exists a nonzero K-fixed vector $v_j \in V_j$. Consider the direct sum $\rho = \bigoplus \rho_j$ of representations ρ_j and take the vector $w = \bigoplus v_j \in \bigoplus V_j$. Let $\mathcal{O} \simeq G/K$ be the G-orbit of the vector $w \in \mathbb{P}(\mathcal{O})$ and $\overline{\mathcal{O}}$ the closure of $\mathcal{O}$ in the projective space $\mathbb{P}(\bigoplus V_j)$. The G-spaces $\overline{\mathcal{O}}$ are called the *projective compactifications* of G/K.

Now we construct a mapping $\pi \colon \mathfrak{R} = G/K \cup \mathfrak{G} \to \overline{\mathcal{O}}$. The mapping $G/K \to \mathcal{O}$ is natural. Let us consider a geodesic $\rho(s) \in \mathfrak{G}$. We can readily prove that the limit $\lim_{s \to \infty} \pi(\gamma(s))$ in $\mathbb{P}(\bigoplus V_j)$ exists. By definition, $\pi(\gamma)$ is this limit.

PROPOSITION 14.4. a) *The mapping $\pi \colon \mathfrak{R} \to \overline{\mathcal{O}}$ is surjective.*
b) *Moreover, the π-image of the sea urchin $\mathfrak{R}^{\mathrm{rat}}$ is the entire space $\overline{\mathcal{O}}$.*

§15. The boundary of a Bruhat–Tits building

An analog of a Riemannian symmetric space in the p-adic case is the so-called Bruhat–Tits building. We discuss an example.

15.1. An analog of the Satake–Furstenberg boundary.
Let $\mathbb{Q}_p$ be p-adic field, $\mathbb{Z}_p$ be the ring of p-adic integers, and let $\mathbb{Q}_p^* = \mathbb{Q}_p \setminus 0$ be the multiplicative group of $\mathbb{Q}_p$.

By a *lattice* in the linear space $\mathbb{Q}_p^n$ we mean an open compact $\mathbb{Z}_p$-submodule of $\mathbb{Q}_p^n$ (any lattice has the form $\mathbb{Z}_p v_1 \oplus \cdots \oplus \mathbb{Z}_p v_n$, where $v_1, \ldots, v_n$ are linearly independent vectors). The group $\mathbb{Q}_p^*$ acts by dilations on the set Lat_n of all lattices in $\mathbb{Q}_p$. The *Bruhat–Tits building* Ens_n is the quotient space $Ens_n = Lat_n/\mathbb{Q}_p^*$.

The space Ens_n possesses a beautiful geometric structure, which is not essential for our purposes. We only note that the group $PGL_n(\mathbb{Q}_p)$ acts on Ens_n in an obvious way and the stabilizer of the lattice $(\mathbb{Z}_p)^n \subset \mathbb{Q}_p^n$ is the group $PGL_n(\mathbb{Z}_p)$, that is, $Ens_n = PGL_n(\mathbb{Q}_p)/PGL_n(\mathbb{Z}_p)$.

We shall describe the analog of the Satake boundary for Ens_n.

To this end we embed Lat_n into the space Sub_n of all $\mathbb{Z}_p$-submodules in $\mathbb{Q}_p^n$. Introduce some natural metric on Sub_n. For example, any submodule under consideration can be regarded as a closed subset of $\mathbb{P}\mathbb{Q}_p^n$, and thus the Hausdorff metric on $[\mathbb{P}\mathbb{Q}_p^n]$ induces a metric on Sub_n.

Furthermore, we apply the construction of the Hausdorff quotient space to the compact space Sub_n, its open subset Lat_n, and the orbits of the group $\mathbb{Q}_p^*$. Then the limit sets in $Sub_n/\mathbb{Q}_p^*$ are as follows: they are the collections

$$0 = L_0 \subset M_1 \subset L_1 \subset \cdots \subset L_{k-1} \subset M_k \subset L_k = \mathbb{Q}_p^n,$$

where for each j, the L_j is a linear subspace and M_j/L_{j-1} is a lattice in L_j/L_{j-1} (defined up to a scalar factor). We call these collections *filtered flags*.

15.2. Velocity compactification. A pair of lattices $R, T \subset \mathbb{Q}_p^n$ can be written in the form of the direct sums

$$R = \mathbb{Z}_p v_1 \oplus \mathbb{Z}_p v_2 \oplus \cdots \oplus \mathbb{Z}_p v_n, \qquad T = p^{k_1}\mathbb{Z}_p v_1 \oplus p^{k_2}\mathbb{Z}_p v_2 \oplus \cdots \oplus p^{k_n}\mathbb{Z}_p v_n,$$

where $k_1 \geqslant \cdots \geqslant k_n$. The set $D(R,T) = (k_1, \ldots, k_n)$ is called the *complex distance* between R and T. The complex distance is an invariant of a pair of lattices with respect to the action of the group $GL(n, \mathbb{Q}_p)$ on the space of lattices, that is, $D(g \cdot R, g \cdot T) = D(R, T)$ for $R, T \subset \mathbb{Q}_p^n$ and $g \in GL(n, \mathbb{Q}_p)$. The vertices of the buildings are lattices defined up to a factor, and hence for a Bruhat–Tits building, the complex distance is defined up to an additive constant. Now we can repeat both constructions of §11.

15.3. Hybrids. We can repeat both constructions of §13 as well. A point of the *analog of the Dynkin–Olshanetsky boundary* of a building Ens_n is given by the following data:

0^* $s \in \{1, \ldots, n-1\}$;
1^* a filtered flag $0 = L_0 \subset M_1 \subset L_1 \subset M_2 \subset \cdots \subset L_s \subset M_s \subset L_{s+1} = \mathbb{Q}_p^n$, where L_j are subspaces and M_j are $\mathbb{Z}_p$-submodules (defined up to a factor) such that M_j/L_j are lattices in L_{j+1}/L_j; and
2^* a point of the simplex Δ_s (see subsection 11.1).

A point of the *analog of the Karpelevich boundary* is given by similar data, namely, by data of the form 0^* and 1^* together with data of the form

2^{**} a point of the polyhedron $\mathcal{K}(1, s)$ of Karpelevich velocities.

15.4. The lattice sky. A point of the *lattice sky* is given by the following data:

$0^\dagger$ $s \in \{1, \ldots, n-1\}$;
$1^\dagger$ a flag $0 = L_0 \subset L_1 \subset \cdots \subset L_{s+1} = \mathbb{Q}_p^n$; and
$2^\dagger$ a point of the simplex Δ_s.

There exist a natural tiling of the lattice sky into simplexes. The simplexes $\Delta_{\mathcal{L}}$ are enumerated by the complete flags

$$\mathcal{L}: \ 0 = L_0 \subset L_1 \subset \cdots \subset L_n = \mathbb{Q}_p^n.$$

Let $\mathcal{A}(\mathcal{L})$ be the set of all nontrivial subflags of the flag $\mathcal{L}$. Then the simplex $\Delta_{\mathcal{L}}$ consists of all elements of the lattice sky such that the data $1^\dagger$ are elements of $\mathcal{A}(\mathcal{L})$ (the faces of the simplex $\Delta_{\mathcal{L}}$ are enumerated by the subflags of the flag $\mathcal{L}$).

§16. Remarks

16.0. Remarks on §0. α) The five conics problem was formulated and (incorrectly) solved by Steiner in 1848; in 0.2 we followed this solution. The correct answer and a correct solution were obtained by Chasles in 1864 [**Chas**]. A similar problem on the number of quadrics in $\mathbb{C}P^3$ that are tangent to nine generic quadrics was solved by Schubert [**Schu**]; this number is equal to

(16.1) $\qquad\qquad\qquad\qquad 666841088.$

For the history of the problem, see [**Kle**].

β) The variety of "complete conics" was introduced by Study [**Stu**]. For the discussion on this variety, see [**Sev**]. A similar completion of $PGL_n(\mathbb{C})/PO_n(\mathbb{C})$ (i.e., of the space of quadrics in $\mathbb{C}P^{n-1}$) was constructed by Semple [**Sem1, Sem3**]. This variety is called that of "complete quadrics", see our §8.

For a discussion of the five conics problem from another point of view, see [**GH, VI.1**].

For the calculation of the number (16.1) and for other problems of the same type, see [**SR,** Chapter XI] and also [**DCP1**].

16.1. Remarks on §1. The category GA and its fundamental representation are defined in [**Ner1**]; see also [**Ner2, Ner3**]. For the categories B, C, and GD and for the theory of representations for the categories GA, B, C, and GD, see [**Ner2, Ner3, Ner6**].

16.2. Remarks on §2. α) This section contains an exposition of the preprint [**Ner5**].

β) The Hausdorff metric was introduced in [**Pom, Hau**].

γ) The standard compactifications of $\mathbb{R}$ are those by means of the points $+\infty$ and $-\infty$ and by means of the singleton ∞ (or $\pm\infty$). Some other compactifications of $\mathbb{R}$ can be obtained as follows: we can embed $\mathbb{R}$ into a compact metric space and take the closure of the image (for instance, we can take a winding of a torus or the curve $r = (\pi/2 + \arctan t)$, $\varphi = t$ on $\mathbb{R}^2$).

An example of a nontrivial and interesting compactification of $\mathbb{R}^n$ is given by the closure of the set of diagonal matrices in the model of §13 for the Karpelevich compactification.

δ) To 2.6. Certainly, the space $(\mathbb{C}^*)^k$ has many nontrivial compactifications. By a *toric variety* we mean a complex algebraic variety L, possibly singular, on which the group $(\mathbb{C}^*)^k$ (the "complex torus") acts in such a way that one of the orbits of $(\mathbb{C}^*)^k$ on L is open.

Let the torus $(\mathbb{C}^*)^k$ act by linear transformations on the space $\mathbb{C}^N$. Then the closures of orbits of $(\mathbb{C}^*)^k$ in $\mathbb{C}P^{N-1}$ are toric varieties, and all toric varieties have this form.

Our space $\overline{\mathbb{T}^{n-1}}$ is an example of such toric variety. For details on toric varieties, see the review [**Dan**].

Certainly, in the construction of 2.6, we can replace the group $\mathbb{C}^*$ by $\mathbb{R}^* = \mathbb{R} \setminus 0$ (in this case we obtain a smooth analytic real variety) or by the multiplicative group of positive reals.

ε) Instead of the words "the closure in the Hausdorff metric", it is customary to speak of "the closure in the Hilbert scheme or Chow scheme". The latter operation is more refined but harder to visualize [**BiS, Kap**]; see [**Kap**] for the construction of a separated quotient of the Grassmannian in $\mathbb{C}^n$ by the action of the torus $\mathbb{C}^* n$.

16.3. **Remarks on §3.** α) The author believes that the language of 3.3 was first used in [**Alg**]. Hinges (under the title "resolving sequences") were introduced in [**Ner4**].

β) The description of the closure of a conjugacy class in $PGL_n(\mathbb{C})$ in the variety Hinge_n seems to be of interest.

16.4. **Remarks on §4.** α) The smooth algebraic variety Hinge_n was first constructed by Semple [**Sem2**] in 1951 as a natural generalization of "complete quadrics". The Semple construction is as follows.

To any operator $A \in PGL_n(\mathbb{C})$ one assigns the collection of operators

$$(16.2) \qquad A = \Lambda^1 A, \Lambda^2 A, \ldots, \Lambda^{n-1} A,$$

that are defined up to a factor, in the exterior powers $\Lambda^j \mathbb{C}^n$ of the space $\mathbb{C}^n$. This collection can be regarded as a point of the product of projective spaces

$$(16.3) \qquad \prod_{j=1}^{n-1} \mathbb{P}(\Lambda^j V) = \prod_{j=1}^{n-1} \mathbb{CP}^{C_n^j - 1}.$$

Semple defined the variety S_n of "complete collineations" as the closure of the image of the group $PGL_n(\mathbb{C})$ in (16.3). Points of this closure are called *Semple complexes*. In our language, Semple complexes are collections of operators of the form $(\lambda_1(\mathcal{P}), \ldots, \lambda_{n-1}(\mathcal{P}))$, where $\mathcal{P}$ is a hinge. In this connection, the varieties S_n and Hinge_n coincide. For the variety S_n, see also [**Tyr, Lak, DCP1**].

β) The papers of Semple [**Sem1, Sem2, Sem3**] are among the first applications of the construction of the *closure of an equivariant embedding*. Let a group G act on a (noncompact) space M and on a compact space N. Let an embedding $\tau \colon M \to N$ be given such that $\tau(g \cdot m) = g \cdot (\tau m)$ (such embeddings are said to be *equivariant*). Furthermore, consider the closure $\overline{\tau(M)}$ of the image of the mapping τ in N. It turns out that a number of meaningful and nontrivial objects can be obtained in this way.

In the case of "complete collineations," the group G is $PGL_n(\mathbb{C}) \times PGL_n(\mathbb{C})$, the space M is the group $PGL_n(\mathbb{C})$ (the group $G = PGL_n(\mathbb{C}) \times PGL_n(\mathbb{C})$ actually acts on $PGL_n(\mathbb{C})$ by left and right multiplications), and N is given by (16.3). The equivariant embedding is defined by formula (16.2).

Let us present some examples of this construction that differ from the examples above and from the examples below.

1^*. *Bohr compacta* (see [**Dix, Rup**]). Consider the product N of a continual set of circles $S^1 = \{z : |z| = 1\}$ endowed with the Tychonoff topology. It is natural to regard the elements of this space as functions $f \colon \mathbb{R} \to S^1$ (without any conditions of continuity and measurability on the functions f). Clearly, N is a group with respect to the pointwise multiplication. Furthermore, the group $\mathbb{R}$ is embedded into N by the rule $s \mapsto F_s(t) = e^{ist}$, $t \in \mathbb{R}$. The *Bohr compactum* is the closure of the image of the group $\mathbb{R}$ in N. It seems that this object can hardly be described in reasonable language, and it is pathological in this sense. On the other hand, sometimes it plays an auxiliary role in very useful and reasonable discussions; see [**Rup**].

2*. *Thurston boundary* (see [**Thu, MS**]). Let Γ_g be the fundamental group of a sphere C_g with g handles. Let Δ_g be the set of homomorphisms from Γ_g into $PSL_2(\mathbb{R})$ whose images are lattices in $PSL_2(\mathbb{R})$ (defined up to conjugation in $PSL_2(\mathbb{R})$) (this set is called the Teichmüller space). Let G be the group of automorphisms of the group Γ_g. Let N be the space of real functions on Γ_g defined up to a factor. We endow this space with the topology of pointwise convergence. Further, to any homomorphism $\Pi \colon \Gamma_g \to PSL_2(\mathbb{R})$ we assign the function $\alpha_\pi(\gamma) = \log \operatorname{tr} \pi(\gamma)$, $\gamma \in \Gamma_g$, regarded as an element of N.

The Thurston compactification of the Teichmüller space is the closure of the image of Δ_g under this mapping in the projective space N.

3*. *Olshanski mantles* [**Ols, Ner3, Ner6**]. Let $M = G$ be an infinite-dimensional group. Let $N = \mathcal{B}(H)$ be the set of linear operators in a Hilbert space H with norm $\leqslant 1$. The set $\mathcal{B}(H)$ is compact with respect to the weak operator topology.

Let ρ be a unitary representation of the group G in the space H. Then ρ can be regarded as a mapping of G into $\mathcal{B}(H)$. Let $\Gamma(G, \rho)$ (a mantle of the group G) be the closure of the image of this mapping. We can readily see that $\Gamma(G, \rho)$ is a semigroup. It turns out that in many cases this semigroup can be described explicitly, and the answer is often nontrivial.

γ) The semigroup $\widetilde{\operatorname{Hinge}}_n$ is defined in [**Ner4**]. Apparently, the semigroup $\widetilde{\operatorname{Hinge}}_n$ coincides with that constructed in [**Vin**]; see also [**Pop**].

16.5. Remarks on §5. None.

16.6. Remarks on §6. α) The construction of 6.2 is a special case of that introduced by de Concini–Procesi [**DCP1**]; see 16.8. It is an example of the closure of an equivariant embedding. Namely, the $PGL_n \times PGL_n$-homogeneous space PGL_n can be embedded in the projectivized space of operators in $V(a)$, and we take the closure of the image.

The construction of 6.3 was obtained in [**Ner4**].

β) Putcha and Renner (see [**Put, Ren**]) studied the following problem. Let G be a complex semisimple Lie group and let ρ be its finite-dimensional holomorphic representation in a space V (in general, reducible). Consider the set $\mathbb{C} \cdot \rho(G)$ of operators of the form $\lambda \cdot \rho(g)$, where $g \in G$ and $\lambda \in \mathbb{C}$. Let $\Gamma(\rho)$ be the closure of the set $\mathbb{C} \cdot \rho(G)$ in the space of all operators on the space V. Clearly, $\Gamma(\rho)$ is a semigroup. It turns out that the semigroups $\Gamma(\rho)$ can be nonisomorphic for different ρ. To be more precise, $\Gamma(\rho) \simeq \Gamma(\rho')$ if the convex hull $Conv(\rho)$ of the weights of the representation ρ can be obtained by a dilation from the convex hull $Conv(\rho')$ of the weights of the representation ρ'.

For the case $G = GL_n$, the semigroup $\widetilde{\operatorname{Hinge}}_n$ is a universal object for almost the same problem. Namely, for all *irreducible* representations ρ, the quotient semigroup $\Gamma(\rho)/\mathbb{C}^*$ is a homomorphic image of the semigroup $\widetilde{\operatorname{Hinge}}_n$.

On the other hand, it seems probable (I do not know the proof) that for any representation ρ (not necessarily irreducible), the semigroup $\Gamma(\rho)$ coincides with the image of the semigroup $\widehat{\text{Hinge}}_n$ (in all events, a semigroup $\Gamma \supset GL_n$ with this property is constructed in [**Vin**]).

16.7. Remarks on §7. We follow [**Ner4**]. The described construction is apparently a special case of some general construction that holds for all homogeneous spaces; see [**Vin, Pop**].

16.8. Remarks on §8. α) *De Concini–Procesi construction.* Complete symmetric varieties were introduced by de Concini and Procesi [**DCP1**] in 1983. Their construction (which is the same for all complex symmetric spaces) is as follows.

Let G/H be a symmetric space, where G and H are complex groups. Let π be a spherical representation of G, that is, an irreducible representation of G that has a nonzero H-stable vector, say, v. The description of all these representations is given by the well-known Helgason theorem ([**Hel1**]; see also [**Hel2**, V.4.1]).

Let $\mathbb{P}_\pi$ be the projectivized representation space of π. Let $\overline{(G/H)_\pi}$ be the closure of the G-orbit of the vector v in $\mathbb{P}_\pi$. Then $\mathbb{P}_\pi$ is a compactification of the symmetric space G/H.

This construction *a priori* depends on π. We say that π is nondegenerate if all numerical labels of π on the Dynkin diagram that can be nonzero (for spherical representations) are nonzero. It turns out that for all nondegenerate irreducible representations π, the varieties $\overline{(G/H)_\pi}$ are isomorphic. Moreover, for all G/H these are smooth complex varieties.

The complete collineations (i.e., elements of the completion of the space $PGL_n \times PGL_n/PGL_n$) fit within the framework of the construction above as follows. Let ρ be a holomorphic representation of GL_n and let ρ' be the dual representation. Consider the representation $\rho \otimes \rho'$ of the group $GL_n \times GL_n$. This is just the representation of $GL_n \times GL_n$ in the space of linear operators in the representation space of ρ (see §6).

For the properties of complete symmetric varieties, see [**DCP1, DCP2, DCS**].

β) Hinge constructions react rather painfully to modifications of G/H that seem to be unessential from any other point of view. For example, the completion of the group SO_n [**Ner4**] can be described in the language of hinges in a much more cumbersome way than the completion $\overline{O_n}$ described in §8.

γ) *Projective compactifications.* Let $\rho_1, \ldots, \rho_k$ be irreducible H-spherical representations of the group G in some spaces $V_1, \ldots, V_k$, and let $v_j \in V_j$ be their spherical vectors. Let $\rho = \bigoplus \rho_j$, $V = \bigoplus V_j$, and $v = \bigoplus v_j \in V$. Consider the G-orbit of the vector v in the projective space $\mathbb{P}V$. Denote by $\mathbb{P}_\rho$ the closure of the orbit Gv in $\mathbb{P}V$.

It turns out that for different representations ρ, the varieties $\mathcal{P}_\rho$ are, in general, different; as a rule, they are not smooth.

Such objects were studied quite intensively, for example, see [**Vus, CX, Kus3**]. We also note that the constructions of this type are possible for any (in general, not symmetric) subgroup H, see [**LV**].

δ) *Closure in the Grassmannian in the adjoint representation.* Consider an irreducible representation ρ of a simple complex linear group in a space V. For the sake of simplicity we assume that on the space V, a nondegenerate symmetric bilinear form $B(\cdot, \cdot)$ is given. Consider the symmetric bilinear form $\widetilde{B}((v,w); (v',w')) = B(v,v') - B(w,w')$, $v, v', w, w' \in V$, on $V \oplus V$. Denote by $\text{Gr} = \text{Gr}(V \oplus V)$ the

Grassmannian of maximal isotropic subspaces of $V \oplus V$. To any element $g \in G$ we assign the graph of the operator $\rho(g)$. Thus, we obtain an embedding of G in Gr. The problem of describing the closure of G in Gr naturally arises.

Now we show that this problem can be reduced to that of item γ. By [**Ner6**, Chap. 2], the spinor representation Spin of the group $O(V)$ can be extended to a continuous embedding of $\mathrm{Gr}(V \oplus V)$ into the projectivized space of operators. Therefore, our problem is reduced to the problem of describing the closure of the group G in the representation Spin $\circ \rho$.

Now we consider the case where ρ is the adjoint representation Ad of the group G. According to [**Kos1, Kos2**], for any simple Lie group G, the representation Spin $\circ$ Ad is the sum of equal representations of the group G whose highest weight is the half-sum of the positive roots. Therefore, the closure of G in the representation Spin $\circ$ Ad and in the Grassmannian $\mathrm{Gr}(\mathfrak{g} \oplus \mathfrak{g})$ coincides with the De Concini–Procesi compactification (with the complete symmetric variety) of the group G.

16.9. Remarks on §9. α) The Satake construction [**Sat**] for the compactification of a Riemannian symmetric space G/K (where G is a simple Lie group and K is a compact subgroup) given in 1960 is precisely the real version of the construction of complete symmetric varieties. We consider a finite-dimensional K-spherical irreducible representation ρ of the group G, and then take the closure of the orbit of a spherical vector in the projective space. Let ρ be nondegenerate in the same sense as in 16.8, α). It turns out that, in this case, the obtained compactification does not depend on ρ.

If ρ can be degenerate, then in this way we can obtain $2^{\mathrm{rk}(G/K)} - 1$ different compactifications, which are also called Satake compactifications (as in our Remark 6.6). The ordinary compactifications, for example, the matrix ball [**Pya**, Chapter II; **Ner2**] belong to the family of Satake compactifications.

For pseudo-Riemannian symmetric spaces (see the list in [**Ber**]), we can repeat the Satake construction.

Oshima and Sekiguchi [**Osh, OS, Schl**] also studied the gluing of compactifications of this type into smooth analytic varieties. In [**Osh**] (see also [**Schl**]) for a Riemannian symmetric space G/K, $2^{\mathrm{rk}(G/K)}$ counterparts of Satake compactifications are glued together. In [**OS**] different real forms G/K_α of the same symmetric space are glued together. The construction [**OS**] is much more involved than that of §9, and the relationship between these constructions remains unclear.

β) The same compactifications of Riemannian symmetric spaces, by a substantially different method, were obtained by Furstenberg [**Fur1, Fur2**].

16.10. Remarks on §10. A more general "hinge superstructure" over the category GA was discussed in [**Ner4**]. Similar "hinge superstructures" exist for the categories B, C, and GD from 1.7.

16.11. Remarks on §11. α) I have not seen these constructions in the literature. For the discussion of the closures of Weyl chambers, see [**Tay**]. There exist many other compactifications of the simplicial cone Σ_n, and hence there exist many other velocity compactifications of symmetric spaces. Some examples are given below in β) and γ). An analog of the set $(\ln a_1, \ldots, \ln a_n)$ for an arbitrary noncompact Riemannian symmetric space is the complex distance (for example, see [**Ner6**, 6.3]).

β) *Satake velocities.* Here we will discuss the closure of a Weyl chamber in the Satake compactification (see also §2). Consider a partition $\mathfrak{t}$ of $I_{1,n}$, where $I_{1,n} =$

$I_{1,\alpha_1} \cup I_{\alpha_1+1,\alpha_2} \cup \cdots \cup I_{\alpha_{s-1},n}$, and assign to $\mathbf{t}$ the *face* $G(\mathbf{t}) = \prod_m \Sigma(I_{\alpha_m+1,\alpha_{m+1}})$. We define the polyhedron of Satake velocities by the relation $\mathcal{S}(n) = \cup_\mathbf{t} G(\mathbf{t})$. Consider a sequence $h^j = \{h_1^{(j)} \geqslant \cdots \geqslant h_n^{(j)} = 0\} \in \Sigma(I_{1,n})$. The sequence $h^{(j)}$ is convergent to a point
$$u = (u_1, \ldots, u_s) \in \prod_m \Sigma(I_{\alpha_m+1,\alpha_{m+1}}) = G(\mathbf{t})$$
if there exist sequences $p_1^{(j)}, p_2^{(j)}, \ldots$ such that
1) for all σ we have $\lim_{j \to \infty}(p_\sigma^{(j)} - p_{\sigma+1}^{(j)}) = +\infty$, and
2) for all σ, the sequences
$$h_{[\alpha]}^{(j)} := (h_{\alpha_{\sigma-1}+1}^{(j)} - p_\sigma^{(j)}, \ldots, h_{\alpha_\sigma}^{(j)} - p_\sigma^{(j)}) \in \Sigma(I_{\alpha_{\sigma-1}+1,\alpha_{\sigma+1}})$$

are convergent, with the limit $u_\sigma \in \Sigma(I_{\alpha_{\sigma-1}+1,\alpha_{\sigma+1}})$.
The polyhedron $\mathcal{S}(n)$ is a compactification of Σ_n.

The projection of the polyhedron of Karpelevich velocities to the polyhedron of Satake velocities. For any tree partition we can delete all its reducible elements, and thus obtain an ordinary partition. Then in the product (11.7) we can omit the second factor.

γ) *Toric velocities.* Consider the space L formed by the sets $(t_1, \ldots, t_m) \in \mathbb{R}^n$ that are defined up to an additive constant. Let $Q \subset L$ be a convex polyhedron with rational vertices contained in the hyperplane $t_1 + \cdots + t_n = 0$. Let Q be invariant with respect to the action of the symmetric group S_n on L. Starting from this polyhedron, we shall construct a compactification of Σ_n. Let $M_j = (m_1^j, \ldots, m_n^j)$, $j = 1, \ldots, N$, be the vertices of Q. For any vertex we introduce the expression

$$\chi_j(t_1, \ldots, t_n) = \exp\left(\sum_{k=1}^n m_k^j t_k\right).$$

Consider the embedding of L in the projective space $\mathbb{RP}^N$ given by the formula

$$\pi \colon (t_1, \ldots, t_n) \to (\chi_1(t_1, \ldots, t_n), \ldots, \chi_N(t_1, \ldots, t_n)).$$

Let $\overline{\pi(L)}$ be the closure of $\pi(L)$ in $\mathbb{RP}^N$. The structure of these closures is well known [**Dan**], and they have a nice and simple description in terms of the geometry of the polyhedron Q. The quotient space $\overline{\pi(L)}/S_n$ is a compactification of Σ_n.

REMARK. The polyhedron of Satake velocities has the form above. Apparently, this is not the case for the polyhedron of Karpelevich velocities.

REMARK. Apparently, projective compactifications (see 14.4) are hybrids of Satake compactifications and toric velocity compactifications.

16.12. **Remarks on §12.** The heroes of our paper are mainly very exotic objects from the point of view of "ordinary" differential geometry. The construction of 12.1 is an exception. The corresponding object of differential geometry is more or less standard, see [**EO, BGS, VEL**]. The Tits metrics on the boundary can also be defined for a general Cartan–Hadamard manifold, see [**BGS**] ([**BGS**] also contains the reference [**ImH**]). Nevertheless, it seems that now it is known only for a few interesting examples.

16.13. Remarks on §13. α) The Dynkin–Olshanetsky boundary [**Dyn, Olse1, Olse2**] is the Martin boundary for the ordinary diffusion on a symmetric space (see 16.16,β).

β) The Karpelevich boundary was constructed in [**Kar**].

16.14. Remarks on §14. α) I have not seen this construction in the literature. Kushner [**Kus1, Kus2**] constructed a compactification for a noncompact Riemannian symmetric space that is universal in the sense of 14.3. Our space $\mathfrak{R}$ is not compact.

β) *Examples of sea-urchin-type constructions*

EXAMPLE (blowing up of cusps). Consider the subgroup $PSL(2,\mathbb{Z}) \subset PSL(2,\mathbb{R})$ and its action on the Lobachevsky plane $\mathcal{L} = PSL(2,\mathbb{R})/SO(n)$ as an upper half-plane. Denote by R the space of oriented geodesics on $\mathcal{L}$ whose limits are rational points on $\overline{\mathbb{R}} = \mathbb{R} \cup \infty$ (it is essential for our purposes that any rational point of the absolute of the Lobachevsky plane is a fixed point for some cusp element of $PSL(2,\mathbb{Z})$). The group $PSL(2,\mathbb{Z})$ has a natural discrete action on the spaces $\mathcal{L}$ and R, and hence we can construct the quotient space $(\mathcal{L} \cup R)/PSL(2,\mathbb{Z})$. This space is a compactification of the space $\mathcal{L}/PSL(2,\mathbb{Z})$. For more general constructions of this type, see [**BS1**].

EXAMPLE (universal toric variety). Consider the torus $(\mathbb{C}^*)^n$. For any

$$L = (l_1, \ldots, l_n) \in \mathbb{Z}^n$$

we consider the one-parameter subgroup $\gamma_L \subset \mathbb{C}^n$ that consists of the elements $(z^{l_1}, \ldots, z^{l_n}) \in (\mathbb{C}^*)^n$ (where $z \in (\mathbb{C}^*)^n$). A point of the boundary $\mathfrak{Q}$ of $(\mathbb{C}^*)^n$ can be defined by the following data:

1* an element $L = (l_1, \ldots, l_n) \in \mathbb{Z}^n$, and

2* an element q of the quotient group $(\mathbb{C}^*)^n/\gamma_L$.

A sequence $u^{(j)} = (u_1^{(j)}, \ldots, u_n^{(j)}) \in (\mathbb{C}^*)^n$ converges to an element $(L,q) \in \mathfrak{Q}$ whenever it satisfies the following conditions:

(i) there exists a real sequence $b^{(j)} \to +\infty$ such that for any $m = 1, \ldots, n$, there exists a finite limit $\lim_{j\to\infty} |u_m^{(j)}/(b^{(j)})^{l_m}|$;

(ii) let $\widetilde{u^{(j)}}$ be the image of $u^{(j)}$ in the quotient group $(\mathbb{C}^*)^n/\gamma_L$. Then $\widetilde{u^{(j)}}$ converges to $q \in (\mathbb{C}^*)^n/\gamma_L$.

EXAMPLE (the complex sea urchin). Consider the space $PSL(n,\mathbb{C})/SO(n,\mathbb{C})$. Choose $s \in \{2, 3, \ldots, n\}$. Choose a set of positive integers $K = (k_1, \ldots, k_s)$ such that $\sum k_\sigma = n$. Choose a set of integers $L : \{l_1 \geqslant \cdots \geqslant l_s\}$ that is defined up to an additive constant. Let $\mathcal{B}(K)$ be the group of $(k_1 + k_2 + \ldots) \times (k_1 + k_2 + \ldots)$ block upper triangular matrices B. Consider the complex curve $\gamma[K,L]$ given by

$$\gamma[K,L](t) = \begin{pmatrix} t^{l_1} E_{k_1} & & \\ & \ddots & \\ & & t^{l_s} E_{k_s} \end{pmatrix},$$

where $t \in \mathbb{C}^*$ and E_ψ is the $\psi \times \psi$ identity matrix. The curves $C\gamma[K,L](t)C^t$, where $C \in GL(n,\mathbb{C})$, are called *geodesic curves*. Choose an element $A \in PSL(n,\mathbb{C})$. Define the family (a *pencil*) $P(K,L|A)$ of all curves $\mathbb{C}^* \to PSL_n/SO_n$ that can be represented in the form $A^t B^t \gamma[K,L] BA$, where $B \in \mathcal{B}(K)$. We can readily see that the curves of a pencil either coincide or are disjoint. The set

$$S(K,A) = \bigcup_{B \in \mathcal{B}(K)} A^t B^t \gamma[K,L] BA$$

is open and dense in $PSL(n,\mathbb{C})/SO(n,\mathbb{C})$.

Now we consider the space

$$PSL(n,\mathbb{C})/SO(n,\mathbb{C}) \cup \bigcup_{K,L} \left(\bigcup_{A \in GL(n,\mathbb{C})/\mathcal{B}(K)} P(K,L\,|\,A) \right)$$

endowed with the topology for which a sequence $x_j \in PSL(n,\mathbb{C})/SO(n,\mathbb{C})$ converges to $\lambda \in P(K,L\,|\,A)$ whenever it satisfies the following conditions.

1. $x_j \in S(K,A)$ for large j.
2. Let $\gamma_j \in P(K,L\,|\,A)$ be a geodesic curve that contains x_j. Then γ_j converges to γ in the natural topology of the pencil $P(K,L\,|\,A)$.
3. Let $B_j \in \mathcal{B}(K)$ be a sequence such that $B_j \to E$ and $B_j \gamma_j = \gamma$. Then the sequence $B_j x_j \in \gamma$ converges to $+\infty$ in γ.

This topology is natural on sets of the form $\bigcup_{A \in GL(n,\mathbb{C})/\mathcal{B}(K)} P(K,L\,|\,A)$ and coincides with the topology of disjoint union on the set

$$\bigcup_{K,L} \left(\bigcup_{A \in GL(n,\mathbb{C})/\mathcal{B}(K)} P(K,L\,|\,A) \right).$$

16.15. Remarks on §15. For the boundaries of the Bruhat–Tits buildings, see [**BS2, Ger1, Ger2**]. Consider the space of lattices in $\mathbb{R}^n$ defined up to dilatations. This space can be compactified by the same way as in subsection 15.1 (see also [**How**]).

16.16. Some general constructions. Here we briefly discuss several general constructions of compactifications for metric (and topological) spaces.

α) *Maximal ideals.* Let X be a noncompact space. Let $\mathcal{A}$ be an algebra of (continuous, holomorphic, algebraic, almost periodic, etc.) functions on X. Let $\mathcal{A}$ satisfy the following conditions:

a) $\mathcal{A}$ contains the constants; and

b) $\mathcal{A}$ separates the points of X (for any $x_1, x_2 \in X$, there exists $f \in \mathcal{A}$ such that $f(x_1) \neq f(x_2)$).

Denote by $\mathrm{Spec}(\mathcal{A})$ the spectrum (the space of maximal ideals) of $\mathcal{A}$. There is a natural embedding $\lambda \colon X \to \mathrm{Spec}(\mathcal{A})$. If $x \in X$, then the set of functions $f \in \mathcal{A}$ that vanish at the point x is a maximal ideal of $\mathcal{A}$. As a rule, $\mathrm{Spec}(\mathcal{A})$ is a compact space and X is dense in $\leq Spec(\mathcal{A})$. Let $\mathcal{A} \subset \mathcal{A}'$ be two algebras satisfying conditions a) and b). Then there is a natural mapping $\mathrm{Spec}(\mathcal{A}') \to \mathrm{Spec}(\mathcal{A})$ given by the formula $J \mapsto J \cap \mathcal{A}$.

EXAMPLE: STONE-ČECH COMPACTIFICATION. For a topological space X we introduce the algebra $C(X)$ of all continuous functions and the algebra $\mathcal{B}(X)$ of all bounded continuous functions. The space $\mathrm{Spec}(\mathcal{B}(X))$ is the universal compactification of X in the following sense. For any dense embedding i of X into a compact topological space K, there exists a mapping $\pi \colon \mathrm{Spec}(\mathcal{B}(X)) \to K$ such that $i = \pi \circ \lambda$. Indeed, $C(K) \subset \mathcal{B}(X)$, and this embedding induces the mapping

$\pi\colon \operatorname{Spec}(\mathcal{B}(X)) \to K = \operatorname{Spec}(C(K))$. This construction seems to be pathological because no point of $\operatorname{Spec}(\mathcal{B}(X)) \setminus X$ can be described explicitly.

β) *Martin boundary* [**Mar, Doo, Shu, KSK, Kai**] Consider a domain $\Omega \subset \mathbb{R}^n$ and a certain boundary problem for the Laplace operator Δ. Let $G(x,y)$ be its Green function; assume that this function is positive. Consider a family of functions $g_x(y) = G(x,y)$, $x, y \in \Omega$, defined up to a positive factor. We thus obtain an embedding of Ω to the projectivized space of functions on Y. By the *Martin compactification* of Ω we mean the pointwise closure of the image of Ω in the projective space. This definition is not constructive, and to describe the Martin boundary, it is necessary to find the asymptotic behavior of Green's function. The Dynkin–Olshanetsky boundary is the Martin boundary for the equation $\Delta u = \mu u$ on a symmetric space [**Dyn, Olse1, Olse2**]. The Martin boundary for the Laplace equation is only a particular case of the very general construction known under the name of Martin boundary. For simplicity, we consider a discrete countable space Ω and a *Markov transition function* $p(x,y)$ on $\Omega \times \Omega$. Recall that for all x we have $\sum_{y \in \Omega} p(x,y) = 1$. A function $f\colon \Omega \to \mathbb{R}$ is said to be λ-harmonic if

$$\lambda f(x) = \sum_{y \in \Omega} p_n(x,y) f(y).$$

Now introduce the functions $p_1(x,y) = p(x,y)$, $p_2(x,y)$, $p_3(x,y)$, ... such that

$$p_n(x,y) = \sum_{y \in \Omega} p(x,z) p_{n-1}(z,y)$$

(i.e., consider the powers of the Markov transition matrix) and define the λ-*Green function* $G_\lambda(x,y) = \sum_{n=1}^\infty \lambda^{-n} p_n(x,y)$. If $G_\lambda(x,y)$ is finite, then we can repeat the arguments above.

γ) *Distance functions* [**Gro, BGS**]. Consider a noncompact metric space X with distance function $d(x,y)$. Let $C(X)$ be the space of continuous functions on X. For any $x \in X$ we introduce the function $a_x(y) = d(x,y)$, $y \in X$. The relation $x \mapsto a_x$, $x \in X$, defines an embedding $X \to C(X)$. Consider the quotient space $C(X)/\mathbb{R}$, i.e., the space of continuous functions on X defined up to an additive constant. Now we take the closure of the image of X with respect to the uniform-on-compacta topology. For symmetric spaces, this construction yields the visibility boundary.

16.17. **Formal geometric constructions.** α) *Gluing of the quotient space.* Consider a noncompact metric space X and a compact set $K \subset X$. Let Y be a compact space and $\sigma\colon X \setminus K \to Y$ a surjective map. Introduce the space $X \cup Y$ endowed with a topology that satisfies the following condition: let $x_j \in X$ be a sequence that has no limit points in X; then $\lim x_j = y \in Y$ whenever the sequence $\sigma(x_j)$ converges to $y \in Y$.

EXAMPLE. In 11.1, $X = SL(n,\mathbb{R})/SO(n)$, K is the point E, and Y is Δ. The mapping σ is the composition of the mappings $X \setminus E \to \Sigma \setminus 0 \to \Delta$.

EXAMPLE. In 12.1, X is a symmetric space, K is the point x_0, and $Y = Sk$.

EXAMPLE. See the construction of the universal toric variety in 16.14.

β) *Gluing of the boundary of the quotient space* (velocity compactifications). Consider a noncompact metric space X and a compact set $K \subset X$. Let Z be a compact space and let $Y \subset Z$ be a compact subset. Let a mapping $\tau\colon X\backslash K \to Z\backslash Y$ be given. Define a topology on $X \cup Y$ by the following condition: a sequence $x_j \in X$ without limit points in X converges to $y \in Y$ whenever $\tau(x_j)$ converges to Y with respect to the topology of Z.

EXAMPLE. Let X_1 and X_2 be two metric spaces. We construct another compactification of $X_1 \times X_2$. Choose points $a_1 \in X_1$ and $a_2 \in X_2$. Let $A = [0, \infty) \times [0, \infty)$. Consider the mapping $X_1 \times X_2 \to A$ given by the rule $(x_1, x_2) \mapsto (d(x_1, a_1), d(x_2, a_2))$. Take a certain boundary of A. For instance, let us consider the mapping $A \to [0, \infty]$ given by the formula $(d_1, d_2) \mapsto d_1/d_2$. Now we can glue the quotient space $[0, \infty]$ to A, and then we can glue $[0, \infty]$ to $X_1 \times X_2$.

γ) *The projective limit of hypersurfaces.* Let us consider a noncompact metric space X and a compact set $K \subset X$. Assume that we have a family S_t, $t > 0$, of closed subsets of X such that
1) $S_t \cup S_\tau = \varnothing$ for $t \neq \tau$,
2) $X \setminus K = \cup_{t>0} S_t$, and
3) for any $t > \tau$, there exists a continuous mapping $\pi_{t\tau}\colon S_t \to S_\tau$ such that for $t > \tau > \sigma$ we have $\pi_{t\sigma} = \pi_{\tau\sigma}\pi_{t\tau}$.

Let S_∞ be the projective limit of the family S_t and let $\pi_{\infty,t}$ be the natural projection of S_∞ onto S_t. We shall define a topology on $X \cup S_\infty$ by the following condition: for a sequence $x_j \in X$ we define $t = t(x_j) \in (0, \infty)$ by the condition $x_j \in S_{t(x_j)}$, and assume that, as $t(x_j) \to \infty$, x_j converges to $y \in S_\infty$ whenever for any $\sigma > 0$, the sequence $\pi_{t(x_j)\sigma}(x_j) \in S_\sigma$ converges to $\pi_{\infty\sigma}(y)$ with respect to the topology of S_σ.

EXAMPLE (visibility boundary). Let $a \in X$. We assume that for any $x \in X$, there exists a unique shortest curve γ_x joining a and x. Let $l = l(x)$ be the length of γ_x. Let $s > 0$. Consider the set S_t given by the equation $l(x) = t$. Let $t > \tau$ and $x \in S_t$. Define the point $\pi_{t\tau}(x) \in S_\tau$ as $S_\tau \cap \gamma_x$. Now we can apply our construction.

EXAMPLE. *Tits-type metrics.* In the situation of the preceding example we denote by ρ the metric on X. Let $p, q \in S_\infty$. Then

$$d(p,q) = \limsup_{t\to\infty} \frac{1}{t} \rho(\pi_{\infty t}(p), \pi_{\infty t}(q))$$

is a metric on S_∞.

References

[Alg]   A. R. Alguneid, *Complete quadrics primal in four-dimensional space*, Proc. Math. Phys. Soc. Egypt **4** (1952), 93–104.

[BGS]   W. Ballmann, M. Gromov, and V. Schroeder, *Manifolds of nonpositive curvature*, Birkhäuser, Boston, 1985.

[Ber]   M. Berger, *Les espaces symétriques non compacts*, Ann. Sci. École Norm. Sup. (3) **74** (1957), 85–177.

[BiS]   Bialynicki–Birulya and A. J. Somerse, *A conjecture about compact quotients by tori*, Adv. Stud. Pure Math. **8** (1987), 59–68.

[BS1]   A. Borel and J.-P. Serre, *Corners of algebraic groups*, Comment. Math. Helv. **48** (1973), 436–491.

[BS2]   ———, *Cohomologie d'immobiles et groupes S-arithmétiques*, Topology **15** (1976), 211–232.

[CX]   E. Casas-Alvero and S. Xambo-Descamps, *The enumerative theory of conics after Halphen*, Lecture Notes in Math., vol. 1196, Springer-Verlag, Heidelberg and Berlin, 1986.

[Chas] M. Chasles, *Détermination du nombre des sections coniques qui doivent toucher cinq courbes données d'ordre quelconque, ou satisfaire à diverses autres conditions*, C. R. Acad. Sci. Paris **59** (1864, janvier–june), 222–226, 297–308, 425–431; **60** (1864, jullet–decembre), 7–15, 93–97, 209–218.

[Dan] V. I. Danilov, *Geometry of toric varieties*, Uspekhi Mat. Nauk **32** (1978), no. 2, 85–134; English transl., Russian Math. Surveys **33** (1978), no. 2, 97–154.

[DCP1] C. De Concini and C. Procesi, *Complete symmetric varieties*, Lecture Notes in Math., vol. 996, Springer-Verlag, Heidelberg and Berlin, 1983, pp. 1–44.

[DCP2] _____, *Complete symmetric varieties II. Intersection theory*, Algebraic Groups and Related Topics (Kyoto/Nagoya 1983), North-Holland, Amsterdam and New York, 1985, pp. 481–513.

[DCS] C. De Concini and T. A. Springer, *Betti numbers of complete symmetric varieties*, Geometry Today (Rome, 1984), Birkhäuser, Boston, 1985, pp. 87–107.

[Dix] J. Dixmier, *Les C^*-algèbres et leurs représentations*, Dunod, Paris, 1969.

[Doo] J. Doob, *Discrete potential theory and boundaries*, J. Math. Mech. **8** (1959), 433–458.

[Dyn] E. B. Dynkin, *Nonnegative eigenfunctions of the Laplace–Beltrami operator and Brownian motion on certain symmetric spaces*, Izv. Akad. Nauk SSSR **30** (1966), 455–478; English transl., Amer. Math. Soc. Transl., vol. 72, Amer. Math. Soc., Providence, RI, 1968, pp. 203–228.

[EO]   P. Eberlein and B. O'Neill, *Visibility manifolds*, Pacific J. Math. **46** (1973), 45–109.

[Fur1] H. A. Furstenberg, *Poisson formula for semisimple Lie groups*, Ann. of Math. (2) **77** (1963), 335–386.

[Fur2] _____, *Boundaries of Riemannian symmetric spaces*, Symmetric Spaces (W. M. Boothby and G. L. Weiss, eds.), Marcel Dekker, 1972, pp. 359–378.

[Ger1] P. Gérardin, *On harmonic functions on symmetric spaces and buildings*, Canadian Math. Soc. Conf. Proc. **1** (1981), 79–92.

[Ger2] _____, *Harmonic functions on buildings of reductive split groups*, Operator Algebras and Group Representations, vol. 1, Pitman, Boston, 1984, pp. 208–221.

[GM]   I. Ya. Goldsheid and G. A. Margulis, *Lyapunov exponents of products of random matrices*, Uspekhi Mat. Nauk **44** (1989), no. 5 (269), 13–60; English transl., Russian Math. Surveys **44** (1989), no. 5, 11–71.

[GH]   P. Griffits and J. Harris, *Principles of algebraic geometry*, Wiley-Interscience, New York, 1978.

[Gro] M. Gromov, *Hyperbolic groups*, Essays in Group Theory (S. M. Gersten, ed.), MSRI Publ., vol. 8, Springer-Verlag, New York, 1987, pp. 75–263.

[GJT] Y. Guivarch, L. Ji, and J. Taylor, *Compactifications of symmetric spaces*, C. R. Acad. Sci. Paris **317** (1993), 1103–1108.

[Hau] F. Hausdorff, *Grundzüge der Mengenlehre*, Leipzig, 1914; English transl., *Set theory*, 4th ed., Chelsea, 1991.

[Hel1] S. Helgason, *Radon–Fourier transform on symmetric spaces and related group representations*, Bull. Amer. Math. Soc. **71** (1965), 757–763.

[Hel2] _____, *Groups and geometric analysis*, Academic Press, Orlando, FL, 1984.

[How] R. Howe, *Some analytic geometry*, Preprint.

[ImH] Im Hof H. C., *Die Geometrie der Weilkammern in symmetrischen Raumen von nichtkompakten Typ*, Preprint, Bonn (1979).

[Kai] V. Kaimanovich, *Boundaries of invariant Markov operators* (to appear).

[Kap] M. M. Kapranov, *Quotients of Grassmannians*, I. M. Gelfand Seminar (S. I. Gelfand and S. G. Gindikin, ed.), Adv. in Sov. Math., vol. 16, Part 2, Amer. Math. Soc., Providence, RI, 1991, pp. 29–110.

[Kar] F. I. Karpelevich, *The geometry of geodesics and the eigenfunctions of the Laplace–Beltrami operator on symmetric spaces*, Trudy Moskov. Matem. Obshch. **14** (1 965), 48–185; English transl., Trans. Moscow Math. Soc. **14** (1965), 51–199.

[KSK] J. G. Kemeny, J. L. Snell, and A. W. Knapp, *Denumerable Markov chains*, Springer-Verlag, New York, 1976.

[Kle] S. L. Kleiman, *Chasles enumerative theory of conics. A historical introduction*, Studies in Algebraic Geometry. Studies in Math., vol. 20, MAA, Washington, DC, 1980.

[Kos1] B. Kostant, *Lie algebra cohomology and generalized Borel–Weil–Bott theorem*, Ann. in Math. **74** (1961), 329–387.

[Kos2] _____, *Clifford algebra analogue of the Hopf-Koszul-Samelson theorem, the ρ-decomposition, $C(\mathfrak{g}) = EndV_\rho \otimes C(P)$, and $\mathfrak{g}$-module structure of $\bigwedge \mathfrak{g}$*, Preprint (1996).

[Kus1] G. F. Kushner, *On the compactification of noncompact symmetric Riemann spaces*, Dokl. Akad. Nauk SSSR **190** (1970), no. 6; English transl., Soviet Math. Dokl **11** (1970), no. 1, 284–287.

[Kus2] _____, *The compactification of noncompact symmetric spaces*, Trudy Sem. Vektor. Tenzor. Anal. **16** (1972), 99–152. (Russian)

[Kus3] _____, *Irreducible projective compactifications of noncompact symmetric spaces*, Trudy Sem. Vekt. Tenzor. Anal. **21** (1983), 153–190. (Russian)

[Lak] D. Laksov, *The geometry of complete maps*, Ark. Mat. **26** (1988), no. 2, 231–263.

[LV] D. Luna and T. Vust, *Plongements d'espaces homogenes*, Comment. Math. Helv. **58** (1983), 186–245.

[Mar] R. S. Martin, *Minimal positive harmonic functions*, Trans. Amer. Math. Soc. **49** (1941), 137–142.

[MS] J. W. Morgan and P. B. Shalen, *Valuations, trees, and degenerations of hyperbolic structures*, Ann. of Math. (2) **120** (1984), no. 3, 401–476.

[Ner1] Yu. A. Neretin, *Spinor representation of infinite-dimensional orthogonal semigroups and the Virasoro algebra*, Funktsional. Anal. i Prilozhen. **23** (1989), no. 3, 32–44; English transl., Functional. Anal. Appl. **24** (1990), 196–207.

[Ner2] _____, *Extensions of representations of classical groups to representations of categories*, Algebra i Analiz **3** (1991), no. 1, 176–202; English transl. in Leningrad Math. J. **3** (1992).

[Ner3] _____, *Infinite-dimensional groups. Their mantles, trains, and representations*, Topics in Representation Theory (A. A. Kirillov, ed.), Adv. in Soviet Math., vol. 2, Amer. Math. Soc., Providence, RI, 1991, pp. 103–171.

[Ner4] _____, *Universal completions of classical groups*, Funktsional. Anal. i Prilozhen. **26** (1992), no. 4, 30–44; English transl., Functional. Anal. Appl. **26** (1993), 254–266.

[Ner5] _____, *One remark on the construction of the separated quotient space*, Preprint MPI-96-26 (1996).

[Ner6] _____, *Categories of symmetries and infinite dimensional groups*, Oxford University Press, Oxford, 1996.

[Ner7] _____, *Hinges and geometric constructions of boundaries of symmetric spaces*, Preprint, MPI-96-78, 1996.

[Olse1] M. A. Olshanetsky, *The Martin boundaries of symmetric spaces with nonpositive curvature*, Uspekhi Mat. Nauk **24** (1969), no. 6, 189–190. (Russian)

[Olse2] _____, *Martin boundaries for real semisimple groups*, J. Funct. Anal. **126** (1994), 169–216.

[Ols] G. I. Olshanskiĭ, *On semigroups related to infinite-dimensional groups*, Topics in Representation Theory (A. A. Kirillov, ed.), Adv. in Soviet Math., vol. 2, Amer. Math. Soc., Providence, RI, 1991, pp. 67–101.

[Osh] T. Oshima, *A realization of Riemannian symmetric spaces*, J. Math. Soc. Japan **30** (1978), 117–132.

[OS] T. Oshima and J. Sekiguchi, *Eigenspaces of invariant differential operators on an affine symmetric space*, Invent. Math. **57** (1980), 1–81.

[Pom] D. Pompeiu, *Sur la continuité des fonctions des variables complexes*, Ann. Fac. Sci. Toulouse **7** (1905), 265–315.

[Pop] V. L. Popov, *Contraction of actions of algebraic groups*, Mat. Sb. **130** (1986), no. 3, 310–334; English transl., Math. USSR-Sb. **58** (1987), 311–335.

[Put] M. S. Putcha, *On linear algebraic semigroups* I, II, Trans. Amer. Math. Soc. **259** (1980), 457–469, 471–491.

[Pya] I. I. Pyatetskii-Shapiro, *Automorphic functions and the geometry of classical domains*, Fizmatgiz, Moscow, 1961; English transl., Gordon and Breach, London, 1969.

[Ren] L. E. Renner, *Classification of semisimple algebraic monoids*, Trans. Amer. Math. Soc. **292** (1985), no. 1, 193–223.

[Rup] W. Ruppert, *Compact semitopological semigroups: an intrinsic theory*, Lecture Notes in Math., vol. 1079, Springer-Verlag, Heidelberg and Berlin, 1984.

[Sat] I. Satake, *On representations and compactifications of symmetric Riemannian spaces*, Ann. of Math. (2) **71** (1960), 77–110.

[Schl] H. Schlichtkrull, *Hyperfunctions and harmonic analysis on symmetric spaces*, Birkhäuser, Boston, 1984.

[Schu] H. Schubert, *Kalkül der Abzählenden Geometrie*, Leipzig, 1879; 2nd ed., Springer-Verlag, Heidelberg and Berlin, 1979.

[Sem1] J. G. Semple, *On complete quadrics* (I), J. London Math. Soc. **26** (1951), 122–125.

[Sem2] _____, *The variety whose points represent complete collineations of S_r on S'_r*, Rend. Mat. Univ. Roma (5) **10** (1951), 201–208.

[Sem3] _____, *On complete quadrics* (II), J. London Math. Soc. **27** (1952), 280–287.

[SR] J. G. Semple and H. Roth, *Introduction to algebraic geometry*, 1st ed., Oxford University Press, Oxford, 1949; 2nd ed., 1985.

[Sev] F. Severi, *I fundamenti della geometria numerativa*, Ann. Mat. (4) **19** (1940), 151–242.

[Shu] M. G. Shur, *The Martin boundary for a linear elliptic second order operator*, Izv. Akad. Nauk SSSR **27** (1963), 45–60. (Russian)

[Stu] E. Study, *Über die Geometrie der Kogelschnitte, insbesondere deren Charakteristikeproblem*, Math. Ann. **27** (1886), 58–101.

[Tay] J. Taylor, *Compactifications defined by a polyhedral cone decompositions of $\mathbb{R}^n$*, Harmonic analysis and discrete potential theory (M. A. Picardello, ed.), Plenum, New York, 1991, pp. 1–14.

[Thu] W. Thurston, *On the geometry and dynamics of diffeomorphisms of surfaces*, Bull. Amer. Math. Soc. **19** (1988), no. 2, 417–431.

[Tit] J. Tits, *Buildings of spherical type and finite BN-pairs*, Lecture Notes in Math., vol. 386, Springer-Verlag, Heidelberg and Berlin, 1974.

[Tyr] J. A. Tyrrell, *Complete quadrics and complete collineations in S_n*, Matematika **3** (1956), 69–79.

[VEL] A. G. Vainstein, V. A. Efremovich, and E. A. Loginov, *Points at infinity of metric spaces*, Trudy Sem. Vektor. Tenzor. Anal **18** (1978), 129–139. (Russian)

[Vin] É. B. Vinberg, *On reductive algebraic semigroups*, Lie Groups and Lie Algebras: E. B. Dynkin's seminar (S. G. Gindikin and E. B. Vinberg, eds.), Transl. Amer. Math. Soc. Ser. 2, vol. 169, Amer. Math. Soc., Providence, RI, 1995, pp. 145–182.

[Vus] T. Vust, *Plongement d'espaces symmetriques algebriques: une classification*, Ann. Scuola Norm. Sup. Pisa (4) **17** (1990), 165–195.

MOSCOW STATE INSTITUTE OF ELECTRONICS AND MATHEMATICS, MOSCOW, RUSSIA
E-mail address: neretin@main.mccme.rssi.ru

Translated by A. I. SHTERN

Multiplicities and Newton Polytopes

Andreĭ Okounkov

Dedicated to A. A. Kirillov on the occasion of his 60th birthday

§1. Introduction

In this paper we consider two applications of the methods used in [**Ok2**] to the analysis of reduction multiplicities.

The first one is about the reduction

$$Sp(2n, \mathbb{C}) \downarrow Sp(2n-2, \mathbb{C}) \times \mathbb{C}^\times.$$

By a theorem of Zhelobenko [**Zh**], the multiplicity corresponding to the two dominant weights

$$\Lambda = (\Lambda_1, \ldots, \Lambda_n) \in Sp(2n)^\wedge, \qquad \lambda = (\lambda_1, \ldots, \lambda_{n-1}) \in Sp(2n-2)^\wedge$$

and to the number $k \in \mathbb{Z} = (\mathbb{C}^\times)^\wedge$ equals the number of the following Gelfand–Zetlin patterns:

(1.1)
$$\begin{array}{ccccccc} \Lambda_1 & & \Lambda_2 & & \ldots & & \Lambda_n & & 0 \\ & \eta_1 & & \eta_2 & & \ldots & & \eta_n & \\ & & \lambda_1 & & \ldots & & \lambda_{n-1} & & 0 \end{array}$$

such that

$$k = 2\sum \eta_i - \sum \Lambda_i - \sum \lambda_i.$$

Here $\eta_1, \ldots, \eta_n$ are auxiliary integer variables.

Here the question of the representation-theoretical meaning of the middle line in (1.1) naturally arises. It was conjectured by Kirillov [**K**] that this middle line corresponds to the intermediate subgroup

$$Sp(2n) \supset \text{``}Sp(2n-1)\text{''} \supset Sp(2n-2),$$

which is the stabilizer of a vector in $\mathbb{C}^{2n}$. This group is isomorphic to the semidirect product of $Sp(2n-2)$ and the $(2n-1)$-dimensional Heisenberg group. This conjecture was partially confirmed by Shtepin in [**Sh1, Sh2**].

1991 *Mathematics Subject Classification.* Primary 20G05.

The author was supported by the NSF under grant DMS 9304580 and the Russian Foundation for Basic Research under grant 95-01-00814.

For all classical groups

$$X(n) = GL(n), \ SO(2n+1), \ Sp(2n), \ SO(2n), \qquad n = 1, 2, \ldots,$$

the reduction multiplicities for the reduction

$$X(n) \downarrow X(n-k) \times (\mathbb{C}^\times)^k$$

are described [**BZ**] by certain Gelfand–Zetlin patterns such that the weight of a pattern is a linear function of its entries. A similar description of the tensor product multiplicities was conjectured in [**BZ**].

In §§2–5 we give a geometric interpretation of the Gelfand–Zetlin patterns for the symplectic group (one can consider the orthogonal cases similarly [**Ok3**]). It is based on the following general construction. The details of the construction can be found in [**Ok2**] or (in a particular example) below.

Consider the following general situation. Suppose we have a connected reductive group representation

(1.2) $$\mathcal{G} \to GL(V)$$

and a $\mathcal{G}$-stable irreducible subvariety

(1.3) $$X \subset \mathbb{P}(V).$$

Denote by $F[X]$ the algebra of polynomials on X and by $F[X]_d \subset F[X]$ the subspace of polynomials of degree d on X. We are interested in the multiplicities of irreducible $\mathcal{G}$-modules in $F[X]_d$.

Let $\mathcal{B} \subset \mathcal{G}$ be a Borel subgroup. Choose a $\mathcal{B}$-stable chain of closed irreducible subvarieties of X

(1.4) $$X = X_0 \supset X_1 \supset \cdots \supset X_{\dim X}, \qquad \operatorname{codim} X_i = i,$$

which always exists. This chain gives rise to a $\mathcal{B}$-invariant valuation

$$v \colon F[X] \to \mathbb{Z}^{\dim X},$$

where the group $\mathbb{Z}^{\dim X}$ is ordered lexicographically. Each quotient

$$F[X]_{d,a} = \{f \in F[X]_d, \ v(f) \geqslant a\}/\{f \in F[X]_d, \ v(f) > a\}$$

is at most 1-dimensional. Its $\mathcal{B}$-weight ν is a linear function $\nu(d,a)$ of d and a.

Denote by $\mathcal{N} \subset \mathcal{B}$ the maximal unipotent subgroup. The invariants $f \in F[X]^{\mathcal{N}}$ are precisely the $\mathcal{G}$-highest vectors in $F[X]$. Therefore, the description of the $\mathcal{G}$-spectrum of $F[X]$ is reduced to the description of the set

(1.5) $$S = \{(d,a) \in \mathbb{Z} \oplus \mathbb{Z}^{\dim X} \mid \exists f \in (F[X]_d)^{\mathcal{N}}, \ v(f) = a\}.$$

The multiplicity of a highest weight λ in $F[X]_d$ then equals

$$\#\{(d,a) \in S \mid \nu(d,a) = \lambda\}.$$

The set

(1.6) $$S(1) = S \cap \{d = 1\} \subset \mathbb{Z}^{2n-1},$$

which describes the $\mathcal{G}$-spectrum in terms of linear functions on X, is of special interest.

If $\mathcal{N}$ happens to have an open orbit on some X_i, $i = 0, 1, \ldots$, then the coordinates $a_{i+1}, a_{i+2}, \ldots$ are identically zero on S, and one can forget them as well as the subvarieties $X_{i+1}, X_{i+2}, \ldots$ themselves.

Now the interpretation of the pattern (1.1) and the role of the intermediate group $Sp(2n-1)$ is the following. Let

$$X = G/B$$

be the flag variety for $G = Sp(2n)$. Let V^Λ be the irreducible $Sp(2n)$-module with highest weight Λ. Consider the unique $Sp(2n)$-equivariant map

$$X \to \mathbb{P}(V^\Lambda)^*.$$

Let $F[X]$ be the algebra of polynomials on the image of X. The space $F[X]_d$ is isomorphic to $V^{d\Lambda}$ as an $Sp(2n)$-module. Finally, let

$$\mathcal{G} = Sp(2n-2) \times \mathbb{C}^\times.$$

There are precisely $2n$ closed $Sp(2n-1)$-stable subvarieties in X. These Schubert varieties form a chain

$$X = X_0 \supset X_1 \supset \cdots \supset X_{2n-1},$$

in which X_{2n-1} is isomorphic to the flag variety for $Sp(2n-2)$. Note that $\mathcal{N}$ has an open orbit on X_{2n-1}.

Consider the corresponding valuation

$$v \colon F[X] \to \mathbb{Z}^{2n-1}$$

and the set $S(1)$ defined in (1.6).

In §5 we prove

THEOREM 1. *For $F[X]$ and $\mathcal{G}$ as above we have*

$$S(1) = \{(a_1, \ldots, a_{2n-1}) \in \mathbb{Z}^{2n-1}, a_i \geq 0,$$
$$a_1 \leq \Lambda_1 - \Lambda_2, a_2 \leq \Lambda_2 - \Lambda_3, \ldots, a_n \leq \Lambda_n,$$
$$a_{i-1} - a_i + a_{2n-i+1} \leq \Lambda_{i-1} - \Lambda_i, i = 2, \ldots, n\}.$$

The $\mathcal{B}$-weight ν of a point $a \in S(1)$ equals

$$\nu(a) = \left(\Lambda_2 - a_2 + a_{2n-1}, \ldots, \Lambda_n - a_n + a_{n+1}, \Lambda_1 - 2a_1 - \sum_2^n a_i\right).$$

In the new coordinates

$$\eta_i = \Lambda_i - a_i, \qquad\qquad i = 1, \ldots, n,$$
$$\lambda_i = \Lambda_{i+1} - a_{i+1} + a_{2n-i}, \quad i = 1, \ldots, n-1,$$

the inequalities which define $S(1)$ and the function ν become precisely those from the Zhelobenko theorem.

Since the variety X_{2n-1} here is isomorphic to the flag variety for $Sp(2n-2)$, one can construct a complete chain (1.4) inductively. Then, using the above theorem, one easily describes the the sets $S(1)$ for

$$\mathcal{G} = Sp(2n-2k) \times (\mathbb{C}^\times)^k, \qquad k = 1, \ldots, n.$$

In particular, in the maximal torus case $k = n$ one obtains Gelfand–Zetlin patterns of the form shown in Theorem 2 below.

Our interpretation of Gelfand–Zetlin patterns appears to be new and is obtained by direct application of the general procedure of [**Ok2**]; however, nothing could be more classical than G/B, and there should exist close connections with other approaches (see, for example, [**AFS, D, La**]).

The other issue we shall consider in this paper is a sharpening of the results of [**Ok2**] about the log-concavity of reduction multiplicities. I realized that this can be done following a discussion in the Kirillov seminar at the University of Pennsylvania.

Let $G \supset \mathcal{G}$ be a group and a subgroup and let

$$\mu(\Lambda, \lambda), \qquad \Lambda \in G^\wedge, \, \lambda \in \mathcal{G}^\wedge,$$

be the corresponding reduction multiplicities. According to a quantum mechanics analogy, this function of Λ and λ corresponds to the number of quantum states with given values of the two "energies" Λ and λ. Hence $\log \mu(\Lambda, \lambda)$ is an analog of entropy, and it is to be expected (see for example [**LL**, §21]) that this function is concave in Λ and λ, at least in some suitable limit.

Using the combinatorics of Gelfand–Zetlin patterns, it was shown in [**Ok1**] that for the pairs

$$\begin{aligned} G &= GL(n), & \mathcal{G} &= GL(k), \\ G &= SO(2n+1), & \mathcal{G} &= SO(2k+1), \\ G &= Sp(2n), & \mathcal{G} &= Sp(2k), \qquad n > k, \end{aligned}$$

with standard embeddings this is indeed the case. However, this log-concavity fails for general pairs of reductive groups.

On the other hand, the following semi-classical limit of reduction multiplicities is always log-concave in λ. It is known that for fixed Λ and λ the function

$$\mu(m\Lambda, m\lambda), \qquad m = 1, 2, \ldots,$$

is a polynomial in m for large m. Let r be the degree of this polynomial for general λ (or for general Λ and λ if we want to let both Λ and λ vary). By definition, set

$$(1.7) \qquad \mu_\infty(\Lambda, \lambda) = \lim \frac{\mu(m\Lambda, m\lambda)}{m^r}, \qquad m \to \infty.$$

If m is sufficiently divisible, for example, if

$$m = m_0!, \qquad m_0 = 1, 2, \ldots,$$

then this definition works for rational Λ and λ as well.

It is well known [**H, GS**] that the number (1.7) has a very natural interpretation in the orbit method of Kirillov [**K1**].

The log-concavity of (1.7) is a particular case of the following general statement. Let the group $\mathcal{G}$ and the variety X be as in (1.2), (1.3). Denote by $\mu(d, \lambda)$ the multiplicity of the highest weight λ in $F[X]_d$; the multiplicity $\mu(m, m\lambda)$ grows as a polynomial in m as $m \to \infty$, say as m^r. The semi-classical multiplicity

$$(1.8) \qquad \mu_\infty(\lambda) = \lim \frac{\mu(m, m\lambda)}{m^r}, \qquad m \to \infty,$$

is always log-concave in λ (see [**Gr**, **Ok2**]). If X is a flag variety of G, then we obtain the log-concavity of $\mu_\infty(\Lambda, \lambda)$ in λ.

The proof of the log-concavity of (1.8) given in [**Ok2**] used the Newton polytope of the projective variety X. Recall the definition of the Newton polytope for a *toric* projective variety. For $\mathcal{G}$ and $X \in \mathbb{P}(V)$ as above suppose that $\mathcal{G} = (\mathbb{C}^\times)^{\dim X}$ and $\mathcal{G}$ has an open orbit on X. Then each subspace $F[X]_d$ is a multiplicity-free direct sum of some irreducible $\mathcal{G}$-modules

$$F[X]_d = \bigoplus V_d^m, \qquad m \in \mathbb{Z}^{\dim X} = \mathcal{G}^\wedge.$$

By the irreducibility of X,

(1.9) $$V_{d_1}^{m_1} \cdot V_{d_2}^{m_2} = V_{d_1+d_2}^{m_1+m_2},$$

provided $V_{d_i}^{m_i} \neq 0$ for $i = 1, 2$. The set

$$\Delta^{\text{toric}}(X) = \{m/d \mid V_d^m \neq 0\} \in \mathbb{Q}^{\dim X}$$

or its closure in $\mathbb{R}^{\dim X}$ is called the *Newton polytope* of the toric variety X. By (1.9) this set is convex.

For general X there is no such torus action and no such grading in the algebra $F[X]$. However, there is always a valuation v and the corresponding filtration of each $F[X]_d$ by subspaces

$$F[X]_{d,a} = \{f \in F[X]_d, \, v(f) \geqslant a\}.$$

The quotients $F[X]_{d,a}$ of this filtration are at most one-dimensional, and

$$F[X]_{d_1,a_1} \cdot F[X]_{d_2,a_2} = F[X]_{d_1+d_2,a_1+a_2}$$

provided $F[X]_{d_i,a_i} \neq 0$ for $i = 1, 2$. In particular, the set S defined in (1.5) is a subsemigroup of $\mathbb{Z}^{1+\dim X}$, and hence the set

(1.10) $$\Delta(X) = \text{Closure of } \{a/d \mid (d,a) \in S\} \in \mathbb{R}^{\dim X}$$

is convex. We call it the *Newton polytope*[1] of X. It depends also on the group $\mathcal{G}$ and the choice of the chain (1.4).

Consider the function $\nu_1(a) = \nu(1, a)$ and the affine-linear map

$$\nu_1 \colon \Delta(X) \to \mathcal{G}^\wedge \otimes_{\mathbb{Z}} \mathbb{R}.$$

One checks that

$$\mu_\infty(\lambda) = \text{vol}(\nu_1^{-1}(\lambda) \cap \Delta(X)).$$

The log-concavity of $\mu_\infty(\lambda)$ follows at once from the classical Brunn–Minkowski inequality [**Le**]. If $\mathcal{G}$ is a torus and X is toric, then ν_1 maps $\Delta(X)$ isomorphically onto the closure of $\Delta^{\text{toric}}(X)$.

In this paper we prove that (1.7) is log-concave in both Λ and λ. Moreover, we prove that (1.8) is log-concave in both λ and the $\mathcal{G}$-linearized sheaf which gives the projective embedding of X. To this end we construct a bigger convex cone $\widetilde{\Delta}(X)$ that contains the polytopes $\Delta(X)$ for all projective embeddings of X. This is done in the Appendix.

---

[1] Perhaps the name "Newton convex set" would be better, as it is likely that $\Delta(X)$ may fail to be a polytope. It is a polytope in important examples, though.

I would like to thank my teacher, A. A. Kirillov, for many, many enlightening discussions both in Moscow and Philadelphia. I am deeply grateful to all participants of his seminar, especially V. Ginzburg and G. Olshanski. I thank the Institute for Advanced Study in Princeton for hospitality.

§2. Flag variety G/B for $G = Sp(2n)$

In this section we fix the notation and recall some well-known facts. Let $G = Sp(2n, \mathbb{C})$. Choose a basis $e_1, \ldots, e_{2n}$ of $\mathbb{C}^{2n}$ in which the matrix of the symplectic form is

$$\begin{pmatrix} & & & & 1 \\ 0 & & & \cdots & \\ & & 1 & & \\ & -1 & & & \\ & \cdots & & 0 & \\ -1 & & & & \end{pmatrix}.$$

Let x_{ij} be the matrix elements in this basis. Let $H, B_+, B_- \subset G$ be the subgroups of diagonal, upper triangular, and lower triangular elements of $Sp(2n)$. We use $x_{11}, \ldots, x_{nn}$ as coordinates in H and use the dual coordinates

$$g^\Lambda = x_{11}^{\Lambda_1} \cdots x_{nn}^{\Lambda_n}, \qquad g \in H, \, \Lambda \in H^\wedge,$$

for weights. The weights

(2.1) $$\Lambda_1 \geqslant \Lambda_2 \geqslant \cdots \geqslant \Lambda_n \geqslant 0$$

are dominant for B_+.

The group B_- is the stabilizer in G of the complete flag

$$0 = L_0^- \subset L_1^- \subset \cdots \subset L_{2n-1}^- \subset L_{2n}^- = \mathbb{C}^{2n},$$

where L_i^- is the linear span of the last i basis vectors, $L_i^- = \langle e_{2n-i+1}, \ldots, e_{2n} \rangle$. The quotient G/B is isomorphic to the variety of all complete flags

(2.2) $$0 = L_0 \subset L_1 \subset \cdots \subset L_{2n-1} \subset L_{2n} = \mathbb{C}^{2n}$$

such that $L_i^\perp = L_{2n-i}$, $i = 0, \ldots, 2n$. Here $L_i^\perp$ is the skew-orthogonal complement. Similarly, we set $L_i^+ = \langle e_1, \ldots, e_i \rangle$.

The set $O = \{L_i \cap L_{2n-i}^+ = 0, \, i = 0, \ldots, 2n\}$ is the open Bruhat cell in G/B. The map $N_+ \to N_+ B_-$ sends the maximal unipotent subgroup N_+ of B_+ isomorphically onto O. We use $x_{ij}, i < j, \, i+j \leqslant 2n+1$, as coordinates in N_+ and O.

The complement of O in G/B is the union of divisors

$$D_i = \{L_i \cap L_{2n-i}^+ \neq 0\}, \qquad i = 1, \ldots, n.$$

Given a dominant weight (2.1), we form the divisor

$$D_\Lambda = (\Lambda_1 - \Lambda_2)D_1 + \cdots + (\Lambda_{n-1} - \Lambda_n)D_{n-1} + \Lambda_n D_n.$$

The space associated with D_Λ $\{f \in \mathbb{C}(G/B) \mid (f) + D_\Lambda \geqslant 0\}$ is one of the appearances of the irreducible G-module V^Λ with highest weight Λ (see, for example, [**BGG**]). We have

$$V^\Lambda \subset \mathbb{C}[O] \subset \mathbb{C}(G/B).$$

The action of G on elements of $V^\Lambda \subset \mathbb{C}[O]$ is the following:

(2.3) $$[g \cdot f](n) = f(n_g) b_{n,g}^{-\Lambda}, \qquad f \in V^\Lambda, \, g \in G, \, n \in N_+,$$

where $n_g \in N_+$ and $b_{n,g} \in B_-$ are found from the decomposition $g^{-1}n = n_g b_{n,g}$.

§3. The $Sp(2n-2)$-highest vectors in V^Λ

Let $Sp(2n-1)$ be the stabilizer of the vector e_{2n} in $Sp(2n)$. Consider the action of this group on G/B. This is a particular case of the classical description of the double cosets $P \backslash G/B$ for a parabolic subgroup $P \subset G$. One easily checks that the Schubert varieties

$$X_i = \{L_i \subset L_{2n-1}^-\}, \qquad i = 1, \ldots, 2n-1,$$

form the unique chain of $Sp(2n-1)$-stable closed subvarieties of G/B

(3.1) $$G/B = X_0 \supset X_1 \supset \cdots \supset X_{2n-1}, \qquad \operatorname{codim} X_i = i.$$

Order $\mathbb{Z}^{2n-1}$ lexicographically. The valuation

$$v \colon \mathbb{C}(G/B) \to \mathbb{Z}^{2n-1}$$

corresponds to the chain (3.1):

(3.2) $$v_1(f) = \operatorname{ord}_{X_1} f, \quad v_2(f) = \operatorname{ord}_{X_2}((x_{1,2n})^{-v_1(f)} f|_{X_1}), \quad \ldots.$$

Here $\operatorname{ord}_{X_1} f$ denotes the order of zero of f on X_1 and $(x_{1,2n})^{-v_1(f)} f|_{X_1}$ denotes the restriction to X_1, which is a well-defined rational function by definition of $v_1(f)$.

Let us set

$$\mathcal{G} = Sp(2n-2) \times \mathbb{C}^\times, \qquad \mathcal{B} = \mathcal{G} \cap B_+, \qquad \mathcal{N} = \mathcal{G} \cap N_+.$$

Observe that $\mathcal{N}$ has an open orbit in X_{2n-1}. Our aim is to describe the set

$$S(1) = \{a \in \mathbb{Z}^{2n-1} \mid \exists f \in (V^\Lambda)^\mathcal{N}, v(f) = a\}.$$

It is clear that

$$(V^\Lambda)^\mathcal{N} \subset \mathbb{C}[O]^\mathcal{N} = \mathbb{C}[x_{12}, \ldots, x_{1,2n}].$$

On O the subvarieties (3.1) are given by the equations

(3.3) $$\begin{aligned} X_1 &= \{x_{1,2n} = 0\}, \\ X_2 &= \{x_{1,2n} = 0, \, x_{1,2n-1} = 0\}, \\ &\ldots\ldots\ldots\ldots\ldots\ldots\ldots\ldots\ldots\ldots\ldots, \\ X_{2n-1} &= \{x_{1,2n} = 0, \, x_{1,2n-1} = 0, \, \ldots, \, x_{12} = 0\}. \end{aligned}$$

By definition of the valuation v, the equality

$$v(f) = a, \qquad f \in \mathbb{C}[O]^\mathcal{N},$$

is equivalent to $x_{1,2n}^{a_1} \cdots x_{12}^{a_{2n-1}}$ being the lowest monomial of f with respect to the following lexicographic ordering of monomials:

$$\prod x_{1i}^{p_i} < \prod x_{1i}^{q_i}$$

if $p_{2n} < q_{2n}$, or if $p_{2n} = q_{2n}$ and $p_{2n-1} < q_{2n-1}$, and so on. Therefore, we must determine which monomials in $x_{12}, \ldots, x_{1,2n}$ can possibly occur as the lowest monomials of a $\mathcal{N}$-fixed vector in V^Λ. Note the analogy with the Gröbner basis construction.

Given a monomial

(3.4) $$x_{1,2n}^{a_1} \cdots x_{12}^{a_{2n-1}}$$

set

(3.5) $$\begin{aligned} \eta_i &= \Lambda_i - a_i, & i &= 1, \ldots, n, \\ \lambda_i &= \Lambda_{i+1} - a_{i+1} + a_{2n-i}, & i &= 1, \ldots, n-1. \end{aligned}$$

These definitions can be depicted as follows:

$$\begin{array}{cccccc}
\Lambda_1 \xrightarrow{a_1} & \Lambda_2 \xrightarrow{a_2} & & \Lambda_i \xrightarrow{a_i} & \\
 \searrow \eta_1 & \searrow \eta_2 & \cdots & \searrow \eta_i & \cdots \\
\lambda_1 \xrightarrow{a_{2n-1}} & & & \lambda_{i-1} \xrightarrow{a_{2n-i+1}} &
\end{array}$$

In §5 we prove the following theorem.

THEOREM 1. *A point $a = (a_1, \ldots, a_{2n-1})$ lies in the set $S(1)$ (or, equivalently, (3.4) is the lowest monomial of a $\mathcal{B}$-semiinvariant polynomial $f \in V^\Lambda$) if and only if the numbers (3.5) satisfy the following interlacing conditions:*

(3.6) $$\begin{array}{ccccccc}
\Lambda_1 & & \Lambda_2 & \cdots & \Lambda_n & & 0 \\
& \eta_1 & & \eta_2 & \cdots & \eta_n & \\
& & \lambda_1 & \cdots & \lambda_{n-1} & & 0
\end{array}$$

The $\mathcal{B}$-weight of f equals

$$\left(\lambda_1, \ldots, \lambda_{n-1}, 2\sum \eta_i - \sum \Lambda_i - \sum \lambda_i\right).$$

In (3.6) the notation

$$\begin{array}{ccccccc}
\Lambda_1 & & \Lambda_2 & \cdots & \Lambda_n & & 0 \\
& \eta_1 & & \eta_2 & \cdots & \eta_n &
\end{array}$$

means

$$\Lambda_1 \geqslant \eta_1 \geqslant \Lambda_2 \geqslant \cdots \geqslant \Lambda_n \geqslant \eta_n \geqslant 0.$$

It will be clear from the proof that it is possible to write the corresponding f explicitly.

§4. Filtration in V^Λ

Observe that X_{2n-1} is isomorphic to the flag variety for $Sp(2n-2)$. Continue the chain (3.1) inductively as follows:

(4.1) $$G/B = X_0 \supset X_1 \supset \cdots \supset X_{n^2} = B_-, \qquad \operatorname{codim} X_i = i,$$

where

$$X_{2n} = \{L_{2n-1} = L^-_{2n-1}, L_2 \subset L^-_{2n-2}\},$$
$$X_{2n+1} = \{L_{2n-1} = L^-_{2n-1}, L_3 \subset L^-_{2n-2}\},$$
$$\dots\dots\dots\dots\dots\dots\dots\dots\dots\dots\dots\dots,$$
$$X_{4n-4} = \{L_{2n-1} = L^-_{2n-1}, L_{2n-2} = L^-_{2n-2}\},$$
$$\dots\dots\dots\dots\dots\dots\dots\dots\dots\dots\dots\dots,$$
$$X_{6n-9} = \{L_{2n-1} = L^-_{2n-1}, L_{2n-2} = L^-_{2n-2}, L_{2n-3} = L^-_{2n-3}\},$$
$$\dots\dots\dots\dots\dots\dots\dots\dots\dots\dots\dots\dots.$$

The chain (4.1) is clearly B_--stable. Note that all X_i are smooth, because they are B_--stable and smooth at the unique B_--fixed point $X_{n^2} \in G/B$.

Extend the valuation v to the valuation

$$v \colon \mathbb{C}(X) \to \mathbb{Z}^{n^2}$$

in the same way as in (3.2). Now

$$v(f) = (v_1, \ldots, v_{n^2}), \qquad f \in \mathbb{C}[O] \cong \mathbb{C}[x_{ij}]_{i<j, i+j \leqslant 2n+1},$$

means that $x_{1,2n}^{v_1} x_{1,2n-1}^{v_2} \cdots$ is the lowest monomial in f with respect to the lexicographic ordering of monomials in which

$$\prod x_{ij}^{p_{ij}} < \prod x_{ij}^{q_{ij}}$$

if $p_{1,2n} < q_{1,2n}$, or if $p_{1,2n} = q_{1,2n}$ and $p_{1,2n-1} < q_{1,2n-1}$, and so on. Note that in particular

$$x_{1,2n} > x_{1,2n-1} > \cdots > x_{12} > x_{2,2n-1} > \cdots > x_{23} > \cdots > x_{n,n+1},$$

which is exactly the ordering of positive roots induced by the standard lexicographic order in $\mathbb{R}^n$.

Let $\mathcal{G}$ be the the maximal torus H of G, i.e., $\mathcal{G} = \mathcal{B} = (\mathbb{C}^\times)^n$. Since $\mathcal{N}$ is trivial, we have $(V^\Lambda)^{\mathcal{N}} = V^\Lambda$. Therefore, to describe the set $S(1)$ we must find out which monomials can occur as the lowest monomial of a polynomial $f \in V^\Lambda$.

Given a monomial

(4.2)
$$\prod_{i<j;\, i+j \leqslant 2n+1} x_{ij}^{p_{ij}},$$

put

(4.3)
$$\begin{aligned}
\eta_i &= \Lambda_i - p_{1,2n-i+1}, & i &= 1, \ldots, n, \\
\lambda_i &= \eta_{i+1} + p_{1,i+1}, & i &= 1, \ldots, n-1, \\
\eta'_i &= \lambda_i - p_{2,2n-i}, & i &= 1, \ldots, n-1, \\
\lambda'_i &= \eta'_{i+1} + p_{2,i+1}, & i &= 1, \ldots, n-2,
\end{aligned}$$
$$\dots\dots\dots\dots\dots.$$

In §5 we deduce the following result from Theorem 1.

THEOREM 2. *The monomial* (4.2) *is the lowest monomial of a polynomial* $f \in V^\lambda$ *if and only if the numbers* (4.3) *satisfy the following conditions of interlacing*:

(4.4)
$$\begin{array}{ccccccc} \Lambda_1 & & \Lambda_2 & & \cdots & \Lambda_n & & 0 \\ & \eta_1 & & \eta_2 & \cdots & & \eta_n & \\ & & \lambda_1 & & \cdots & \lambda_{n-1} & & 0 \\ & & & \eta'_1 & \cdots & & \eta'_{n-1} & \\ & & & & \cdots & \cdots & & \end{array}$$

The weight of the monomial (4.2) *equals*

$$\left(2\sum \eta_i - \sum \Lambda_i - \sum \lambda_i,\ 2\sum \eta'_i - \sum \lambda_i - \sum \lambda'_i,\ \dots \right).$$

§5. Proof of the theorems

Proof of Theorem 1. To simplify the notation, we put

$$x_i = x_{1,i}, \qquad i = 2, \dots, 2n.$$

Let us examine the poles of $x_2, \dots, x_{2n}$ on the divisors D_i. Suppose $i < n$ and let s_i be the Weyl group element permuting x_{ii} and $x_{i+1,i+1}$. On $s_i O \subset G/B$ we have

$$D_i = s_i O \setminus O.$$

Local coordinates in $s_i O \subset G/B$ are

$$\begin{pmatrix} \cdots & y_{1,i} & y_{1,i+1} & \cdots & y_{1,2n-i} & y_{1,2n-i+1} & \cdots \\ \ddots & \vdots & \vdots & & \vdots & \vdots & \\ & 0 & 1 & & & & \\ & 1 & y_{i,i+1} & & & & \\ & & & \ddots & \vdots & \vdots & \\ & & & & 0 & 1 & \\ & & & & 1 & -y_{i,i+1} & \\ & & & & & & \ddots \end{pmatrix} B_-.$$

In these coordinates the divisor D_i is defined by the equation $y_{i,i+1} = 0$. Using Gauss elimination, we get

$$x_i = y_{1,i+1} - y_{1i} y_{i,i+1}, \qquad x_{i+1} = y_{1,i+1}/y_{i,i+1},$$
$$x_{2n-i} = y_{1,2n-i+1} + y_{1,2n-i} y_{i,i+1}, \qquad x_{2n-i+1} = -y_{1,2n-i+1}/y_{i,i+1},$$
$$x_j = y_{1,j}, \qquad j \neq i, i+1, 2n-i, 2n-i+1.$$

Therefore

$$\operatorname{ord}_{D_i}(x_j) = \begin{cases} -1, & j = i+1, 2n-i+1, \\ 0, & \text{otherwise}. \end{cases}$$

Note that no polynomials from $\mathbb{C}[O]$ may have zeros on the divisor D_i. Indeed, such polynomials form an N_+-stable subspace which (provided it is nonempty) contains an N_+-fixed vector different from $f \equiv 1$. Similarly,

$$\operatorname{ord}_{D_n}(x_j) = \begin{cases} -1, & j = n+1, \\ 0, & \text{otherwise}. \end{cases}$$

Now suppose (4.2) is the lowest monomial of f. Since f is divisible by $x_{2n}^{p_{2n}}$, we have $\operatorname{ord}_{D_1}(f) \leqslant -p_{2n}$, whence $p_{2n} \leqslant \Lambda_1 - \Lambda_2$.

Expand the polynomial f in powers of x_{2n}:

$$f = \sum_{k \geqslant p_{2n}} x_{2n}^k f_k, \qquad f_k \in \mathbb{C}[x_2, \ldots, x_{2n-1}].$$

On $s_2 O$ we have $x_{2n} = y_{1,2n}$ and

$$f = \sum_{k \geqslant p_{2n}} y_{1,2n}^k \tilde{f}_k, \qquad \tilde{f}_k \in \mathbb{C}[y_{1,2}, \ldots, y_{1,2n-1}][y_{23}, y_{23}^{-1}].$$

Because of the factors $y_{1,2n}^k$, the poles of polynomials $\tilde{f}_k$ cannot cancel; therefore, for each polynomial f_k, we have $\operatorname{ord}_{D_2}(f_k) \geqslant \operatorname{ord}_{D_2}(f)$. In particular, $f_{p_{2n}}$ is divisible by $x_{2n-1}^{p_{2n-1}}$. Hence

$$p_{2n-1} \leqslant -\operatorname{ord}_{D_2}(f) \leqslant \Lambda_2 - \Lambda_3.$$

Continuing in the same way, we get

(5.1) $\qquad p_{2n-2} \leqslant -\operatorname{ord}_{D_3}(f) \leqslant \Lambda_3 - \Lambda_4, \ \ldots, \ p_{n+1} \leqslant -\operatorname{ord}_{D_n}(f) \leqslant \Lambda_n,$

which means that $\Lambda_1 \geqslant \eta_1 \geqslant \Lambda_2 \geqslant \cdots \geqslant \Lambda_n \geqslant \eta_n \geqslant 0$.

Now let $\prod x_i^{q_i}$ be any monomial from f. Since f is a weight vector for $\mathcal{B}$, we have

$$p_i - p_{2n-i+1} = q_i - q_{2n-i+1}, \qquad i = 2, \ldots, n.$$

Suppose $p_i - p_{2n-i+1} = s > 0$ for some i. Then f is divisible by x_i^s and by (5.1) we have

$$\operatorname{ord}_{D_{i-1}}(f) \leqslant -\max(s,0) - p_{2n-(i-1)+1}.$$

This implies

$$\max(p_i - p_{2n-i+1}, 0) + p_{2n-i+2} \leqslant \Lambda_{i-1} - \Lambda_i \quad \text{and} \quad \max(\Lambda_i, \lambda_{i-1}) \leqslant \eta_{i-1},$$

which is exactly (3.6).

To conclude the proof of the theorem, consider the fundamental representations of $Sp(2n)$

$$\pi_k : Sp(2n) \to GL(\wedge^k \mathbb{C}^{2n}), \qquad k = 1, \ldots, n.$$

Their matrix coefficients

$$\begin{array}{ll} 1, x_2, x_{2n}, & k = 1, \\ 1, x_{k+1}, x_{2n-k+1}, x_1 x_{2n} + \cdots + x_k x_{2n-k+1}, & 1 < k < n, \\ 1, x_{n+1}, x_1 x_{2n} + \cdots + x_n x_{n+1}, & k = n, \end{array}$$

clearly have only poles of order $\leqslant 1$ on D_k. Consider the lowest monomials of these polynomials:

$$\begin{array}{ll} 1, x_2, x_{2n}, & k = 1, \\ 1, x_{k+1}, x_{2n-k+1}, x_k x_{2n-k+1}, & 1 < k < n, \\ 1, x_{n+1}, x_n x_{n+1}, & k = n. \end{array}$$

It is easy to check that by multiplying these monomials we get all the lowest monomials (3.4) in V^Λ satisfying (3.6).

The weight of each monomial (3.4) can be easily found from (2.3). $\qquad \square$

Proof of Theorem 2. Set $\mathbb{K} = \mathbb{C}[x_{ij}]_{2 \leq i < j, i+j \leq 2n+1}$. Consider the $\mathbb{Z}^{2n-1}$-filtration of
$$\mathbb{C}[O] = \mathbb{K}[x_2, \ldots, x_{2n}],$$
which corresponds to the chain $X_1 \supset \cdots \supset X_{2n-1}$. This filtration is $Sp(2n-2)$-invariant and hence trivial on all irreducible $Sp(2n-2)$-submodules. Hence by Theorem 1
$$V^\lambda = \bigoplus_a V_a$$
where $a \in \mathbb{Z}^{2n-1}$ runs over the set described in theorem 1 and V_a is the $Sp(2n-2)$-irreducible submodule consisting of polynomials of the form
$$g = g' x_{12}^{a_{2n-1}} \cdots x_{1,2n}^{a_1} + \text{higher terms}, \qquad g' \in \mathbb{K}, \quad g' \neq 0.$$

Since $g' \neq 0$ for all $g \in V_a$, the polynomial $g' \in \mathbb{K}$ determines $g \in V_a$ uniquely. It is easy to see that the group $Sp(2n-2)$ acts on g' exactly as on elements of the irreducible $Sp(2n-2)$-module with highest weight $\lambda = (\lambda_1, \ldots, \lambda_{n-1})$, the polynomial $g' \equiv 1$ being the highest vector. Thus we can use the induction hypothesis to obtain all possible lowest monomials in V^Λ. The weight of each monomial is readily seen from (2.3). □

Appendix. Remark on the log-concavity of multiplicities

Consider a representation of a connected reductive group $\mathcal{G} \to GL(V)$ and a $\mathcal{G}$-stable irreducible subvariety

(A.1) $$X \subset \mathbb{P}(V).$$

For fixed $\mathcal{G}$ and X, all embeddings (A.1) form a semigroup with respect to the tensor product of $\mathcal{G}$-modules V (in other words, with respect to the product of the corresponding very ample sheaves in the group

(A.2) $$\text{Pic}^{\mathcal{G}}(X)$$

of $\mathcal{G}$-linearized invertible sheaves). Let $\mathcal{O}$ be the sheaf corresponding to (A.1). For simplicity, assume that the group (A.2) is finitely generated. This assumption is not essential. It is clearly satisfied for flag varieties.

Consider the $\mathcal{G}$-module
$$H^0(X, \mathcal{O}) = \bigoplus_\lambda \mu(\mathcal{O}, \lambda) V^\lambda,$$
where $\mu(\mathcal{O}, \lambda)$ is the multiplicity of the irreducible $\mathcal{G}$-module V^λ. Suppose that m is sufficiently divisible,
$$m = m_0!, \qquad m_0 = 1, 2, \ldots.$$

The number $\mu(\mathcal{O}^{\otimes m}, m\lambda)$ grows as a polynomial in m, say as m^r, for general $\mathcal{O}$ and λ. By definition, let us put

(A.3) $$\mu_\infty(\mathcal{O}, \lambda) = \lim \frac{\mu(\mathcal{O}^{\otimes m}, m\lambda)}{m^r}, \qquad m \to \infty.$$

By construction, (A.3) is a well-defined function on the cone
$$A \subset \text{Pic}^{\mathcal{G}}(X) \otimes_\mathbb{Z} \mathbb{Q}$$

generated by the ample sheaves in the $\mathbb{Q}$-vector space $\text{Pic}^{\mathcal{G}}(X) \otimes_{\mathbb{Z}} \mathbb{Q}$. By our assumption this vector space is finite-dimensional.

Let $\mathcal{B} \subset G$ be a Borel subgroup. There always exists a $\mathcal{B}$-stable chain of closed irreducible subvarieties of X

$$X = X_0 \supset X_1 \supset \cdots \supset X_{\dim X}, \quad \text{codim } X_i = i.$$

Denote by v the corresponding valuation $\mathbb{C}(X) \to \mathbb{Z}^{\dim X}$. This valuation is also defined for sections of any line bundle on X.

Recall the following definition of the Newton polytope of X corresponding to the sheaf $\mathcal{O}$:

$$\Delta(X, \mathcal{O}) = \text{Closure of } \{a \mid \exists m \, \exists f \in H^0(X, \mathcal{O}^{\otimes m})^{\mathcal{N}}, v(f)/m = a\} \subset \mathbb{R}^{\dim X}.$$

Here $(\cdots)^{\mathcal{N}}$ denotes invariants of the maximal unipotent subgroup $\mathcal{N} \subset \mathcal{B}$. By construction, $\Delta(X, \mathcal{O})$ is well defined for all $\mathcal{O} \in A$.

LEMMA. *For any $r \in \mathbb{Q}$ and $\mathcal{O} \in A$ we have $\Delta(X, \mathcal{O}^{\otimes r}) = r\Delta(X, \mathcal{O})$.*

PROOF. It suffices to prove that $\Delta(X, \mathcal{O}^{\otimes r}) \supset r\Delta(X, \mathcal{O})$. Suppose $r = p/q$ and $f \in H^0(X, \mathcal{O}^{\otimes m})^{\mathcal{N}}$, $v(f)/m = a$. Then $f^{\otimes p} \in H^0(X, (\mathcal{O}^{\otimes p/q})^{\otimes mq})$ and

$$v(f^{\otimes p})/mq = pv(f)/mq = ra. \quad \square$$

Consider the following set:

$$\widetilde{\Delta}(X) = \text{Closure of } \{(\mathcal{O}, \Delta(X, \mathcal{O})) \mid \mathcal{O} \in A\} \subset \bar{A} \oplus \mathbb{R}^{\dim X},$$

where $\bar{A}$ is the closure of A. By the lemma, $\widetilde{\Delta}(X)$ is a cone.

PROPOSITION. *The cone $\widetilde{\Delta}(X)$ is convex.*

PROOF. Suppose $\mathcal{O}_1, \mathcal{O}_2 \in A$ and

$$f_i \in H^0(X, \mathcal{O}_i^{\otimes m_i})^{\mathcal{N}}, \quad v(f_i)/m_i = a_i, \quad i = 1, 2.$$

Take two arbitrary numbers $k_1, k_2 \in \mathbb{Z}_+$, $k_1 + k_2 > 0$. Then $f_1^{\otimes k_1} \otimes f_2^{\otimes k_2}$ is a nonzero (because X is irreducible) global $\mathcal{N}$-invariant section of

$$(\mathcal{O}_1^{\otimes k_1 m_1/(k_1 m_1 + k_2 m_2)} \otimes \mathcal{O}_2^{\otimes k_2 m_2/(k_1 m_1 + k_2 m_2)})^{\otimes (k_1 m_1 + k_2 m_2)}$$

and

$$\frac{v(f_1^{\otimes k_1} \otimes f_2^{\otimes k_2})}{k_1 m_1 + k_2 m_2} = \frac{k_1 m_1}{(k_1 m_1 + k_2 m_2)} \frac{v(f_1)}{m_1} + \frac{k_2 m_2}{(k_1 m_1 + k_2 m_2)} \frac{v(f_2)}{m_2}.$$

It is clear that the corresponding points of $\widetilde{\Delta}(X)$ are dense in the segment

$$[(\mathcal{O}_1, v(f_1)/m_1), (\mathcal{O}_2, v(f_2)/m_2)],$$

which, therefore, lies in $\widetilde{\Delta}(X)$. $\square$

Now suppose that $f \in H^0(X, \mathcal{O}^{\otimes m})^{\mathcal{N}}$, $v(f)/m = a$, is $\mathcal{B}$-semi-invariant and denote by $\nu_{\mathcal{O}}(a)$ the $\mathcal{B}$-weight of f. As in [**Ok2**], it is easy to see that it is a well-defined function, linear in $\mathcal{O}$ and a. In [**Ok2**] we proved that

$$\mu_{\infty}(\mathcal{O}, \lambda) = \text{vol}(\nu_{\mathcal{O}}^{-1}(\lambda) \cap \Delta(X, \mathcal{O})).$$

Since the map $(\mathcal{O}, a) \mapsto (\mathcal{O}, \nu_{\mathcal{O}}(a))$ is linear, we have the following theorem.

THEOREM 3. *The function $\log \mu_\infty(\mathcal{O}, \lambda)$ is concave in $\mathcal{O}$ and λ.*

Taking for X the flag variety G/B of G, we obtain the following:

COROLLARY. *The function $\log \mu_\infty(\Lambda, \lambda)$ defined in (1.7) is concave in Λ and λ.*

References

[AFS] A. Alekseev, L. Faddeev, and S. Shatashvili, *Quantization of symplectic orbits of compact Lie groups by means of the functional integral*, J. Geom. Phys. **5** (1988), 391–406.

[BGG] I. N. Bernstein, I. M. Gelfand, and S. I. Gelfand, *Schubert cells and cohomology of the spaces G/P*, Uspekhi Mat. Nauk **28** (1973), no. 3, 3–26; English transl. in Russian Math. Surveys **28** (1973).

[BZ] A. Berenstein and A. Zelevinsky, *Tensor product multiplicities and convex polytopes in partition space*, J. Geom. Phys. **5** (1988), 453–472.

[D] C. de Concini, *Symplectic standard tableaux*, Adv. Math. **34** (1979), 1–27.

[Gr] W. Graham, *Logarithmic convexity of push-forward measures*, Invent. Math. **123** (1996), 315–322.

[GS] V. Guillemin and S. Sternberg, *Geometric quantization and multiplicities of group representations*, Invent. Math **67** (1982), 515–538.

[H] G. J. Heckmann, *Projections of orbits and asymptotic behavior of multiplicities for compact connected Lie groups*, Invent. Math **67** (1982), 333–356.

[La] V. Lakshmibai, *Geometry of G/P. VII. The symplectic group and the involution σ*, J. Algebra **108** (1987), 403–434.

[Le] K. Leichtweiss, *Konvexe Mengen*, Springer-Verlag, Berlin–Heidelberg–New York, 1980.

[LL] L. D. Landau and E. M. Lifshitz, *Statistical physics*, Pergamon Press, Oxford–New York, 1969.

[K1] A. A. Kirillov, *Elements of the theory of representations*, Grundlehren der Mathematischen Wissenschaften, Band 220, Springer-Verlag, Berlin–New York, 1976.

[K2] _____, *A remark on the Gelfand–Tsetlin patterns for symplectic groups*, J. Geom. Phys. **5** (1988), 473–482.

[Ok1] A. Okounkov, *Log-concavity of multiplicities with application to characters of $U(\infty)$*, Adv. Math. (to appear).

[Ok2] _____, *Brunn–Minkowski inequality for multiplicities*, Invent. Math. **125** (1996), 405–411.

[Ok3] _____, *Newton polytopes for spherical varieties*, Preprint (1994).

[Sh1] V. V. Shtepin, *Splitting of multiple points of the spectrum in the reduction $Sp(2n) \downarrow Sp(2n-2)$*, Funktsional. Anal. i Prilozhen. **20** (1986), no. 4, 93–95; English transl. in Functional Anal. Appl. **20** (1986).

[Sh2] _____, *On a class of finite dimensional $Sp(2n-1)$-modules*, Uspekhi Mat. Nauk **41** (1986), no. 3, 207–208; English transl. in Russian Math. Surveys **41** (1986).

[Zh] D. P. Zhelobenko, *Classical groups. The spectral analysis of finite-dimensional representations*, Uspekhi Mat. Nauk **17** (1962), no. 1, 27–120; English transl. in Russian Math. Surveys **17** (1962).

DEPARTMENT OF MATHEMATICS, UNIVERSITY OF CHICAGO, 5734 SOUTH UNIVERSITY AVENUE, CHICAGO, ILLINOIS 60637-1546

E-mail address: okounkov@msri.org

Shifted Schur functions II.
The Binomial Formula for Characters of Classical Groups and its Applications

Andreĭ Okounkov and Grigori Olshanski

To A. A. Kirillov on his 60th birthday

ABSTRACT. Let G be any of the complex classical groups $GL(n)$, $SO(2n+1)$, $Sp(2n)$, $O(2n)$, $\mathfrak{g}$ the Lie algebra of G, and $Z(\mathfrak{g})$ the subalgebra of G-invariants in the universal enveloping algebra $U(\mathfrak{g})$. We derive a Taylor-type expansion for finite-dimensional characters of G (the binomial formula) and use it to specify a distinguished linear basis in $Z(\mathfrak{g})$. The eigenvalues of the basis elements in highest weight $\mathfrak{g}$-modules are certain shifted (or factorial) analogs of Schur functions. We also study the associated homogeneous basis in $I(\mathfrak{g})$, the subalgebra of G-invariants in the symmetric algebra $S(\mathfrak{g})$. Finally, we show that both bases are related by a G-equivariant linear isomorphism $\sigma\colon I(\mathfrak{g}) \to Z(\mathfrak{g})$, called special symmetrization.

Introduction

The present paper is a continuation of our previous work [OO1] but can be read independently. Here we aim to carry over to orthogonal and symplectic groups some of the results obtained in [OO1] for the general linear group.

Let G be one of the groups $SO(2n+1,\mathbb{C})$, $Sp(2n,\mathbb{C})$,[1] $\mathfrak{g}$ the Lie algebra of G, $U(\mathfrak{g})$ the universal enveloping algebra of $\mathfrak{g}$, and $Z(\mathfrak{g})$ the center of $U(\mathfrak{g})$. The commutative algebra $Z(\mathfrak{g})$ is also known as the algebra of bi-invariant differential operators on the group G, or the algebra of Laplace operators.

Irreducible finite-dimensional representations of the group G are indexed by partitions λ of length $\leqslant n$; we denote them by V_λ, and we view each V_λ also as a $U(\mathfrak{g})$-module.

Our main object is a distinguished linear basis $\{\mathbb{T}_\mu\} \subset Z(\mathfrak{g})$. Here the index μ ranges over partitions of length $\leqslant n$, and each basis element $\mathbb{T}_\mu$ can be characterized,

---

1991 *Mathematics Subject Classification*. Primary 17B10, 17B35; Secondary 05E15, 22E46.

The authors were supported by the Russian Foundation for Basic Research (grant 95-01-00814). The first author was supported by the NSF, grant DMS 9304580.

[1]To simplify the discussion, in the present Introduction we exclude the even orthogonal group $SO(2n,\mathbb{C})$. However, after minor modifications, all our constructions hold for this group as well.

up to a scalar multiple, as the unique element in $Z(\mathfrak{g})$ of degree $2|\mu|$ satisfying the following *vanishing condition*: the eigenvalue of $\mathbb{T}_\mu$ in V_λ equals 0 if $|\lambda| \leqslant |\mu|$ and $\lambda \neq \mu$ (moreover, it turns out that the eigenvalue is 0 if $\lambda_i < \mu_i$ for at least one i).

For a general λ, denote the eigenvalue of $\mathbb{T}_\mu$ in V_λ by $t_\mu^*(\lambda) = t_\mu^*(\lambda_1, \ldots, \lambda_n)$. The functions t_μ^* are inhomogeneous polynomials with a certain kind of symmetry; they can be identified with certain *factorial* analogs of the Schur polynomials. Note that the top homogeneous component t_μ of the inhomogeneous polynomial t_μ^* coincides with a Schur polynomial in the squares of the arguments:

$$t_\mu(x_1, \ldots, x_n) = s_\mu(x_1^2, \ldots, x_n^2).$$

The polynomials t_μ^* arise in the *binomial formula* for the characters of G. Denote by $\chi_\lambda = \chi_\lambda(z_1, \ldots, z_n)$ the character of V_λ; here $z_1, \ldots, z_n$ are natural coordinates of a maximal torus of G. Assume that the 'additive' variables $x_1, \ldots, x_n$ are connected with the 'multiplicative' variables $z_1, \ldots, z_n$ by the relation

$$x_i = z^{1/2} - z^{-1/2}, \quad \text{i.e.,} \quad x_i^2 = z + z^{-1} - 2.$$

The binomial formula is written as

$$\frac{\chi_\lambda(z_1, \ldots, z_n)}{\chi_\lambda(1, \ldots, 1)} = \sum_\mu \frac{t_\mu^*(\lambda_1, \ldots, \lambda_n) t_\mu(x_1, \ldots, x_n)}{c_\pm(n, \mu)},$$

where $c_\pm(n, \mu)$ are simple normalization factors. This is a Taylor-type expansion for the character near the identity element of the group.[2]

One more important property of the basis elements $\mathbb{T}_\mu$ is the existence of a relation between the bases corresponding to different values of the parameter n. We call this the *coherence property* of the canonical basis.

Then we study a counterpart of $\{\mathbb{T}_\mu\}$ for the algebra $I(\mathfrak{g})$, the subalgebra of G-invariants in the symmetric algebra $S(\mathfrak{g})$. Note that $I(\mathfrak{g})$ is a graded algebra which is canonically isomorphic to $\operatorname{gr} Z(\mathfrak{g})$, the graded algebra associated to the natural filtration in $Z(\mathfrak{g})$.

To the basis $\{\mathbb{T}_\mu\}$ corresponds a homogeneous basis $\{T_\mu\}$ of $I(\mathfrak{g})$: each T_μ coincides with the leading term of $\mathbb{T}_\mu$ (with respect to the identification $I(\mathfrak{g}) = \operatorname{gr} Z(\mathfrak{g})$). On the other hand, the basis $\{T_\mu\}$ can be described without reference to $Z(\mathfrak{g})$, in intrinsic terms of the G-module $S(\mathfrak{g})$. Note that the elements T_μ admit an explicit expression as polynomials in the natural generators of the Lie algebra $\mathfrak{g}$.

Finally, we show that there exists a G-equivariant linear isomorphism

$$\sigma \colon S(\mathfrak{g}) \to U(\mathfrak{g})$$

which preserves leading terms and takes T_μ to $\mathbb{T}_\mu$ for any μ. We call σ the *special symmetrization* for the orthogonal/symplectic Lie algebras.

The special symmetrization σ is contained in a family of 'generalized symmetrizations' studied in the paper [**O2**]. The results of [**O2**] provide explicit combinatorial formulas both for σ and for its inverse σ^{-1}. Since we have at our disposal an explicit expression for T_μ, this yields a certain formula for the elements $\mathbb{T}_\mu = \sigma(T_\mu)$.

We conclude this Introduction with a brief discussion of some related works.

---

[2] Note that the binomial formula for characters of classical groups of type B, C, D is a particular case of an expansion of type BC_n Jacobi polynomials (see Lassalle [**Lass**]).

For the general linear group $GL(n,\mathbb{C})$, the binomial theorem and distinguished bases $\{\mathbb{S}_\mu\} \subset Z(\mathfrak{gl}(n,\mathbb{C}))$ and $\{S_\mu\} \subset I(\mathfrak{gl}(n,\mathbb{C}))$ with similar properties were considered earlier in [**OO1**]. The elements $\mathbb{S}_\mu$, which are called *quantum immanants*, appear in a higher version of the classical Capelli identity and admit a remarkable explicit expression (see [**Ok1, N, Ok2**]). The eigenvalues of the elements $\mathbb{S}_\mu$ in highest weight modules are described by certain polynomials $s^*_\mu(\lambda_1,\ldots,\lambda_n)$, called *shifted Schur polynomials* in [**OO1**]. Both families of polynomials, $\{s^*_\mu\}$ and $\{t^*_\mu\}$, have similar properties.[3]

Let μ be of the form (1^m) for the orthogonal group or of the form (m) for the symplectic group ($m = 1, 2, \ldots$). Then the corresponding elements $\mathbb{T}_\mu$ coincide with Laplace operators considered by Molev and Nazarov in [**MN**]. As shown in [**MN**], these elements also occur in a Capelli-type identity and admit an explicit expression in terms of the generators of the Lie algebra. An open problem is to generalize the results of [**MN**] to arbitrary μ. Note that the explicit formulas found in [**MN, Ok1, N, Ok2**] differ from the formulas obtained via special symmetrization.

Many of the results of [**OO1**] and the present paper should have counterparts for classical Lie superalgebras $\mathfrak{gl}(p|q)$, $\mathfrak{q}(n)$, and $\mathfrak{osp}(n|2m)$. Some work in this direction was done by Borodin and Rozhkovskaya [**BR**], Molev [**Mo**], Ivanov and Okounkov (see [**I**]).

In the present work (as in [**OO1**]), we use very elementary tools, such as explicit formulas for the (ordinary and factorial) Schur polynomials and the Weyl character formula. We believe our approach can be developed further by making use of finer techniques. In particular, there exist more involved versions of the binomial formula (see [**OO2**] and [**Ok3**]).

§0. Main notation

Unless otherwise stated, the symbol $G(n)$ or G will denote any of the complex classical groups of rank n

$$GL(n,\mathbb{C}), \quad Sp(2n,\mathbb{C}), \quad SO(2n+1,\mathbb{C}), \quad SO(2n,\mathbb{C}),$$

which constitute the series A, C, B, D, respectively. By $\mathfrak{g}(n)$ or $\mathfrak{g}$ we denote the corresponding complex Lie algebras. In the case of the series D, we also consider the nonconnected groups $G' = G'(n) = O(2n,\mathbb{C})$.

By $U(\mathfrak{g})$ we denote the universal enveloping algebra of $\mathfrak{g}$. For the series A, C, B, we denote by $Z(\mathfrak{g})$ the center of $U(\mathfrak{g})$ (which coincides with the subalgebra of G-invariants in $U(\mathfrak{g})$); for the series D the same symbol will denote the subalgebra of G'-invariants in $U(\mathfrak{g})$ (which is a proper subalgebra of the center).

Similarly, we define $I(\mathfrak{g}) \subset S(\mathfrak{g})$ as the subalgebra of G-invariants (or G'-invariants, for the series D) in the symmetric algebra of $\mathfrak{g}$.

V_λ or $V_{\lambda|n}$ is the irreducible finite-dimensional complex-analytic representation of G with highest weight λ; it is also viewed as a $U(\mathfrak{g})$-module.

$\mathbb{C}[G]$ is the space of regular functions on G, where G is viewed as an algebraic group over $\mathbb{C}$; in other words, $\mathbb{C}[G]$ is the linear span of matrix elements of all representations V_λ.

By μ we always denote a partition of length $l(\mu) \leq n$; it is also viewed as a Young diagram; $|\mu|$ denotes the number of boxes of the diagram μ.

---

[3]However, in contrast to the polynomials s^*_μ, the polynomials t^*_μ are not stable as $n \to \infty$.

§1. Binomial formula

Here we derive a multidimensional analog of the binomial formula

$$(1+x)^k = \sum_{m=0}^{k} \frac{1}{m!} k(k-1)\cdots(k-m+1) x^m,$$

where the powers of a variable will be replaced by characters of G.

Let us agree about the choice of a Borel subgroup $B \subset G$ and a maximal torus $H \subset B$ (thus positive roots and dominant weights will be specified):

• For the series A, B is the subgroup of upper triangular matrices and H is the subgroup of diagonal matrices,

$$H = \{\mathrm{diag}(z_1,\ldots,z_n)\}, \qquad z_1,\ldots,z_n \in \mathbb{C}^*.$$

• For the series C–B–D we identify G with a subgroup in $GL(N,\mathbb{C})$, where $N = 2n$ or $N = 2n+1$,

(1.1) $$G = \{g \in GL(N,\mathbb{C}) \mid g'Mg = M\},$$

where M stands for the following symmetric (case B–D) or antisymmetric (case C) matrix of order N:

(1.2) $$M = \begin{bmatrix} & & & & 1 \\ & 0 & & \cdot\cdot\cdot & \\ & & & 1 & \\ & & \pm 1 & & \\ & \cdot\cdot\cdot & & 0 & \\ \pm 1 & & & & \end{bmatrix}.$$

Then for B and H we take the subgroups of upper triangular or diagonal matrices in G. Thus

$$H = \{\mathrm{diag}(z_1,\ldots,z_n,z_n^{-1},\ldots,z_1^{-1})\}, \qquad \text{cases C–D},$$
$$H = \{\mathrm{diag}(z_1,\ldots,z_n,1,z_n^{-1},\ldots,z_1^{-1})\}, \qquad \text{case B}.$$

For all the series, the weights of H are of the form

$$z^\lambda = z_1^{\lambda_1}\cdots z_n^{\lambda_n}, \quad \text{where } \lambda = (\lambda_1,\ldots,\lambda_n) \in \mathbb{Z}^n,$$

while the dominant weights are distinguished by the supplementary conditions

$$\lambda_1 \geq \cdots \geq \lambda_n, \qquad \text{case A},$$
$$\lambda_1 \geq \cdots \geq \lambda_n \geq 0, \qquad \text{cases C–B},$$
$$\lambda_1 \geq \cdots \geq \lambda_{n-1} \geq |\lambda_n|, \qquad \text{case D}.$$

The dominant weights for the series A are called the *signatures* and those for the series C–B are called the *positive signatures* (=partitions of length $\leq n$).

The irreducible character of G corresponding to a dominant weight λ (i.e., the character of $V_\lambda = V_{\lambda|n}$) will be denoted by the symbol

$$\chi_\lambda = \chi_\lambda^{gl(n)}, \quad \chi_\lambda^{sp(2n)}, \quad \chi_\lambda^{so(2n+1)} \quad \text{or} \quad \chi_\lambda^{so(2n)}.$$

For the series D, it is more convenient to deal with the *reducible* character
$$\chi_\lambda^{o(2n)} = \chi_{\lambda_1,\ldots,\lambda_n}^{so(2n)} + \chi_{\lambda_1,\ldots,\lambda_{n-1},-\lambda_n}^{so(2n)}, \quad \text{where } \lambda_1 \geqslant \cdots \geqslant \lambda_n \geqslant 0.$$

When $\lambda_n = 0$, this is simply $2\chi_\lambda^{so(2n)}$, and when $\lambda_n > 0$, this coincides with the restriction to the group $G = SO(2n, \mathbb{C})$ of an irreducible character of the group $G' = O(2n, \mathbb{C})$.

Thus, we are working with the characters
$$\chi_\lambda^{gl(n)}, \quad \chi_\lambda^{sp(2n)}, \quad \chi_\lambda^{so(2n+1)}, \quad \chi_\lambda^{o(2n)},$$
where λ is either a signature (case A) or a positive signature (case C–B–D).

To each λ we assign another weight $l = (l_1, \ldots, l_n)$ defined as

(1.3) $\quad l = (\lambda_1 + n - 1, \lambda_2 + n - 2, \ldots, \lambda_n), \qquad$ case A,

(1.4) $\quad l = (\lambda_1 + n - 1 + \varepsilon, \lambda_2 + n - 2 + \varepsilon, \ldots, \lambda_n + \varepsilon), \qquad$ case C–B–D,

where

(1.5) $$\varepsilon = \begin{cases} 1, & \text{case C}, \\ 1/2, & \text{case B}, \\ 0, & \text{case D}. \end{cases}$$

For the series C–B–D, l simply equals λ plus the half-sum of positive roots, and for the series A, the same holds after restriction to $SL(n, \mathbb{C}) \subset GL(n, \mathbb{C})$.

Unless otherwise stated, we shall regard characters of G as functions on the torus H, i.e., as functions of the coordinates $z_1, \ldots, z_n$. The following formulas follow from the general Weyl character formula (the determinants below are of order n):

$$\chi_\lambda^{gl(n)}(z_1, \ldots, z_n) = \frac{\det[z_j^{l_i}]}{\prod_{i<j}(z_i - z_j)},$$

$$\chi_\lambda^{sp(2n)}(z_1, \ldots, z_n) = \frac{\det[(z_j^{l_i} - z_j^{-l_i})/(z_j - z_j^{-1})]}{\prod_{i<j}(z_i + z_i^{-1} - z_j - z_j^{-1})},$$

$$\chi_\lambda^{so(2n+1)}(z_1, \ldots, z_n) = \frac{\det[(z_j^{l_i} - z_j^{-l_i})/(z_j^{1/2} - z_j^{-1/2})]}{\prod_{i<j}(z_i + z_i^{-1} - z_j - z_j^{-1})},$$

$$\chi_\lambda^{o(2n)}(z_1, \ldots, z_n) = \frac{\det[z_j^{l_i} + z_j^{-l_i}]}{\prod_{i<j}(z_i + z_i^{-1} - z_j - z_j^{-1})}.$$

Note that if λ is a partition, then $\chi_\lambda^{gl(n)}(z_p, \ldots, z_n) = s_\lambda(z_1, \ldots, z_n)$, the Schur polynomial in n variables. The characters $\chi_\lambda^{gl(n)}$ are symmetric Laurent polynomials in $z_1, \ldots, z_n$, whereas the other characters are symmetric polynomial functions of the variables $z_i + z_i^{-1}$, $i = 1, \ldots, n$. Note that the last claim fails when $\chi_\lambda^{o(2n)}$ is replaced by $\chi_\lambda^{so(2n)}$; this is the reason why we deal with the characters $\chi_\lambda^{o(2n)}$.

Let $a = (a_1, a_2, \ldots)$ be an arbitrary number sequence. The *generalized factorial powers* of a variable x are defined as
$$(x|a)^k = \begin{cases} (x - a_1) \cdots (x - a_k), & k \geqslant 1, \\ 1, & k = 0. \end{cases}$$

When $a \equiv 0$ these are the ordinary powers, when $a = (0, 1, 2, \dots)$ these are the *falling factorial powers*, and when $a = (0, -1, -2, \dots)$ these are the *rising factorial powers* (or the Pohgammer symbol).

Further, the *generalized factorial Schur polynomial* in n variables, indexed by a partition μ with $l(\mu) \leqslant n$, is defined as

$$s_\mu(z_1, \dots, x_n \,|\, a) = \frac{\det\left[(x_j\,|\,a)^{\mu_i+n-i}\right]}{\det\left[(x_j\,|\,a)^{n-i}\right]} = \frac{\det\left[(x_j\,|\,a)^{\mu_i+n-i}\right]}{\prod_{i<j}(x_i - x_j)}.$$

When $a \equiv 0$ this turns into the standard formula for the ordinary Schur polynomial $s_\mu(x_1, \dots, x_n)$, and in the general case $s_\mu(x_1, \dots, x_n \,|\, a)$ is an *inhomogeneous* symmetric polynomial of degree $|\mu| = \mu_1 + \cdots + \mu_n$ with the top homogeneous component equal to $s_\mu(x_1, \dots, x_n)$:

$$s_\mu(x_1, \dots, x_n \,|\, a) = s_\mu(x_1, \dots, x_n) + \text{lower terms}.$$

It follows that the polynomials $s_\mu(x_1, \dots, x_n\,|\,a)$ form a basis in the algebra of symmetric polynomials in n variables.

The polynomials $s_\mu(x_1, \dots, x_n\,|\,0, 1, \dots)$ were introduced by Biedenharn and Louck and called the *factorial Schur polynomials* (see [**BL1, BL2**]). The general definition is due to Macdonald (see [**M1**] and [**M2**, I, §3, ex. 20]; note that Macdonald uses the notation $(x\,|\,a)^k = (x + a_1) \cdots (x + a_k)$).

THEOREM 1.1 (Binomial formula for $GL(n)$). *Assume that*

$$z_1 = 1 + x_1, \quad \dots, \quad z_n = 1 + x_n.$$

Then

$$(1.6) \quad \frac{\chi_\lambda^{gl(n)}(z_1, \dots, z_n)}{\chi_\lambda^{gl(n)}(1, \dots, 1)} = \sum_\mu \frac{s_\mu(l_1, \dots, l_n\,|\,0, 1, 2, \dots)\, s_\mu(x_1, \dots, x_n)}{c(n, \mu)}.$$

Here $\lambda = (\lambda_1, \dots, \lambda_n)$ *is an arbitrary signature,* $l_i = \lambda_i + n - i$ $(i = 1, \dots, n)$, μ *ranges over partitions of length* $\leqslant n$, *and*

$$(1.7) \quad c(n, \mu) = \prod_{(i,j)\in\mu} (n + j - i) = \prod_{i=1}^n \frac{(\mu_i + n - i)!}{(n - i)!},$$

where the first product is taken over all boxes (i, j) *of the Young diagram representing the partition* μ.

COMMENT. This version of the binomial formula was proposed by the authors (see [**OO1**, Theorem 5.1]). It is equivalent to the expansion

$$s_\lambda(1 + x_1, \dots, 1 + x_n) = \sum_\mu d_{\lambda\mu} s_\mu(x_1, \dots, x_n),$$

where the coefficients $d_{\lambda\mu}$ are given by

$$d_{\lambda\mu} = \det\left[\binom{\lambda_i + n - i}{\mu_j + n - j}\right]_{1 \leqslant i, j \leqslant n}$$

(see Lascoux [**Lasc**], Macdonald [**M2**, I, §3, ex. 10]).

Although the latter formula looks simpler than (1.6), formula (1.6) has a number of advantages, as explained in our paper [**OO1**]. E.g., an important idea is to consider $l_1, \ldots, l_n$ as variables rather than parameters, and then formula (1.6) makes evident the fact that the roles of $l_1, \ldots, l_n$ and of $x_1, \ldots, x_n$ are almost symmetric.

THEOREM 1.2 (Binomial formula for the series C–B–D). *Assume the variables z_i and x_i are subject to the relations*

$$x_i^2 = z_i + z_i^{-1} - 2, \quad i.e., \quad x_i = \pm(z_i^{1/2} - z_i^{-1/2}), \qquad i = 1, \ldots, n.$$

Let G be any of the groups of the series C, B, D, and let us use the abbreviation χ_λ for $\chi_\lambda^{sp(2n)}$, $\chi_\lambda^{so(2n+1)}$ or $\chi_\lambda^{o(2n)}$, where λ is a partition of length $\leq n$. Then we have

$$(1.8) \qquad \frac{\chi_\lambda(z_1, \ldots, z_n)}{\chi_\lambda(1, \ldots, 1)} = \sum_\mu \frac{s_\mu(l_1^2, \ldots, l_n^2 \mid \varepsilon^2, (\varepsilon+1)^2, \ldots) s_\mu(x_1^2, \ldots, x_n^2)}{c_\pm(n, \mu)}.$$

Here $l_i = \lambda_i + n - i + \varepsilon$, ε is defined by (1.5), μ ranges over partitions of length $\leq n$, and

$$(1.9) \qquad c_\pm(n, \mu) = \prod_{(i,j) \in \mu} 4(n + j - i)(n \pm 1/2 + j - i),$$

where the sign "+" is taken for the series C, B, while the sign "−" is taken for the series D.

Formula (1.8) is a particular case of the expansion of a type BC_n Jacobi polynomial with "Jack parameter" $\alpha = 1$ into a series of Schur polynomials. This expansion probably is well known to experts (see, e.g., Lassalle [**Lass**, Théorème 9]). For completeness we present a detailed proof.

PROOF. Consider the classical Jacobi polynomials $P_k^{(\alpha, \beta)}$ and put

$$\widetilde{P}_k^{(\alpha,\beta)} = P_k^{(\alpha,\beta)} / P_k^{(\alpha,\beta)}(1).$$

Let T_k and U_k be the Chebyshev polynomials of the first and second kind, respectively. Then for any $k = 0, 1, \ldots$

$$\frac{z^{k+1} - z^{-k-1}}{z - z^{-1}} = U_k\left(\frac{z + z^{-1}}{2}\right) \sim \widetilde{P}_k^{(1/2, 1/2)}\left(\frac{z + z^{-1}}{2}\right),$$

$$\frac{z^{k+1/2} - z^{-k-1/2}}{z^{1/2} - z^{-1/2}} \sim \widetilde{P}_k^{(1/2, -1/2)}\left(\frac{z + z^{-1}}{2}\right),$$

$$z^k + z^{-k} = 2T_k\left(\frac{z + z^{-1}}{2}\right) \sim \widetilde{P}_k^{(-1/2, -1/2)}\left(\frac{z + z^{-1}}{2}\right),$$

where the symbol "$\sim$" means equality up to a numerical factor depending on k but independent of z (see Szegö [**S**, (4.1.7) and (4.1.8)]).

Substituting these expressions into the formulas for the characters χ_λ given above, we obtain

$$(1.10) \qquad \chi_\lambda(z_1, \ldots, z_n) \sim \frac{\det[\widetilde{P}_{\lambda_i+n-i}^{(\alpha,\beta)}(1 + t_j)]_{1 \leq i,j \leq n}}{\prod_{i<j}(t_i - t_j)},$$

where
$$t_i = \frac{z_i + z_i^{-1}}{2} - 1, \quad i = 1, \ldots, n, \qquad (\alpha, \beta) = \begin{cases} (1/2, 1/2), & \text{case C,} \\ (1/2, -1/2), & \text{case B,} \\ (-1/2, -1/2), & \text{case D,} \end{cases}$$

and the symbol "$\sim$" means equality up to a factor not depending on $t_1, \ldots, t_n$.

Recall a well-known identity: for any power series
$$f_i(t) = \sum_{m=0}^{\infty} a_m^{(i)} t^m, \qquad i = 1, \ldots, n,$$

we have

(1.11) $$\frac{\det[f_i(t_j)]_{1 \leq i,j \leq n}}{\prod_{i<j}(t_i - t_j)} = \sum_{\mu} \det[a_{\mu_j+n-j}^{(i)}] s_\mu(t_1, \ldots, t_n),$$

where μ ranges over partitions of length $\leq n$ (see, e.g., [**Hua**, Theorem 1.2.1]).

We shall apply this identity to
$$f_i(t) = \widetilde{P}_{\lambda_i+n-i}^{(\alpha,\beta)}(1+t), \qquad i = 1, \ldots, n,$$

and we shall use the well-known expansion
$$\widetilde{P}_k^{(\alpha,\beta)}(1+t) = \sum_{m \geq 0} \frac{k(k-1)\cdots(k-m+1)(k+\alpha+\beta+1)\cdots(k+\alpha+\beta+m)}{2^m m!\,(\alpha+1)\cdots(\alpha+m)} t^m$$

(see, e.g., Szegö [**S**, (4.21.2)]).

Note that in our situation $\alpha + \beta + 1 = 2\varepsilon$, so that
$$(k-i)(k+\alpha+\beta+1+i) = (k+\varepsilon)^2 - (\varepsilon+i)^2, \qquad i = 0, 1, \ldots.$$

Further, note that $\alpha = \pm 1/2$ according to the assumption on the sign made in the statement of the theorem. It follows that the above expansion can be rewritten as

(1.12) $$\widetilde{P}_k^{(\alpha,\beta)}(1+t) = \sum_{m \geq 0} \frac{((k+\varepsilon)^2 \mid \varepsilon^2, (\varepsilon+1)^2, \ldots)^m t^m}{2^m m!\,(\pm 1/2+1)\cdots(\pm 1/2+m)}.$$

Formulas (1.10), (1.11), and (1.12) imply that
$$\chi_\lambda(z_1, \ldots, z_n)$$
$$\sim \sum_{\mu} \frac{\det[(l_i^2 \mid \varepsilon^2, (\varepsilon+1)^2, \ldots)^{\mu_j+n-j}] s_\mu(t_1, \ldots, t_n)}{\prod_{i=1}^{n}(2^{\mu_i+n-i}(\mu_i+n-i)!\,(\pm 1/2+1)\cdots(\pm 1/2+\mu_i+n-i))}.$$

The normalizing factor in this formula is determined by the condition that the constant term (i.e., the coefficient of $s_\varnothing(t_1, \ldots, t_n) \equiv 1$) should be equal to 1. After simple transformations, using the relations
$$s_\mu(t_1, \ldots, t_n) = 2^{-|\mu|} s_\mu(x_1^2, \ldots, x_n^2),$$
$$\frac{\det[(l_i^2 \mid \varepsilon^2, (\varepsilon+1)^2, \ldots)^{\mu_j+n-j}]}{\det[(l_i^2 \mid \varepsilon^2, (\varepsilon+1)^2, \ldots)^{n-j}]} = s_\mu(l_1^2, \ldots, l_n^2 \mid \varepsilon^2, (\varepsilon+1)^2, \ldots),$$

we obtain the desired formula. $\square$

§2. A distinguished basis in $Z(\mathfrak{g})$

The aim of this section is to construct and characterize a linear basis in $Z(\mathfrak{g})$. Our construction is suggested by the binomial formula of §1.

Let $\mathcal{O}_e(G)$ be the algebra of germs of holomorphic functions at the unity e of the group G and let $\mathcal{M}_e(G)$ be the maximal ideal of $\mathcal{O}_e(G)$ formed by germs vanishing at e. We may identify $U(\mathfrak{g})$ with the space of the linear functionals on $\mathcal{O}_e(G)$ that vanish on $\mathcal{M}_e(G)^m$, where m is large enough. This defines a nondegenerate pairing $\langle \cdot, \cdot \rangle$ between $U(\mathfrak{g})$ and $\mathbb{C}[G]$ (see §0 for the definition of $\mathbb{C}[G]$). If V is a finite-dimensional G-module (also viewed as a $U(\mathfrak{g})$-module), V^* is the dual module, $\xi \in V$ and $\eta \in V^*$ are arbitrary vectors, and $f_{\xi\eta}(g) = \eta(g\xi)$ is the corresponding matrix coefficient, then we have

$$\langle X, f_{\xi\eta} \rangle = \eta(X\xi) \quad \text{for any } X \in U(\mathfrak{g}).$$

Let $I(G) \subset \mathbb{C}[G]$ be the subalgebra of invariants of the group G (for series A, C, B) or the group G' (for series D) with respect to its action on G by conjugations; note that this definition is parallel to that of $Z(\mathfrak{g}) \subset U(\mathfrak{g})$, see §0.

Note that both $U(\mathfrak{g})$ and $\mathbb{C}[G]$ are semisimple modules over G or G'. It follows that there are canonical projections

(2.1) $\qquad\qquad\qquad \#\colon U(\mathfrak{g}) \to Z(\mathfrak{g}),$

(2.2) $\qquad\qquad\qquad \#\colon \mathbb{C}[G] \to I(G);$

these projections also can be defined as averaging over a compact form of the group G or G'. The pairing $\langle \cdot, \cdot \rangle$ over $U(\mathfrak{g})$ and $\mathbb{C}[G]$ is invariant over G or G' and so defines a nondegenerate pairing between $Z(\mathfrak{g})$ and $I(G)$.

LEMMA-DEFINITION 2.1. *Let μ range over the set of partitions of length $\leqslant n$. There exist central elements*

$$\mathbb{S}_\mu \in Z(\mathfrak{g}), \quad \deg \mathbb{S}_\mu \leqslant |\mu| \qquad (\text{series A}),$$
$$\mathbb{T}_\mu \in Z(\mathfrak{g}), \quad \deg \mathbb{T}_\mu \leqslant 2|\mu| \qquad (\text{series C, B, D}),$$

uniquely specified by the condition

$$(f^\# \big|_H)(z_1, \ldots, z_n) = \sum_\mu \frac{\langle \mathbb{S}_\mu, f \rangle s_\mu(x_1, \ldots, x_n)}{c(n, \mu)},$$

$$(f^\# \big|_H)(z_1, \ldots, z_n) = \sum_\mu \frac{\langle \mathbb{T}_\mu, f \rangle s_\mu(x_1^2, \ldots, x_n^2)}{c_\pm(n, \mu)},$$

where $H \subset G$ is the diagonal torus defined in §1, $f \in \mathbb{C}[G]$ stands for an arbitrary test function, $f^\# \in I(G)$ is the image of f under the projection (2.2), and the variables $x_1, \ldots, x_n$ and the normalizing factors $c(n, \mu)$, $c_\pm(n, \mu)$ are given by (1.6), (1.9).

PROOF. The function $f^\# \big|_H$ is a symmetric Laurent polynomial in $z_1, \ldots, z_n$; in the case of the series C, B, D, it is also invariant under the transformations $z_i \mapsto z_i^{-1}$. It follows that $f^\# \big|_H$ can be expanded into a series of polynomials $s_\mu(x_1, \ldots, x_n)$ or $s_\mu(x_1^2, \ldots, x_n^2)$, where $x_i = z_i - 1$ (for series A) or $x_i^2 = z_i + z_i^{-1} - 2$ (for series C, B, D). Given μ, the map assigning to f the μth coefficient in that expansion is a linear functional on $\mathbb{C}[G]$ which depends only on the finite jet of

f at the point $e \in G$ (the order of the jet is equal to $|\mu|$ or $2|\mu|$). Hence this functional is an element of the algebra $U(\mathfrak{g})$ of degree $\leqslant |\mu|$ or $\leqslant 2|\mu|$. This proves the existence of the elements $\mathbb{S}_\mu$ or $\mathbb{T}_\mu$. As will be clear in what follows, we have the exact equalities
$$\deg \mathbb{S}_\mu = |\mu|, \qquad \deg \mathbb{T}_\mu = 2|\mu|.$$
Finally, the elements $\mathbb{S}_\mu$, $\mathbb{T}_\mu$ belong to $Z(\mathfrak{g})$, because their values at any test function $f \in \mathbb{C}[G]$ depend on $f^{\#}$ only. $\square$

Recall that V_λ denotes the irreducible G-module with highest weight $\lambda = (\lambda_1, \ldots, \lambda_n)$, which is also viewed as a $U(\mathfrak{g})$-module. Any central element $X \in U(\mathfrak{g})$ acts in V_λ as a scalar operator $\text{const} \cdot 1$; the corresponding constant is called the *eigenvalue of the central element X in V_λ*.

THEOREM 2.2. *Let $G = GL(n, \mathbb{C})$. The elements $\mathbb{S}_\mu$ defined in Lemma 2.1 form a basis in $Z(\mathfrak{g})$ and are characterized (within a scalar multiple) by the following properties*:
  (i) $\deg \mathbb{S}_\mu \leqslant |\mu|$;
  (ii) *the eigenvalue of $\mathbb{S}_\mu$ in a module V_λ equals 0 for all partitions λ such that $l(\lambda) \leqslant n$, $|\lambda| \leqslant |\mu|$, $\lambda \neq \mu$;*
  (iii) *the eigenvalue of $\mathbb{S}_\mu$ in V_μ is nonzero.*

THEOREM 2.3. *Let G be a classical group of type C,B,D. The elements $\mathbb{T}_\mu$ defined in Theorem 2.1 form a basis in $Z(\mathfrak{g})$ and are characterized (within a scalar multiple) by the following properties*:
  (i) $\deg \mathbb{T}_\mu \leqslant 2|\mu|$;
  (ii) *the eigenvalue of $\mathbb{T}_\mu$ in a module V_λ equals 0 for all partitions λ such that $l(\lambda) \leqslant n$, $|\lambda| \leqslant |\mu|$, $\lambda \neq \mu$;*
  (iii) *the eigenvalue of $\mathbb{T}_\mu$ in V_μ is nonzero.*

COMMENT. Note a difference between these two claims: in Theorem 2.3 we used all irreducible G-modules[4] to characterize our basis, while in Theorem 2.2 we used only a part of the modules (namely, the polynomial ones). This implies that for the series A, our basis $\{\mathbb{S}_\mu\}$ is not invariant under the outer automorphism of $\mathfrak{gl}(n, \mathbb{C})$.

To prove Theorems 2.2–2.3, we shall restate them as purely combinatorial claims. We shall need the Harish-Chandra homomorphism. Recall (see, e.g., Dixmier [**D**, 7.4]) that it is defined for any reductive Lie algebra $\mathfrak{g}$ over $\mathbb{C}$. Fix a triangular decomposition $\mathfrak{g} = \mathfrak{n}_- \oplus \mathfrak{h} \oplus \mathfrak{n}_+$, where $\mathfrak{h}$ is a Cartan subalgebra and $\mathfrak{n}_+$ and $\mathfrak{n}_-$ are spanned by positive and negative root vectors. The Harish-Chandra homomorphism maps $U(\mathfrak{g})^\mathfrak{h}$, the centralizer of $\mathfrak{h}$ in $U(\mathfrak{g})$, onto $U(\mathfrak{h}) = S(\mathfrak{h})$. Let us identify $S(\mathfrak{h})$ with $\mathbb{C}[\mathfrak{h}^*]$, the algebra of polynomial functions on the dual space $\mathfrak{h}^*$. It is known that the restriction of the Harish-Chandra homomorphism to the center of $U(\mathfrak{g})$ is injective and its image in $\mathbb{C}[\mathfrak{h}^*]$ consists of the polynomials $f(\lambda)$ on $\mathfrak{h}^*$ that are invariant under transformations

(2.1) $$\lambda \mapsto w(\lambda + \rho) - \rho, \qquad \lambda \in \mathfrak{h}^*,$$

---

[4] The fact that for the series D there exist dominant highest weights $\lambda = (\lambda_1, \ldots, \lambda_n)$ with $\lambda_n < 0$ is not relevant here, because for any element of $Z(\mathfrak{o}(2n))$, its eigenvalue does not change under the transformation $\lambda_n \mapsto -\lambda_n$.

where w ranges over the Weyl group $W = W(\mathfrak{g}, \mathfrak{h})$ and ρ denotes the half-sum of positive roots. Moreover, if X is a central element of $U(\mathfrak{g})$ and f_X is the corresponding invariant polynomial, then the value of f_X at a point $\lambda \in \mathfrak{h}^*$ coincides with the eigenvalue of X in the irreducible highest weight module over $\mathfrak{g}$ with highest weight λ. Although this is true for all $\lambda \in \mathfrak{h}^*$, we shall deal with dominant weights λ only; note that any polynomial $f(\lambda)$ is uniquely determined by its values on the set of dominant weights. Finally, note that $\deg X = \deg f_X$.

After this digression, let us return to the classical Lie algebras. First we examine the series A. Then the group W coincides with the symmetric group $S(n)$ and the weight ρ satisfies the property $\rho_i - \rho_{i+1} = 1$, $1 \leqslant i \leqslant n-1$. It follows that a polynomial $f(\lambda_1, \ldots, \lambda_n)$ is invariant under the action (2.1) if and only if it satisfies the following symmetry property:

$$(2.2) \quad f(\lambda_1, \ldots, \lambda_n) = f(\lambda_1, \ldots, \lambda_{i-1}, \lambda_{i+1} - 1, \lambda_i + 1, \lambda_{i+2}, \ldots, \lambda_n)$$

for $i = 1, \ldots, n-1$. In our paper [**OO1**], such polynomials were called *shifted symmetric polynomials*, and the algebra of shifted symmetric polynomials in n variables was denoted by $\Lambda^*(n)$. Clearly, $f(\lambda_1, \ldots, \lambda_n)$ is shifted symmetric if and only if it is symmetric in the variables $l_1, \ldots, l_n$ that are related to $\lambda_1, \ldots, \lambda_n$ by (1.3). Note that the top homogeneous component of a shifted symmetric polynomial is a symmetric polynomial. The Harish-Chandra homomorphism induces an algebra isomorphism $Z(\mathfrak{gl}(n)) \to \Lambda^*(n)$.

Next assume G is one of the classical groups of type C, B, D. Then the weight ρ has the form

$$\rho = \rho_\varepsilon = (n - 1 + \varepsilon, n - 2 + \varepsilon, \ldots, \varepsilon), \quad \text{where } \varepsilon = 1, 1/2, 0,$$

and the group W is either the hyperoctahedral group $H(n) = S(n) \ltimes \{\pm 1\}^n$ (C-B case) or its subgroup $H'(n) \subset H(n)$ of index 2 (D case); recall that the action of the generators of the subgroup $\{\pm 1\}^n$ is of the form $\lambda_i \mapsto -\lambda_i$. However, for the series D, we have replaced the center of $U(o(2n, \mathbb{C}))$ by its proper subalgebra $Z(o(2n, \mathbb{C}))$, which is just equivalent to the replacement of $H'(n)$ by the whole group $H(n)$. This allows us to describe the algebras $Z(\mathfrak{g})$ in a uniform way for the three series C, B, D.

Denote by $\mathrm{M}^*(n) = \mathrm{M}^*_\varepsilon(n)$ the algebra of polynomials in $\lambda_1, \ldots, \lambda_n$ that can be written in the form

$$f(\lambda_1, \ldots, \lambda_n) = g(l_1^2, \ldots, l_n^2), \quad l = \lambda + \rho_\varepsilon,$$

where g is an arbitrary symmetric polynomial. Then we obtain an algebra isomorphism $Z(\mathfrak{g}) \to \mathrm{M}^*(n)$ induced by the Harish-Chandra homomorphism.

LEMMA-DEFINITION 2.4. *For the series* A, *put*

$$s^*_\mu(\lambda_1, \ldots, \lambda_n) = s_\mu(l_1, \ldots, l_n \,|\, 0, 1, 2, \ldots)$$

and for the series C, B, D *put*

$$t^*_\mu(\lambda_1, \ldots, \lambda_n) = s_\mu(l_1^2, \ldots, l_n^2 \,|\, \varepsilon^2, (\varepsilon + 1)^2, \ldots),$$

where $\varepsilon = 1, 1/2, 0$, *respectively. The Harish-Chandra homomorphism sends* $\mathbb{S}_\mu$ *to* s^*_μ *and* $\mathbb{T}_\mu$ *to* t^*_μ.

PROOF. This follows at once from the above discussion if one compares the definition of the elements $\mathbb{S}_\mu$, $\mathbb{T}_\mu$ (Lemma-Definition 2.1), the claim of Theorems 1.1–1.2, and the above definition of s_μ^* and t_μ^*. □

Since the map $Z(\mathfrak{gl}(n)) \to \Lambda^*(n)$ preserves degree and since $\deg s_\mu^* = |\mu|$, we have $\deg \mathbb{S}_\mu = |\mu|$. Similarly, as the top homogeneous component of $t_\mu^*(\lambda_1, \ldots, \lambda_n)$ coincides with $s_\mu(\lambda_1^2, \ldots, \lambda_n^2)$, we have $\deg t_\mu^* = 2|\mu|$, so that $\deg \mathbb{T}_\mu = 2|\mu|$.

Now it is clear that Theorems 2.2–2.3 are equivalent to the following claims about the polynomials s_μ^* and t_μ^*.

THEOREM 2.5. *The polynomials $s_\mu^*(\lambda_1, \ldots, \lambda_n)$ and $t_\mu^*(\lambda_1, \ldots, \lambda_n)$ form a basis of the algebras $\Lambda^*(n)$ and $\mathrm{M}^*(n)$, respectively. They can be characterized (up to a scalar multiple) by the following properties*:
  (i) $s_\mu^* \in \Lambda^*(n)$ and $\deg s_\mu^* = |\mu|$, $t_\mu^* \in \mathrm{M}^*(n)$ and $\deg t_\mu^* = 2|\mu|$;
  (ii) $s_\mu^*(\lambda) = 0$ and $t_\mu^*(\lambda) = 0$ *for all partitions λ such that* $l(\lambda) \leqslant n$, $|\lambda| \leqslant |\mu|$, $\lambda \neq \mu$;
  (iii) $s_\mu^*(\mu) \neq 0$ and $t_\mu^*(\mu) \neq 0$.

PROOF. The part of these claims concerning the polynomials s_μ^* was established in [**Ok1**]; see also [**OO1**, Theorem 3.3]. So we shall only consider the case of t_μ^* (which, however, is quite similar to that of s_μ^*).

Let $\mathrm{M}(n)$ denote the algebra of polynomials in n variables $\lambda_1, \ldots, \lambda_n$, the latter being invariant under the natural action of the hyperoctahedral group $H(n)$ (the generators of $\{\pm 1\}^n \subset H(n)$ change the sign of the variables). Clearly, the polynomials

$$t_\mu(\lambda_1, \ldots, \lambda_n) := s_\mu(\lambda_1^2, \ldots, \lambda_n^2)$$

form a homogeneous basis in $\mathrm{M}(n)$.

On the other hand, $\mathrm{M}(n)$ is canonically isomorphic to the graded algebra associated with the filtered algebra $\mathrm{M}^*(n)$. Since t_μ coincides with the top component of t_μ^*, we conclude that the polynomials t_μ^* form a basis in $\mathrm{M}^*(n)$.

Let us check that the polynomials t_μ^* satisfy conditions (i)–(iii). Condition (i) is evident, and conditions (ii), (iii) will be deduced from the explicit formula

$$\begin{aligned} t_\mu^*(\lambda_1, \ldots, \lambda_n) &= s_\mu(l_1^2, \ldots, l_n^2 \mid \varepsilon^2, (\varepsilon+1)^2, \ldots) \\ &= \frac{\det[(l_i^2 \mid \varepsilon^2, (\varepsilon+1)^2, \ldots)^{\mu_j + n - j}]}{\prod_{i<j}(l_i^2 - l_j^2)}. \end{aligned}$$

Note that the denominator is nonzero because $l_1^2, \ldots, l_n^2$ strictly decrease. Thus, we must analyze the vanishing properties of the numerator.

Assume $|\lambda| \leqslant |\mu|$ and $\lambda \neq \mu$. Then there exists $k \in \{1, \ldots, n\}$ such that $\lambda_k < \mu_k$, whence

$$\lambda_i < \mu_j \quad \text{for } j \leqslant k \leqslant i.$$

For such couples (i, j), the (i, j)th entry of the determinant in the numerator is

$$(l_i^2 \mid \varepsilon^2, (\varepsilon+1)^2, \ldots)^{\mu_j + n - j} = \prod_{r=1}^{\mu_j + n - j} ((\lambda_i + n - i + \varepsilon)^2 - (\varepsilon + r - 1)^2) = 0,$$

so that the determinant vanishes.

If $\lambda = \mu$, then a similar argument shows that the matrix in the numerator is strictly upper triangular with nonzero diagonal entries, so that the determinant is nonzero.

Thus, we have verified properties (i)–(iii), and it remains to prove that polynomials with such properties are unique. For $d = 2, 4, 6, \ldots$, consider the subspace $\mathrm{M}_d^*(n) \subset \mathrm{M}^*(n)$ formed by elements of degree $\leqslant d$. The existence property verified above implies that the linear functionals on $\mathrm{M}_d^*(n)$ of the form

$$f \mapsto f(\lambda), \qquad |\lambda| \leqslant d,$$

are linearly independent. On the other hand, their number is exactly equal to $\dim \mathrm{M}_d^*(n)$, which implies the uniqueness claim. □

Note that Theorem 2.5 also can be directly deduced from the binomial formula; cf. the second proof of Theorem 5.1 in [**OO1**].

§3. Coherence property

Until now we fixed a classical Lie algebra $\mathfrak{g}$. In this section we deal with the whole series $\{\mathfrak{g}(n)\}$ of classical Lie algebras (of type A, C, B, or D). We define natural embeddings $Z(\mathfrak{g}(n)) \to Z(\mathfrak{g}(n+1))$ and we show that the elements of the canonical bases are stable (up to scalar multiples) with respect to these embeddings.

Since the Lie algebra will vary (inside a classical series), we shall use more detailed notation for the polynomials in n variables introduced above:

$$s_{\mu|n}, \ s^*_{\mu|n}, \ t_{\mu|n}, \ t^*_{\mu|n},$$

and the elements of the distinguished basis of $Z(\mathfrak{g}(n))$ will be denoted by $\mathbb{S}_{\mu|n}$ or $\mathbb{T}_{\mu|n}$.

For any n, we shall identify $\mathbb{C}^n$ with the subspace of $\mathbb{C}^{n+1}$ spanned by first n basis vectors. Since $\mathfrak{g}(n)$ is realized in $\mathbb{C}^n$ (for series A) or in $\mathbb{C}^n \oplus \mathbb{C}^n$ (for series C, B, D), we obtain a natural Lie algebra embedding $\mathfrak{g}(n) \to \mathfrak{g}(n+1)$, which induces a natural algebra embedding $U(\mathfrak{g}(n)) \to U(\mathfrak{g}(n+1))$.

Next, we define the *averaging operator*

$$\mathrm{Avr}_{n,n+1} \colon Z(\mathfrak{g}(n)) \to Z(\mathfrak{g}(n+1))$$

as the composition

$$Z(\mathfrak{g}(n)) \hookrightarrow U(\mathfrak{g}(n)) \hookrightarrow U(\mathfrak{g}(n+1)) \to Z(\mathfrak{g}(n+1)),$$

where the latter arrow is the projection # defined in (2.1). Similarly, for any couple $n < m$, we define the averaging operator

$$\mathrm{Avr}_{nm} \colon Z(\mathfrak{g}(n)) \to Z(\mathfrak{g}(m)),$$

which also coincides with composition of averagings,

$$Z(\mathfrak{g}(n)) \to Z(\mathfrak{g}(n+1)) \to \cdots \to Z(\mathfrak{g}(m)).$$

THEOREM 3.1. *Let $c(n,\mu)$ and $c_\pm(n,\mu)$ be the normalizing factors occurring in the binomial formula (see (1.7), (1.9)). For any $n < m$ we have*

$$\mathrm{Avr}_{nm}\left(\frac{1}{c(n,\mu)}\mathbb{S}_{\mu|n}\right) = \frac{1}{c(m,\mu)}\mathbb{S}_{\mu|m},$$

$$\mathrm{Avr}_{nm}\left(\frac{1}{c_\pm(n,\mu)}\mathbb{T}_{\mu|n}\right) = \frac{1}{c_\pm(m,\mu)}\mathbb{T}_{\mu|m}.$$

We call this the *coherence property* of the basis $\{\mathbb{S}_\mu\}$ or $\{\mathbb{T}_\mu\}$. For the series A the coherence property was established in our paper [**OO1**, §10], by three different methods which can also be carried over to the series C, B, D. We shall now give one of the proofs, which is based on the binomial formula.

PROOF. We assume $\{\mathfrak{g}(n)\}$ is one the series C, B, D. First of all we restate Theorem 1.2. Let $g \in G$ and $X \in \mathfrak{g}$ be related by

$$g^{1/2} - g^{-1/2} = X,$$

where g is close to $e \in G$ and X is close to $0 \in \mathfrak{g}$. Let $f \in I(G)$ be a test function and let $f' = f'(X)$ be the corresponding function in a neighborhood of $0 \in \mathfrak{g}$. Write $\pm x_1, \pm x_2, \ldots$ for the eigenvalues of the matrix X, and define

$$T_\mu(X) = s_\mu(x_1^2, x_2^2, \ldots), \qquad X \in \mathfrak{g},\ l(\mu) \leqslant \mathrm{rank}\,\mathfrak{g}.$$

One can expand $f'(X)$ into a series of the invariant polynomials $T_\mu(X)$, which converges in a neighborhood of the origin. We claim that this expansion can be written as

(3.1) $$f'(x) = \sum_{l(\mu) \leqslant \mathrm{rank}\,\mathfrak{g}} \langle \widetilde{\mathbb{T}}_\mu, f \rangle T_\mu(X),$$

where

$$\widetilde{\mathbb{T}}_\mu := \frac{1}{c_\pm(\mathrm{rank}\,\mathfrak{g},\mu)}\,\mathbb{T}_\mu.$$

Indeed, since $I(G)$ is spanned by the characters χ_λ (here we use the same abbreviation as in Theorem 1.2), we may assume $f = \chi_\lambda/\chi_\lambda(e)$. But then our claim is just a restatement of the binomial formula. To see this, we must recall the definition of the elements $\mathbb{T}_\mu$ and use the fact that for any element $A \in Z(\mathfrak{g})$ (in particular, for $A = \mathbb{T}_\mu$), $\langle A, \chi/\chi(e)\rangle$ equals the eigenvalue of A in the $U(\mathfrak{g})$-module corresponding to χ_λ.

Now let $n < m$ and let $G(n) \hookrightarrow G(m)$ be the natural embedding. Let $f \in I(G(m))$ be a test function and $f|G(n)$ be its restriction to $G(n)$, which is clearly an element of $I(G(n))$. Let us write the expansion (3.1) for f and $f|G(n)$:

$$f'(X) = \sum_{l(\mu)\leqslant m} \langle \widetilde{\mathbb{T}}_{\mu|m}, f\rangle T_{\mu|m}(X), \qquad X \in \mathfrak{g}(m),$$

$$(f|G(n))'(Y) = \sum_{l(\mu)\leqslant n} \langle \widetilde{\mathbb{T}}_{\mu|n}, f|G(n)\rangle T_{\mu|n}(Y), \qquad Y \in \mathfrak{g}(n).$$

Assume $X = Y \in \mathfrak{g}(n)$. Then the left-hand sides of both expansions coincide. On the other hand, by the definition of the polynomial functions T_μ and the stability property of the Schur functions, we have

$$T_{\mu|m}(Y) = \begin{cases} T_{\mu|n}(Y), & \text{if } l(\mu) \leq n, \\ 0, & \text{if } l(\mu) > n. \end{cases}$$

It follows that

$$\langle \widetilde{\mathbb{T}}_{\mu|n}, f | G(n) \rangle = \langle \widetilde{\mathbb{T}}_{\mu|m}, f \rangle, \qquad f \in I(G(m)), \ l(\mu) \leq n.$$

More generally, for any $f \in \mathbb{C}[G]$ we have

$$\langle \widetilde{\mathbb{T}}_{\mu|m}, f \rangle = \langle \widetilde{\mathbb{T}}_{\mu|m}, f^\# \rangle = \langle \widetilde{\mathbb{T}}_{\mu|n}, f^\# | G(n) \rangle,$$

which is equivalent to the claim of the theorem. $\square$

REMARK 3.2. It was shown in [**OO1**] that the coherence property of the basis $\{\mathbb{S}_\mu\}$ is equivalent to the following relation satisfied by the polynomials s_μ^*: for any partition μ of length $\leq n$ and any signature Λ of length $n+1$, we have

$$(3.2) \quad \frac{\chi_\Lambda^{gl(n+1)}(e)}{c(n+1,\mu)} s^*_{\mu|n+1}(\Lambda_1, \ldots, \Lambda_{n+1}) = \sum_{\lambda \prec \Lambda} \frac{\chi_\lambda^{gl(n)}(e)}{c(n,\mu)} s^*_{\mu|n}(\lambda_1, \ldots, \lambda_n),$$

where λ denote a signature of length n and $\lambda \prec \Lambda$ denotes the Gelfand–Tsetlin betweenness condition

$$\Lambda_1 \geq \lambda_1 \geq \Lambda_2 \geq \cdots \geq \Lambda_n \geq \lambda_n \geq \Lambda_{n+1}.$$

A similar relation, which is equivalent to the coherence property of $\{\mathbb{T}_\mu\}$, holds for the polynomials t_μ^*:

$$(3.3) \quad \frac{\chi_\Lambda(e)}{c_\pm(n+1,\mu)} t^*_{\mu|n+1}(\Lambda_1, \ldots, \Lambda_{n+1}) = \sum_\lambda [\chi_\Lambda : \chi_\lambda] \frac{\chi_\lambda(e)}{c_\pm(n,\mu)} t^*_{\mu|n}(\lambda_1, \ldots, \lambda_n).$$

Here Λ is a partition of length $\leq n+1$, μ is a partition of length $\leq n$, χ_Λ is one of the characters of $G(n+1)$ (i.e., $\chi_\Lambda^{sp(2n+2)}$, $\chi_\Lambda^{so(2n+3)}$ or $\chi_\Lambda^{o(2n+2)}$), χ_λ has the same meaning (but for the group $G(n)$ of rank n), and $[\chi_\Lambda : \chi_\lambda]$ denotes the multiplicity of χ_λ in the decomposition of χ_Λ as restricted to $G(n) \subset G(n+1)$.[5]

In [**OO1**, §10] we gave a direct derivation of the relation (3.2). The relation (3.3) can be directly verified by a similar (but more complicated) argument. This gives another approach to the coherence property.

§4. A distinguished basis in $I(\mathfrak{g})$

In this section, we study a distinguished basis in $I(\mathfrak{g}) \subset S(\mathfrak{g})$ (see §0 for the definition of $I(\mathfrak{g})$).

We equip $\mathfrak{g}$ with an invariant inner product: for any $X, Y \in \mathfrak{g}$,

$$\langle X, Y \rangle = \begin{cases} \operatorname{tr} XY, & \text{for the series A}, \\ \frac{1}{2} \operatorname{tr} XY, & \text{for the series C, B, D}. \end{cases}$$

---

[5]This multiplicity is given by the well-known 'branching rules' for symplectic and orthogonal groups (see Želobenko [**Z**, §§129–130]).

We shall identify $\mathfrak{g}$ with its dual space $\mathfrak{g}^*$ by making use of the product $\langle\,\cdot\,,\,\cdot\,\rangle$. This will allow us to interpret $S(\mathfrak{g})$ as the algebra of polynomial functions on $\mathfrak{g}$, and then elements of $I(\mathfrak{g})$ will become invariant polynomial functions on $\mathfrak{g}$.

Assume X ranges over $\mathfrak{g}$. In the case of the series A, we shall denote by $x_1,\ldots,x_n$ the eigenvalues of X. In the case of the series C, D, the eigenvalues of X may be written as $\pm x_1,\ldots,\pm x_n$, and for the series B one must add one 0. In this notation we set

$$S_\mu(X) = S_{\mu|n}(X) = s_\mu(x_1,\ldots,x_n),$$
$$T_\mu(X) = T_{\mu|n}(X) = t_\mu(x_1,\ldots,x_n) = s_\mu(x_1^2,\ldots,x_n^2).$$

Clearly, $\{S_\mu\}$ or $\{T_\mu\}$ is a homogeneous basis in $I(\mathfrak{g})$.

Note that $I(\mathfrak{g})$ may be identified with $\operatorname{gr} Z(\mathfrak{g})$, the graded algebra associated with the fibered algebra $Z(\mathfrak{g})$. The next claim follows immediately from the definitions.

PROPOSITION 4.1. *Under the identification $I(\mathfrak{g}) = \operatorname{gr} Z(\mathfrak{g})$, each basis element $S_\mu \in I(\mathfrak{g})$ or $T_\mu \in I(\mathfrak{g})$ coincides with the leading term of the basis element $\mathbb{S}_\mu \in Z(\mathfrak{g})$ or $\mathbb{T}_\mu \in Z(\mathfrak{g})$, respectively.* □

Now write $\mathfrak{g}(n)$ instead of $\mathfrak{g}$ and consider the chain $\cdots \hookrightarrow \mathfrak{g}(n) \hookrightarrow \mathfrak{g}(n+1) \hookrightarrow \cdots$ of natural embeddings that we already discussed in §3. Note that these embeddings are isometric with respect to the inner product $\langle\,\cdot\,,\,\cdot\,\rangle$. Therefore, for each couple $n < m$ there is a natural projection $\mathfrak{g}(m) \to \mathfrak{g}(n)$ and so an algebra morphism $S(\mathfrak{g}(m)) \to S(\mathfrak{g}(n))$, which in turn induces an algebra morphism

$$\operatorname{Proj}_{nm}\colon I(\mathfrak{g}(m)) \to I(\mathfrak{g}(n)), \qquad n < m.$$

PROPOSITION 4.2 (Stability of the bases $\{S_\mu\}$, $\{T_\mu\}$). *Assume $n < m$. If $l(\mu) \leqslant n$ then*

$$\operatorname{Proj}_{nm}(S_{\mu|n}) = S_{\mu|n}, \qquad \operatorname{Proj}_{nm}(T_{\mu|m}) = T_{\mu|n}.$$

If $n < l(\mu) \leqslant m$ then the result is zero.

PROOF. This follows at once from the stability property of the Schur polynomials. □

On the other hand, for any $n < m$ we can define linear maps

$$\operatorname{Avr}_{nm}\colon I(\mathfrak{g}(n)) \to I(\mathfrak{g}(m))$$

(the *averaging operators*) in exactly the same way as for the invariants in the enveloping algebras (see §3).

PROPOSITION 4.3 (Coherence property). *Assume $n < m$ and $l(\mu) \leqslant n$. Then*

$$\operatorname{Avr}_{nm}(S_{\mu|n}) = \frac{c(n,\mu)}{c(m,\mu)} S_{\mu|m}, \qquad \operatorname{Avr}_{nm}(T_{\mu|n}) = \frac{c_\pm(n,\mu)}{c_\pm(m,\mu)} T_{\mu|m}.$$

In particular, the averaging operators are injective.

PROOF. This follows from Theorem 3.1 and Proposition 4.1. □

THEOREM 4.4 (Characterization of the basis). *Let n be fixed and μ range over partitions of length $\leqslant n$. The basis elements $S_{\mu|n}$ (for the series A) or $T_{\mu|n}$ (for the series C, B, D) are the unique, within a scalar multiple, elements of $I(\mathfrak{g}(n))$ that are eigenvectors of all linear mappings*

$$\mathrm{Proj}_{nm} \circ \mathrm{Avr}_{nm} \colon I(\mathfrak{g}(n)) \to I(\mathfrak{g}(n)), \qquad m = n+1, n+2, \ldots.$$

PROOF. By Propositions 4.2 and 4.3, we have for any $m > n$

$$\mathrm{Proj}_{nm}(\mathrm{Avr}_{nm}(S_{\mu|n})) = \frac{c(n,\mu)}{c(m,\mu)} S_{\mu|n},$$

$$\mathrm{Proj}_{nm}(\mathrm{Avr}_{nm}(T_{\mu|n})) = \frac{c_{\pm}(n,\mu)}{c_{\pm}(m,\mu)} T_{\mu|n}.$$

Thus, each basis element is an eigenvector for all the maps $\mathrm{Proj}_{nm} \circ \mathrm{Avr}_{nm}$. It remains to show that the eigenvalues corresponding to two different basis vectors are distinct for certain m. But this follows from the explicit expressions for $c(n,\mu)$ and $c_{\pm}(n,\mu)$ given in §1. □

In [OO1, (2.10)], we obtained an explicit expression of the basis elements $S_{\mu|n} \in I(\mathfrak{gl}(n))$ through the matrix limits $E_{ij} \in \mathfrak{gl}(n)$, $1 \leqslant i,j \leqslant n$:

PROPOSITION 4.5. *Let μ be a partition of length $\leqslant n$, $k = |\mu|$, and χ^{μ} the irreducible character of the symmetric group $\mathfrak{S}(k)$ indexed by μ. Then*

$$S_{\mu|n} = (k!)^{-1} \sum_{i_1,\ldots,i_k=1}^{n} \sum_{s \in \mathfrak{S}(k)} \chi^{\mu}(s) E_{i_1 i_{s(1)}} \cdots E_{i_k i_{s(k)}}.$$

We shall give now a similar formula for the series C, B, D. Assume that $\mathfrak{g} = \mathfrak{g}(n)$ is of type C, B, D and realize it as an involutive subalgebra in $\mathfrak{gl}(N, \mathbb{C})$, where $N = 2n$ or $N = 2n+1$, as indicated in §1. We shall assume the canonical basis in $\mathbb{C}^N$ is labelled by the numbers $i = -n, -n+1, \ldots, n-1, n$, where 0 is included for the series B only, and we put

$$F_{ij} = E_{ij} - \theta_{ij} E_{-j,-i},$$

where $\theta_{ij} \equiv 1$ for the series B, D and $\theta_{ij} = \mathrm{sgn}(i)\,\mathrm{sgn}(j)$ for the series C. Note that for our embedding $\mathfrak{g}(n) \hookrightarrow \mathfrak{gl}(N,\mathbb{C})$ the elements F_{ij} form a basis in $\mathfrak{g}(n)$.

Given a partition $\mu \vdash k$, we define a central function φ^{μ} on the symmetric group $\mathfrak{S}(2k)$ as follows. If a permutation $s \in \mathfrak{S}(2k)$ is such that all its cycles are even (so that the circle type of s may be written as 2ρ for a certain $\rho \vdash k$), then we put

$$\varphi^{\mu}(s) = \chi^{\mu}_{\rho},$$

where χ^{μ}_s denotes the value of the irreducible character χ^{μ} at any permutation with cycle type ρ. Otherwise (i.e., if $s \in \mathfrak{S}(2k)$ has at least one cycle of odd length) $\varphi^{\mu}(s) = 0$.

The next claim is an exact analog of Proposition 4.5.

PROPOSITION 4.6. *Assume $\mathfrak{g} = \mathfrak{g}(n)$ is of type C, B, D. Let μ be an arbitrary partition of length $\leqslant n$ and $k = |\mu|$. Then*

$$T_{\mu|n} = \frac{1}{(2k)!} \sum_{i_1,\ldots,i_{2k}=-n}^{n} \sum_{s \in \mathfrak{S}(2k)} \varphi^\mu(s) F_{i_1 i_{s(1)}} \cdots F_{i_{2k} i_{s(2k)}}.$$

PROOF. Let us view $T_{\mu|n}$ as an invariant polynomial function $T_{\mu|n}(X)$, where X ranges over $\mathfrak{g}(n)$, and denote by $\pm x_1, \ldots, \pm x_n$ the eigenvalues of X (for the series B we exclude the zero eigenvalue). We shall use the standard notation of Macdonald's book [M2]: p_ρ are the power sums and

$$z_\rho = 1^{k_1} k_1! \, 2^{k_2} k_2! \cdots \quad \text{for } \rho = (1^{k_1} 2^{k_2} \cdots).$$

By definition of the basis $\{T_\mu\}$, we have

$$T_{\mu|n}(X) = s_\mu(x_1^2, \ldots, x_n^2) = \sum_{\rho \vdash k} z_\rho^{-1} \chi_\rho^\mu p_\rho(x_1^2, \ldots, x_n^2)$$

$$= \sum_{\rho \vdash k} z_\rho^{-1} 2^{-l(\rho)} \chi_\rho^\mu p_{2\rho}(x_1, -x_1, \ldots, x_n, -x_n)$$

$$= \sum_{\rho \vdash k} z_{2\rho}^{-1} \chi_\rho^\mu p_{2\rho}(x_1, -x_1, \ldots, x_n, -x_n).$$

Let s range over elements of $\mathfrak{S}(2k)$ with even cycle type 2ρ (where $\rho \vdash k$ depends on s). Then the latter expression can be written as

$$T_{\mu|n}(X) = \frac{1}{(2k)!} \sum_s \varphi^\mu(s) p_{2\rho}(x_1, -x_1, \ldots, x_n, -x_n).$$

On the other hand, for any (even) m, the invariant polynomial function

$$p_m(X) = \text{tr}(X^m) = p_m(x_1, -x_1, \ldots, x_n, -x_n)$$

corresponds to the element

$$\sum_{i_1,\ldots,i_m=-n}^{n} F_{i_1 i_2} F_{i_2 i_3} \cdots F_{i_m i_1} \in I(\mathfrak{g}(n)),$$

which also can be written as

$$\sum_{i_1,\ldots,i_m=-n}^{n} F_{i_1 i_{s(1)}} \cdots F_{i_m i_{s(m)}}$$

for any cyclic permutation s of the indices $1, \ldots, m$. This concludes the proof. □

There exists another characterization of the bases $\{S_\mu\}$, $\{T_\mu\}$, which is based on the following observation, which is undoubtedly well known.

PROPOSITION 4.7. *Define the classical group $\widetilde{G}$ containing G as follows.*

If $G = GL(n, \mathbb{C})$, then $\widetilde{G} = GL(n, \mathbb{C}) \times GL(n, \mathbb{C})$ and the embedding $G \hookrightarrow \widetilde{G}$ has the form $g \mapsto (g, (g')^{-1})$.

If G is one of the groups $SO(2n+1, \mathbb{C})$ or $Sp(2n, \mathbb{C})$, $SO(2n, \mathbb{C})$, then $\widetilde{G} = GL(N, \mathbb{C})$, where $N = 2n+1$ or $N = 2n$, respectively.

Then the adjoint action of G in $\mathfrak{g}$ can be extended to a linear action of the group $\widetilde{G} \supset G$ such that the induced representation of $\widetilde{G}$ in the symmetric algebra $S(\mathfrak{g})$ is a multiplicity-free polynomial representation.

PROOF. Assume $G = GL(n, \mathbb{C})$. Using the bijection $E_{ij} \leftrightarrow e_j \otimes e_i$, where $\{e_i\}$ stands for the natural basis of $\mathbb{C}^n$, we identify $\mathfrak{g} = \mathfrak{gl}(n, \mathbb{C})$ with $\mathbb{C}^n \otimes \mathbb{C}^n$. The action of the group $\widetilde{G} = GL(n, \mathbb{C}) \times GL(n, \mathbb{C})$ in the vector space $\mathfrak{g} = \mathbb{C}^n \otimes \mathbb{C}^n$ is the natural one. Clearly, its restriction to the subgroup $G = \{(g, (g')^{-1})\}$ is equivalent to the adjoint action. It is well known that action of the group $GL(n, \mathbb{C}) \times GL(n, \mathbb{C})$ in $S(\mathbb{C}^n \otimes \mathbb{C}^n)$ is a multiplicity-free polynomial representation (see, e.g., Želobenko [**Z**, §56] or Howe [**H**, 2.1]).

Assume $G = Sp(2n, \mathbb{C})$ and put $N = 2n$. One can identify $\mathfrak{g} = sp(2n, \mathbb{C})$ with $S^2(\mathbb{C}^n)$ in such a way that the adjoint action of G in $\mathfrak{g}$ will coincide with the natural action of $\widetilde{G} = GL(N, \mathbb{C})$ in $S^2(\mathbb{C}^N)$, restricted to $G \subset \widetilde{G}$. It is well known [**H**, 3.1] that the natural action of $GL(N, \mathbb{C})$ in $S(S^2(\mathbb{C}^N))$ is a multiplicity-free polynomial representation.

Assume now that $G = SO(N, \mathbb{C})$, where $N = 2n+1$ or $N = 2n$. Then one can identify $\mathfrak{g}$ with $\Lambda^2(\mathbb{C}^N)$ in such a way that the adjoint action of G in $\mathfrak{g}$ will coincide with the restriction to G of the natural action of the group $\widetilde{G} = GL(N, \mathbb{C})$ in $\Lambda^2(\mathbb{C}^N)$. On the other hand, it is well known [**H**, 3.8] that $S(\Lambda^2 \mathbb{C}^N)$ is a multiplicity-free polynomial $GL(N, \mathbb{C})$-module.

Finally, note that in this argument, we could replace the special orthogonal group $SO(N, \mathbb{C})$ by the complete orthogonal group $O(N, \mathbb{C})$. □

The next result provides us with an alternative characterization of the bases $\{S_\mu\}$, $\{T_\mu\}$.

THEOREM 4.8. *Let G be a classical group, and let $\widetilde{G} \supset G$ be as in Proposition 4.7. The basis elements $S_\mu \in I(\mathfrak{g})$ or $T_\mu \in I(\mathfrak{g})$ are the unique (up to a scalar multiple) elements of $I(\mathfrak{g}) \subset S(\mathfrak{g})$ that generate irreducible submodules of $S(\mathfrak{g})$ under the action of $\widetilde{G}$.*

PROOF. Examine the case $G = GL(n, \mathbb{C})$. The representation of the group $\widetilde{G} = GL(n, \mathbb{C}) \times GL(n, \mathbb{C})$ in the space $S(\mathbb{C}^n \otimes \mathbb{C}^n)$ can be decomposed into the direct sum of the irreducibles $V_{\mu|n} \otimes V_{\mu|n}$, where μ ranges over partitions of length $\leq n$ (see [**Z**, §56] or [**H**, 2.1.2]). Each component $V_{\mu|n} \otimes V_{\mu|n}$ clearly contains a unique (up to a scalar multiple) vector S'_μ, invariant under the subgroup $G = \{(g, (g')^{-1})\}$. It is also evident that the elements S'_μ can be characterized as the only elements of $I(\mathfrak{g})$ generating irreducible G-submodules of $S(\mathfrak{g})$. Thus, we must check that $S'_\mu = S_\mu$ (up to a scalar multiple).

By Theorem 4.4, it suffices to show that each S'_μ is an eigenvector for the maps $\operatorname{Proj}_{nm} \circ \operatorname{Avr}_{nm}$, $m = n+1, n+2, \ldots$; but this follows from the fact that

$$\operatorname{Avr}_{nm}(V_{\mu|n}) \subset V_{\mu|m}, \qquad \operatorname{Proj}_{nm}(V_{\mu|m}) = V_{\mu|n},$$

which in turn follows from the very definition of these maps.

For other groups G the arguments are similar.

Let $G = Sp(2n, \mathbb{C})$ and put $N = 2n$. The representation of $\tilde{G} = GL(N, \mathbb{C})$ in $S(S^2(\mathbb{C}^N))$ is the direct sum of the irreducibles $V_{M|N}$, where M is a Young diagram with even rows (the number of rows does not exceed N); on the other hand, an irreducible polynomial $GL(N, \mathbb{C})$-module $V_{M|N}$ contains a vector invariant under $Sp(2n, \mathbb{C})$, where $n = N/2$, if and only if all the columns of M are even, and then such a vector is unique within a scalar multiple (see [**H**, 3.1] or [**M2**, VII, (6.11)]).

Thus, an irreducible component $V_{M|N} \subset S(S^2(\mathbb{C}^{2n}))$ has a nonzero (and then one-dimensional) intersection with $I(\mathfrak{g})$ if and only if M can be written as

$$M = 2\mu \cup 2\mu = (2\mu_1, 2\mu_1, 2\mu_2, 2\mu_2, \ldots, 2\mu_n, 2\mu_n),$$

where μ is a partition of length $\leqslant n$. Let T'_μ be any nonzero element of the one-dimensional space $V_{2\mu \cup 2\mu|2n} \cap I(\mathfrak{g})$. The same argument as above shows that $T'_\mu = T_\mu$, up to a scalar multiple.

(Note, however, a difference with the case $G = GL(n, \mathbb{C})$ examined above. For $G = GL(n, \mathbb{C})$, we saw that each irreducible $\tilde{G}$-submodule contained a G-invariant, while now only a part of components possess G-invariants.)

Now let $G = SO(N, \mathbb{C})$, where $N = 2n+1$ or $N = 2n$, and remark that in both cases $I(\mathfrak{g})$ coincides with the subspace of $S(\mathfrak{g})$ formed by the G'-invariants, where $G' = O(N, \mathbb{C})$. The representation of $\tilde{G} = GL(N, \mathbb{C})$ in the space $S(\Lambda^2(\mathbb{C}^N))$ decomposes into the direct sum of irreducible modules $V_{M|N}$, where M is a diagram with even columns. On the other hand, a G'-invariant in $V_{M|N}$ exists (and then is unique, within a scalar multiple) if and only if M has even rows (see [**H**, 3.8] or [**M2**, VII, (3.14)]). Thus we see again that M has the form $2\mu \cup 2\mu$, and we conclude the proof as above. $\square$

§5. A relationship between the bases in $Z(\mathfrak{g})$ and $I(\mathfrak{g})$

Here we exhibit a linear isomorphism $S(\mathfrak{g}) \to U(\mathfrak{g})$, called special symmetrization, which maps $I(\mathfrak{g})$ onto $Z(\mathfrak{g})$ and takes the canonical basis in $I(\mathfrak{g})$ to that of $Z(\mathfrak{g})$. The main result is Theorem 5.2. We reduce it to Proposition 5.3, which in turn is reduced to Proposition 5.4. These two propositions are of independent interest.

We shall need the concept of generalized symmetrization proposed by Olshanski [**O2**]. Let us identify the algebra $U(\mathfrak{g})$ with a subspace of the dual of $\mathcal{O}_e(G)$ (see the beginning of §2). Similarly, we shall identify $S(\mathfrak{g})$ with a subspace of the dual space to $\mathcal{O}_0(\mathfrak{g})$, the space of germs of holomorphic functions defined at a neighborhood of the origin $0 \in \mathfrak{g}$.

Assume we are given a map $F: \mathfrak{g} \to G$ with the following properties:
  (i) F is holomorphic and defined in a neighborhood of the origin, invariant under the adjoint representation;
  (ii) F takes $0 \in \mathfrak{g}$ to $e \in G$;
  (iii) the differential of F at the origin is the identical map $\mathfrak{g} \to \mathfrak{g}$;
  (iv) F is equivariant with respect to the group G acting by the adjoint representation in $\mathfrak{g}$ and by conjugations on itself.

Then F induces an isomorphism of local rings $\mathcal{O}_0(\mathfrak{g}) \to \mathcal{O}_e(G)$, which by duality determines a linear isomorphism

$$\sigma \colon S(\mathfrak{g}) \to U(\mathfrak{g}).$$

Note that this construction is meaningful for any complex Lie group; in fact it also works for formal groups.

When F is the exponential map, the corresponding map σ coincides with the standard symmetrization map. Following [**O2**], in the general case we call σ a *generalized symmetrization*.

A generalized symmetrization σ shares a number of properties of the standard symmetrization. In particular, σ preserves leading terms and also induces a bijection between invariants of the group.

Thus, for a classical group G, the symmetrization σ induces a linear isomorphism $I(\mathfrak{g}) \to Z(\mathfrak{g})$ (for the series D we shall assume F is invariant under $G' \supset G$).

THEOREM 5.1 ([**OO1**, Theorem 14.1]). *Assume that* $\mathfrak{g} = \mathfrak{gl}(n, \mathbb{C})$ *and* $F(X) = 1 + X$, *and let σ be the corresponding generalized symmetrization. Then*
$$\sigma(S_{\mu|n}) = \mathbb{S}_{\mu|n} \quad \text{for any } \mu, \ \ell(\mu) \leqslant n. \qquad \square$$

This generalized symmetrization was introduced in [**O1**]; then it was used in the note [**KO**] and called *special symmetrization*. We aim to find a similar generalized symmetrization for the series C, B, D.

THEOREM 5.2. *Let* $\mathfrak{g} = \mathfrak{g}(n)$ *be any classical Lie algebra of type C, B, D, realized as an involutive subalgebra of* $\mathfrak{gl}(N, \mathbb{C})$, *where* $N = 2n$ *or* $N = 2n + 1$. *Define* $F \colon \mathfrak{g} \to G$ *by the relation*
$$F(X)^{1/2} - F(X)^{-1/2} = X \quad \text{for } X \in \mathfrak{g},$$
i.e.,
$$F(X) = 1 + X^2/2 + ((1 + X^2/2)^2 - 1)^{1/2} = 1 + X^2/2 + X(1 + X^2/4)^{1/2}$$
$$= 1 + X + \cdots,$$
where the eigenvalues of the matrix X are assumed sufficiently small.

Let σ be the corresponding generalized symmetrization. Then
$$\sigma(T_{\mu|n}) = \mathbb{T}_{\mu|n} \quad \text{for any } \mu, \ l(\mu) \leqslant n.$$

Note that F is indeed a well-defined (local) map from $\mathfrak{g}$ to G. We call σ the *special symmetrization* for the classical Lie algebras of type C, B, D. We refer the reader to [**O2**] for explicit formulas for $\sigma \colon S(\mathfrak{g}) \to U(\mathfrak{g})$ and its inverse $\sigma^{-1} \colon U(\mathfrak{g}) \to S(\mathfrak{g})$.

Note that the proof of the theorem given below differs from the argument used in [**OO1**] for the series A. On the other hand, the argument given below also holds for the series A. We do not know if the approach of [**OO1**, Theorem 14.1], can be carried over to the series C, B, D.

PROOF. Let $\varphi \in \mathcal{O}_e(G)$ stand for a test element. By definition, we must prove the equality
$$\langle \mathbb{T}_{\mu|n}, \varphi \rangle = \langle T_{\mu|n}, \varphi \circ F \rangle,$$
where the brackets on the left-hand side denote the pairing between $U(\mathfrak{g})$ and $\mathcal{O}_e(G)$, while the brackets on the right-hand side denote the pairing between $S(\mathfrak{g})$ and $\mathcal{O}_0(\mathfrak{g})$. Let
$$\mathcal{O}_e(G)^{\text{inv}} \subset \mathcal{O}_e(G) \quad \text{and} \quad \mathcal{O}_0(\mathfrak{g})^{\text{inv}} \subset \mathcal{O}_0(\mathfrak{g})$$

denote the subspaces of invariants with respect to the action of the group G (or the group G', for the series D). Choose a compact form $K \subset G$ (or $K' \subset G'$). Averaging over K (or K') determines an invariant projection

$$\mathcal{O}_e(G) \to \mathcal{O}_e(G)^{\text{inv}}.$$

Using it, we reduce our problem to the case $\varphi \in \mathcal{O}_e(G)^{\text{inv}}$.

We shall apply the binomial formula (Theorem 1.2), which can be conveniently written as

$$(\varphi \circ F)(X) = \sum_{l(\mu) \leqslant n} \frac{\langle \mathbb{T}_{\mu|n}, \varphi \rangle}{c_\pm(n, \mu)} T_{\mu|n}(X),$$

where $\varphi \in \mathcal{O}_e(G)^{\text{inv}}$. Indeed, if φ has the form $\varphi = \chi_\lambda/\chi_\lambda(e)$, where χ_λ is one of the characters $\chi_\lambda^{sp(2n)}$, $\chi_\lambda^{so(2n+1)}$, $\chi_\lambda^{o(2n)}$, then the above relation just coincides with the binomial formula, because of the definition of the map F and the relation

$$\langle \mathbb{T}_{\mu|n}, \chi_\lambda/\chi_\lambda(e) \rangle = t^*_{\mu|n}(\lambda_1, \ldots, \lambda_n).$$

Further, since the linear span of the characters χ_λ is dense in $\mathcal{O}_e(G)^{\text{inv}}$ with respect to the adic topology defined by the unique maximal ideal of $\mathcal{O}_e(G)^{\text{inv}}$, our expansion holds for any $\varphi \in \mathcal{O}_e(G)^{\text{inv}}$.

On the other hand, any element $\psi \in \mathcal{O}_0(\mathfrak{g})^{\text{inv}}$ can be uniquely written as a series

$$\psi = \sum_{l(\mu) \leqslant n} c_\mu(\psi) T_{\mu|n}, \qquad c_\mu(\psi) \in \mathbb{C},$$

converging in the adic topology of $\mathcal{O}_0(\mathfrak{g})^{\text{inv}}$. Taking $\psi = \varphi \circ F$ and comparing the two expansions, we reduce the problem to the following relation:

$$c_\mu(\psi) = \frac{\langle T_{\mu|n}, \psi \rangle}{c_\pm(n, \mu)}, \qquad \psi \in \mathcal{O}_0(\mathfrak{g})^{\text{inv}}.$$

Finally, without loss of generality we may assume $\psi = T_{\nu|n}$ for a certain partition ν of length $\leqslant n$, because $\{T_{\nu|n}\}$ is a topological basis in $\mathcal{O}_0(\mathfrak{g})^{\text{inv}}$.

Thus, we have reduced the problem to the following claim.

PROPOSITION 5.3. *Let μ, ν be partitions of length $\leqslant n$. Then*

$$\langle T_{\mu|n}, T_{\nu|n} \rangle = \delta_{\mu\nu} c_\pm(n, \mu).$$

COMMENT. The brackets in the left-hand side can be understood in two different but equivalent ways. First, they represent the pairing between a distribution supported at the origin and a polynomial function, i.e.,

$$\langle T_{\mu|n}, T_{\nu|n} \rangle = (\partial(T_{\mu|n}) T_{\nu|n})(0),$$

where $\partial(T_{\mu|n})$ stands for the differential operator on $\mathfrak{g}$ with constant coefficients that corresponds to $T_{\mu|n} \in S(\mathfrak{g})$. Second, the brackets may denote the canonical extension to $S(\mathfrak{g})$ of the inner product on $\mathfrak{g}$.

PROOF. Probably, the proposition could be proved starting from the explicit expression for $T_{\mu|n}$ given in Proposition 4.6, but we prefer to use a different approach.

Let us extend the inner product $\langle \cdot, \cdot \rangle$ in $\mathfrak{g}$ to $S(\mathfrak{g})$. Then with $(S(\mathfrak{g}), \langle \cdot, \cdot \rangle)$ one can associate a reproducing kernel $\mathcal{E}(X, Y)$, where $X, Y \in \mathfrak{g}$. By definition,

$$\mathcal{E}(X, Y) = \sum_\alpha \psi_\alpha(X) \psi_\alpha^*(Y),$$

where $\{\psi_\alpha\}$ is an arbitrary *homogeneous* basis in $S(\mathfrak{g})$ and $\{\psi_\alpha^*\}$ is the dual basis. Note that $\mathcal{E}(X, Y)$ does not depend on the choice of the basis.

Similarly, to the inner product space $(I(\mathfrak{g}), \langle \cdot, \cdot \rangle)$ there also corresponds a reproducing kernel $\mathcal{F}(X, Y)$.

We have the following evident relation between these two kernels:

$$\mathcal{F}(X, Y) = \int \mathcal{E}(X, \operatorname{Ad} u \cdot Y) \, du = \int \mathcal{E}(\operatorname{Ad} u \cdot X, Y) \, du,$$

where the integral is taken over the compact form $K \subset G$ (or $K' \subset G'$) equipped with the normalized Haar measure.

Further, taking as $\{\psi_\alpha\}$ the basis of monomials formed from a basis in $\mathfrak{g}$, we obtain

$$\mathcal{E}(X, Y) = e^{\langle X, Y \rangle},$$

whence

$$\mathcal{F}(X, Y) = \int e^{\langle X, \operatorname{Ad} u \cdot Y \rangle} \, du.$$

On the other hand, the claim of the proposition means that

$$\mathcal{F}(X, Y) = \sum_{l(\mu) \leq n} \frac{T_{\mu|n}(X) T_{\mu|n}(Y)}{c_\pm(n, \mu)},$$

because $\{T_{\mu|n}\}$ is a homogeneous basis in $I(\mathfrak{g})$.

Thus, we have reduced our problem to the following claim.

PROPOSITION 5.4. *Let u range over a compact form $K \subset G$ (or $K' \subset G'$, for the series* D*) and let du denote the normalized Haar measure. Then*

(5.1) $$\int e^{\langle X, \operatorname{Ad} u \cdot Y \rangle} \, du = \sum_{l(\mu) \leq n} \frac{T_{\mu|n}(X) T_{\mu|n}(Y)}{c_\pm(n, \mu)}, \qquad X, Y \in \mathfrak{g} = \mathfrak{g}(n).$$

PROOF. We show that this can be obtained from the binomial formula (Theorem 1.2) by passing to a limit.

First of all, without loss of generality we can assume that X and Y are diagonal matrices with diagonal entries $\pm x_i$ and $\pm y_i$, respectively (as usual, we omit the zero entry in the case of the series D). Then in the right-hand side one can replace $T_{\mu|n}(X)$ by $t_{\mu|n}(x_1, \ldots, x_n)$ and $T_{\mu|n}(Y)$ by $t_{\mu|n}(y_1, \ldots, y_n)$. After this the right-hand side becomes very similar to the right-hand side of the binomial formula: the only difference is that in the binomial formula, we have $t_{\mu|n}^*(\lambda_1, \ldots, \lambda_n)$ instead of $t_{\mu|n}(y_1, \ldots, y_n)$.

Now let us compare the left-hand sides of the two formulas (in the discussion below one must replace K by K' for the series D).

The normalized irreducible character $\chi_\lambda/\chi_\lambda(e)$ occurring in the binomial formula (1.8) can be viewed as a spherical function for the Gelfand pair $(K \times K, K)$. The left-hand side of formula (5.1) can also be viewed as a spherical function for a Gelfand pair. Namely, this pair consists of the Cartan motion group $K \ltimes \mathfrak{k}$ (the semidirect product of K and its Lie algebra $\mathfrak{k}$ viewed as a K-module) and its subgroup K. The vector $x = (\pm x_i)$ can be regarded as the argument of the spherical function, and $y = (\pm y_i)$ is the parameter.

Now we use the well-known relation between the spherical functions of a symmetric space (say, of compact type) and the spherical functions of the corresponding Cartan motion group (see, e.g., Dooley and Rice [**DR**]).

In our context, this relation is expressed as follows. Let ε be a small parameter which will then tend to 0. Assume $g \in K$ has the form $g = 1 + \varepsilon X + O(\varepsilon)$, where $X \in \mathfrak{k}$, and the partition λ has the form $\lambda = \varepsilon^{-1} y + O(\varepsilon^{-1})$.[6] Then in the limit as $\varepsilon \to 0$ the normalized character indexed by λ turns into the spherical function for the Cartan motion group, indexed by y.

Since the top homogeneous component of the polynomial $t^*_{\mu|n}$ coincides with $t_{\mu|n}$, the right-hand side of the binomial formula (1.8) will turn, after this passage to the limit, into the right-hand side of formula (5.1).

This concludes the proof of Proposition 5.4, and at the same time those of Proposition 5.3 and of Theorem 5.2.

§6. Appendix: Bispherical functions on $Sp(2n, \mathbb{C}) \setminus GL(2n, \mathbb{C})/O(2n, \mathbb{C})$

In this section we discuss a curious fact suggested by the results of §4. Consider two Gelfand pairs $(\widetilde{G}, G)$:

$$(GL(N, \mathbb{C}), O(N, \mathbb{C})), \qquad N = 2n+1 \text{ or } N = 2n,$$
$$(GL(N, \mathbb{C}), Sp(N, \mathbb{C})), \qquad N = 2n.$$

The spherical functions of these pairs are matrix elements $g \mapsto (V(g)\xi, \xi)$, where g ranges over $\widetilde{G}$, V is an irreducible finite-dimensional $\widetilde{G}$-module, and ξ is a G-invariant vectors. It is well known that these spherical functions can be identified (after an appropriate parametrization of the double G-cosets in $\widetilde{G}$) as Jack polynomials $P_\mu^{(\alpha)}$ in n variables, where the parameter α takes the value $\alpha = 2$ for the former pair and the value $\alpha = 1/2$ for the latter one, and μ ranges over partitions of length $\leq n$.

The aim of this Appendix is to examine what happens when these two pairs are "mixed" in the following sense. We set

$$\widetilde{G} = GL(2n, \mathbb{C}), \quad G_1 = O(2n, \mathbb{C}), \quad G_2 = Sp(2n, \mathbb{C}),$$

and assume V is an irreducible $\widetilde{G}$-module admitting both a G_1-invariant ξ and a G_2-invariant η. Then the dual module V^* also possesses a G_2-invariant η^*, and we form the matrix element

$$\varphi(g) = \langle V(g)\xi, \eta^* \rangle, \qquad g \in \widetilde{G},$$

---

[6]Since both sides of (5.1) are invariant with respect to the action of the hyperoctahedral group on $y_1, \ldots, y_n$, we may assume $y_1 \geq \cdots \geq y_n \geq 0$, so that the approximation of the vectors $\varepsilon^{-1} y$ by partitions does exist.

which we call a *bispherical function*. We shall calculate φ in terms of a natural parametrization of the (G_2, G_1)-cosets in $\widetilde{G}$ and see that φ is a Schur polynomial (note that Schur polynomials are Jack polynomials with $\alpha = 1$).

Without loss of generality we may assume V is a polynomial module. We shall specify the embeddings $G_1 \hookrightarrow \widetilde{G}$ as indicated in §1.

THEOREM 6.1. *Let V be an irreducible polynomial representation of the group $\widetilde{G} = GL(2n, \mathbb{C})$ admitting both a G_1-invariant ξ and a G_2-invariant η (where $G_1 = O(2n, \mathbb{C})$ and $G_2 = Sp(2n, \mathbb{C}))$, i.e.,*

$$V = V_{2\mu \cup 2\mu | 2n},$$

where μ is a partition of length $\leqslant n$ and

$$2\mu \cup 2\mu = (2\mu_1, 2\mu_1, \ldots, 2\mu_n, 2\mu_n).$$

Then the spherical function given by the formula

$$\varphi_\mu(g) = \langle V_{2\mu \cup 2\mu | 2n}(g)\xi, \eta^* \rangle,$$

where $g \in \widetilde{G}$ and η^* stands for a G_2-invariant in V^*, is proportional to the Schur polynomial s_μ in n variables under a suitable parametrization of the double cosets $G_2 g G_1$ in $\widetilde{G}$.

FIRST PROOF. This proof is based on Theorem 4.8. Let $g \mapsto g'$ denote the transposition of $2n \times 2n$ matrices that corresponds to the symmetric form M preserved by the subgroup $G_1 = O(2n, \mathbb{C})$ (see §1 for the choice of M). That is, $g \mapsto g'$ is the transposition with respect to the secondary diagonal. The map $g \mapsto gg'$ defines a bijection between the cosets $gG_1 \subset \widetilde{G}$ and (nondegenerate) $2n \times 2n$ matrices, symmetric relative the secondary diagonal, so one can write

$$\varphi_\mu(g) = \psi_\mu(gg'),$$

where ψ_μ is a polynomial function on symmetric matrices.

The group $\widetilde{G}$ acts on symmetric matrices, and under this action ψ_μ is determined by two properties: first, it transforms according to $V_{2\mu \cup 2\mu}$, and, second, it is G_2-invariant.

Next, identify the space of symmetric matrices with the Lie algebra $sp(2n, \mathbb{C})$ or, better, with its dual space. Under this identification, the functions ψ_μ become elements of the space $I(sp(2n, \mathbb{C}))$, and an application of Theorem 4.8 implies that these are just the basis elements T_μ (as usual, up to a scalar factor).

Given an n-tuple $(x_1, \ldots, x_n)$ of nonzero complex numbers, let us form the diagonal matrix

$$g(x) = \operatorname{diag}(x_1, \ldots, x_n, x_n, \ldots, x_1).$$

Then

$$g(x)(g(x))' = \operatorname{diag}(x_1^2, \ldots, x_n^2, x_n^2, \ldots, x_1^2).$$

Under the above identification between symmetric matrices and elements of the (dual space to the) Lie algebra $sp(2n, \mathbb{C})$, the latter matrix turns into

$$\operatorname{diag}(x_1^2, \ldots, x_n^2, -x_n^2, \ldots, -x_1^2),$$

which is an element of the Cartan subalgebra of $sp(2n, \mathbb{C})$.

It follows that

$$\varphi_\mu(g(x)) = \psi_\mu(g(x)(g(x))') = \text{const } s_\mu(x_1^4, \ldots, x_n^4),$$

by the definition of the elements T_μ for the series C. To conclude the proof, we note that each double (G_2, G_1)-coset in $\widetilde{G}$ has the form $G_2 g(x) G_1$ with a certain $x = (x_1, \ldots, x_n)$. □

SECOND PROOF (sketch). One can choose a compact form $\widetilde{G}^u \subset \widetilde{G}$ (isomorphic to $U(2n)$) such that $G_i^u = \widetilde{G}^u \cap G_i$ is a compact form of G_i for $i = 1, 2$. We also can assume that the matrices $g(x)$ are in $\widetilde{G}^u$ provided $|x_1| = \cdots = |x_n| = 1$. Then each (G_2^u, G_1^u)-coset in $\widetilde{G}^u$ is of the form $G_2^u g(x) G_1^u$, so that we can use $x_1, \ldots, x_n$ as parameters of the (G_2^u, G_1^u)-cosets in $\widetilde{G}^u$.

A direct computation shows that the "radial part" of the Haar measure of the compact group $\widetilde{G}^u$, expressed in these parameters, has the density

$$w(x) = \text{const} \prod_{1 \leq i < j \leq n} |x_i^4 - x_j^4|^2.$$

It follows that the bispherical functions are symmetric orthogonal polynomials in $x_1^4, \ldots, x_n^4$ with weight $w(x)$.

Finally, an analysis of the weights of $V_{2\mu \cup 2\mu | 2n}$ shows that the leading term of the bispherical function (with respect to the lexicographic order on monomials in $x_1^4, \ldots, x_n^4$) is equal to

$$\text{const } x_1^{4\mu_1} \cdots x_n^{4\mu_n}.$$

Therefore our orthogonal polynomials will coincide with the Schur polynomials $s_\mu(x_1^4, \ldots, x_n^4)$. □

References

[BL1] L. C. Biedenharn and J. D. Louck, *A new class of symmetric polynomials defined in terms of tableaux*, Adv. in Appl. Math. **10** (1989), 396–438.

[BL2] _____, *Inhomogeneous basis set of symmetric polynomials defined by tableaux*, Proc. Nat. Acad. Sci. U.S.A. **87** (1990), 1441–1445.

[BR] A. M. Borodin and N. A. Rozhkovskaya, *On a super-analog of the Schur–Weyl duality*, Preprint ESI (1995), Vienna.

[D] J. Dixmier, *Algèbres enveloppantes*, Gauthier-Villars, Paris–Bruxelles–Montréal, 1974.

[Ho] R. Howe, *Perspectives on invariant theory: Schur duality, multiplicity-free actions and beyond*, The Schur Lectures (1992), Israel Math. Conf. Proc., vol. 8, 1995, pp. 1–182.

[Hua] L. K. Hua, *Harmonic analysis of functions of several complex variables in classical domains*, Amer. Math. Soc., Providence, RI, 1963.

[I] V. N. Ivanov, *Dimensions of skew shifted Young diagrams and projective characters of the infinite symmetric group*, Representation Theory, Dynamical Systems, Combinatorial and Algorithmic Methods, II, Zapiski Nauch. Sem. POMI, vol. 240, 1997 (Russian); English transl., J. Math. Sci. (to appear).

[Lasc] A. Lascoux, *Classes de Chern d'un produit tensoriel*, C. R. Acad. Sci. Paris Sér. A **286** (1978), 385–387.

[Lass] M. Lassalle, *Polynômes de Jacobi généralisés*, C. R. Acad. Sci. Paris Sér. I **312** (1991), 425–428.

[M1] I. G. Macdonald, *Schur functions: theme and variations*, Publ. I. R. M. A. Strasbourg 1992, 498/S-27, Actes 28-e Séminaire Lotharingien, pp. 5–39.

[M2] _____, *Symmetric functions and Hall polynomials*, 2nd edition, Oxford University Press, 1995.

[Mo] A. Molev, *Factorial supersymmetric Schur functions and super Capelli identities*, this volume.

[MN] A. Molev and M. Nazarov, *Capelli identities for classical Lie algebras*, Preprint (1997).

[N] M. Nazarov, *Yangians and Capelli identities*, this volume.

[Ok1] A. Yu. Okounkov, *Quantum immanants and higher Capelli identities.*, Transform. Groups **1** (1996), 99–126.

[Ok2] _____, *Young basis, Wick formula, and higher Capelli identities*, Intern. Math. Res. Notices (1996), no. 17, 817–839.

[Ok3] _____, *(Shifted) Macdonald polynomials: q-integral representation and combinatorial formula*, Preprint, April 1996; q-alg/9605013.

[OO1] A. Okounkov and G. Olshanski, *Shifted Schur functions*, Algebra i Analiz **9** (1997), no. 2, 79–147 (Russian); English transl. to appear in St. Petersburg Math. J. **9** (1998).

[OO2] _____, *Shifted Jack polynomials, binomial formula, and applications*, Math. Res. Lett. **4** (1997), 69–78.

[O1] G. I. Olshanskiĭ, *Representations of infinite-dimensional classical groups, limits of enveloping algebras, and Yangians*, Topics in representation theory (A. A. Kirillov, ed.), Advances in Soviet Mathematics, vol. 2, Amer. Math. Soc., Providence, RI, 1991, pp. 1–66.

[O2] _____, *Generalized symmetrization in enveloping algebras*, Transformation Groups **2** (1997), 197–213.

[S] G. Szegö, *Orthogonal polynomials*, Amer. Math. Soc., New York, 1959.

[Z] D. P. Želobenko, *Compact Lie groups and their representations*, "Nauka", Moscow, 1970 (Russian); English transl., Amer. Math. Soc., Providence, RI, 1973.

DEPT. OF MATHEMATICS, UNIVERSITY OF CHICAGO, 5734 SOUTH UNIVERSITY AV., CHICAGO, ILLINOIS 60637-1546
E-mail address: `okounkov@math.uchicago.edu`

INSTITUTE FOR PROBLEMS OF INFORMATION TRANSMISSION, BOLSHOY KARETNY 19, 101447 MOSCOW GSP-4, RUSSIA
E-mail address: `olsh@ippi.ac.msk.su`